Lecture Notes in Networks and Systems 793

The series "Lecture Notes in Networks and Systems" publishes the latest developments in Networks and Systems—quickly, informally and with high quality. Original research reported in proceedings and post-proceedings represents the core of LNNS.

Volumes published in LNNS embrace all aspects and subfields of, as well as new challenges in, Networks and Systems.

The series contains proceedings and edited volumes in systems and networks, spanning the areas of Cyber-Physical Systems, Autonomous Systems, Sensor Networks, Control Systems, Energy Systems, Automotive Systems, Biological Systems, Vehicular Networking and Connected Vehicles, Aerospace Systems, Automation, Manufacturing, Smart Grids, Nonlinear Systems, Power Systems, Robotics, Social Systems, Economic Systems and other. Of particular value to both the contributors and the readership are the short publication timeframe and the worldwide distribution and exposure which enable both a wide and rapid dissemination of research output.

The series covers the theory, applications, and perspectives on the state of the art and future developments relevant to systems and networks, decision making, control, complex processes and related areas, as embedded in the fields of interdisciplinary and applied sciences, engineering, computer science, physics, economics, social, and life sciences, as well as the paradigms and methodologies behind them.

Indexed by SCOPUS, INSPEC, WTI Frankfurt eG, zbMATH, SCImago.

All books published in the series are submitted for consideration in Web of Science.

For proposals from Asia please contact Aninda Bose (aninda.bose@springer.com).

Krassimir T. Atanassov · Vassia Atanassova ·
Janusz Kacprzyk · Andrzej Kałuszko ·
Maciej Krawczak · Jan W. Owsiński ·
Sotir S. Sotirov · Evdokia Sotirova ·
Eulalia Szmidt · Sławomir Zadrożny
Editors

Uncertainty and Imprecision in Decision Making and Decision Support - New Advances, Challenges, and Perspectives

Selected Papers from BOS/SOR-2022, Held on October 13–15, 2022, and IWIFSGN-2022, Held on October 13–14, 2022, in Warsaw, Poland

Springer

Editors
Krassimir T. Atanassov
Department of Bioinformatics
and Mathematical Modelling
Institute of Biophysics and Biomedical
Engineering, Bulgarian Academy of Sciences
Sofia, Bulgaria

Vassia Atanassova
Department of Bioinformatics
and Mathematical Modelling
Institute of Biophysics and Biomedical
Engineering, Bulgarian Academy of Sciences
Sofia, Bulgaria

Janusz Kacprzyk
Polish Academy of Sciences
Systems Research Institute
Warsaw, Poland

Andrzej Kałuszko
Polish Academy of Sciences
Systems Research Institute
Warsaw, Poland

Maciej Krawczak
WIT- Warsaw School of Information
Technology
Warsaw, Poland

Jan W. Owsiński
Polish Academy of Sciences
Systems Research Institute
Warsaw, Poland

Sotir S. Sotirov
Faculty of Technical Sciences
Prof. Assen Zlatarov University
Burgas, Bulgaria

Evdokia Sotirova
Intelligent Systems Laboratory
Prof. Assen Zlatarov University
Burgas, Bulgaria

Eulalia Szmidt
Polish Academy of Sciences
Systems Research Institute
Warsaw, Poland

Sławomir Zadrożny
Polish Academy of Sciences
Systems Research Institute
Warsaw, Poland

ISSN 2367-3370 ISSN 2367-3389 (electronic)
Lecture Notes in Networks and Systems
ISBN 978-3-031-45068-6 ISBN 978-3-031-45069-3 (eBook)
https://doi.org/10.1007/978-3-031-45069-3

This Springer imprint is published by the registered company Springer Nature Switzerland AG
The registered company address is: Gewerbestrasse 11, 6330 Cham, Switzerland

Paper in this product is recyclable.

Preface

This volume contains selected papers from two important conferences held in Warsaw, Poland, on October 13–15, 2022: the BOS/SOR'2022—National Conference on Operational and Systems Research, one of premiere conferences in the field of operational and systems research in Poland and at the European level. and the Twentieth International Workshop on Intuitionistic Fuzzy Sets and Generalized Nets, IWIFSGN-2022, one of premiere conferences on fuzzy logic, notably its extensions, with a part on the Generalized Nets (GNs), an extension of the traditional Petri nets. A joint publication of selected papers from BOS/SOR-2022 and IWIFSGN-2022 follows a long tradition and is well substantially justified.

The scope of the BOS/SOR conferences covers systems modeling, systems analysis, broadly perceived operational research, notably optimization, decision making, and decision support, to just mention a few, often under uncertain and imprecise information. Therefore, in addition to traditional probabilistic and statistical tools and techniques, the use of fuzzy sets and their extensions is also relevant, notably the intuitionistic fuzzy sets which constitute the main topic of the IWIFSGN conferences. These reasons have always made the BOS/SOR and IWIFSGN conferences to be a perfect venue for the exchange of ideas, cross fertilization and mutual inspiration.

The volume is composed of some parts that cover the main areas and challenges of the fields concerned. Part I, "Fuzzy sets, intuitionistic fuzzy sets, rough sets", contains papers that provide a foundational and theoretical basis for the next discussion.

Krassimir Atanassov ("On intuitionistic fuzzy extended modal topological structures") presents the concept and analyses of an intuitionistic fuzzy evaluation of the usability of facts in a database. The access to the facts by the users is evaluated. The most used facts are determined according to the frequency of the users' access, and its intuitionistic fuzzy evaluations are aggregated. Relations to the big data contexts are shown.

Veselina Bureva, Petar Petrov, Velin Andonov, and Krassimir Atanassov ("Intuitionistic Fuzzy Evaluation of User Requests Frequency") present an intuitionistic fuzzy evaluation of the usability of facts in a database evaluating the access to the facts by the users. The most used facts are determined according to the frequency of the users' access. In each step, the intuitionistic fuzzy evaluations are aggregated. An example using relational database with user calls is presented, and a relation to big data is emphasized.

Zofia Matusiewicz ("Balanced and intuitionistic fuzzy systems of equations") considers positive and negative information to interpret human behavior using balanced and intuitionistic sets and relations. Differences and similarities in using sets, relations, and systems of balanced and intuitionistic equations are shown, and a comparison of results of the application of intuitionistic systems of equations with a product (max-min, min-max) and balanced systems of equations with balanced max and min operations are shown, and examples are given.

Stela Todorova ("On the solutions of some equations with intuitionistic fuzzy index matrices") shows the solutions of three equations with intuitionistic fuzzy index matrices. The conditions for their existence are given.

Katarína Čunderlíková ("Convergence of functions of several intuitionistic fuzzy observables") shows connection between the random variables in classical probability space and the intuitionistic fuzzy observables studying a convergence in distribution, a convergence in measure, and an almost everywhere convergence of a function of several intuitionistic fuzzy observables induced by a Borel measurable function.

Dawid Ewald, Jacek M. Czerniak, Jan Baumgart, and Huber Zarzycki ("Review of fuzzification functionals dedicated to OFN") are concerned with the ordered fuzzy numbers (OFNs) which are useful tools to describe the real world. Properties of the OFNs make it possible to more accurately represent the natural human understanding of the world and convert this into a useful code understandable to computers. The authors describe methods for the fuzzification of real numbers to the OFNs and propose theoretical backgrounds of the OFN arithmetics and fuzzification functionals, adding some examples.

Jacek M. Czerniak, Jan Baumgart, Hubert Zarzycki, and Łukasz Apiecionek ("Using modified Canberra distance as OFN numbers comparison operator") deal with the comparison of the fuzzy numbers focusing on a comparison operator that would operate intuitively on fuzzy numbers similarly to operators in other number systems. T use of the Canberra distance metric is proposed. Examples of using the fuzzy numbers in the OFN notation are shown.

Part II, "Advanced data analysis and machine learning", is concerned with various problems and aspects of data analysis and machine learning which are the backbones of virtually all modern information technologies.

Urszula Bentkowska and Marcin Mrukowicz ("Parameterized interval-valued aggregation functions in classification of data with large number of missing values") apply parameterized interval-valued aggregation functions in an algorithm based on the k-nearest neighbors that uses interval modeling to improve the quality of binary classification in the case of large number of missing values in datasets. The authors apply a multiple imputation method. The interval-valued aggregation functions proposed belong to diverse families of aggregations considered with respect to monotonicity conditions based on classical partial order for intervals and also other comparability relations for intervals and are also defined with diverse parameters.

Krzysztof Pancerz and Jaromir Sarzyński ("Rough Set Flow Graphs for Information Systems over Ontological Graphs") present an approach in which information flow over time is represented by Rough Set Flow Graphs with the underlying information consisting of concepts embedded in ontologies. The Rough Set Flow Graph Visualizer module, implemented as a part of the Classification and Prediction Software System (CLAPSS), is shown. An example to model sample data containing sequential information about the educational paths and job careers is discussed.

Jan G. Bazan, Stanisława Bazan-Socha, Urszula Bentkowska, Wojciech Gałka, Marcin Mrukowicz, and Lech Zaręba ("Aggregation functions in researching connections between bio-markers and DNA micro-arrays") consider the use of diverse types of regression models in the case of microarray datasets and propose an ensemble algorithm in which datasets for training constituent regression models are deterministically formed, and aggregation functions are used to combine the output values of the constituent regression models. The regression tree, K-nearest neighbors, and support vector regression are used for tests. Families of aggregation functions are applied and compared, with the arithmetic mean, exponential mean, Olympic aggregation, arithmetic-min average, arithmetic-max average, and median. A comparison with the bagging regression model with the optimized parameters based on grid search is shown. It is shown that the proposed ensemble regression model outperforms significantly the corresponding single models.

Anna Wilbik, Barbara Pękala, Krzysztof Dyczkowski, and Jarosław Szkoła ("A comparison of client weighting schemes in federated learning") extend their previous approach to federated learning with missing information and investigate different weighting schemes that depend on the effectiveness of the local models. The initial results are promising.

Slaviiana Danailova–Veleva, Lyubka Doukovska, and Atanas Dukovski ("InterCriteria Analysis of the Supervisory Statistic Data for Selected 8 EU Countries during the Period 2020-2021") analyze the correlation between some indicators for financial condition of the significant banks in eight European countries using a multi-criteria decision-making method—InterCriteria Analysis which is based on two fundamental concepts: the intuitionistic fuzzy sets (IFSs) and index matrices (IMs).

Part III, "Industrial and business applications", contains accounts of various successful applications of the tools and techniques proposed.

Velin Andonov, Stoyan Poryazov, and Emiliya Saranova ("Generalized Net Models of Traffic Quality Evaluation of a Service Stage") use the Generalized Nets (GNs) as models of a service phase and a service stage in an overall telecommunication system aiming at traffic quality evaluation. The values of some traffic quality indicators are characteristics of tokens of the GNs. The proposed GN models are an important step toward the monitoring, prediction, and management of the Quality of Service (QoS) in overall telecommunication systems.

Velin Andonov, Stoyan Poryazov, and Emiliya Saranovan ("Generalized Net Model of Overall Network Efficiency Evaluation") use the Generalized Nets (GNs) for the modeling of telecommunication systems, specifically for the traffic quality evaluation of a service phase and a service stage in an overall telecommunication system. The model can be used for monitoring, prediction, and management of the QoS in overall telecommunication systems. T

Hristo Blidov and Lyubka Doukovska ("Generalized Net Model of the General Claim Process – Cassation Proceedings before the Supreme Court of Cassation") present an application of the Generalized Nets (GNs) for the modeling the cassation proceedings before the Supreme Court of Cassation. The model can help detect and eliminate unnecessary complications in the process itself, leading to its simplification and optimization, presenting a complex process simpler and more understandable.

Hristo Blidov and Lyubka Doukovska ("Generalized Net Model of the General Claim Process – Annulment Proceedings before the Supreme Court of Cassation") show new methods for the analysis of processes in the administration of justice and, in particular, the application of the Generalized Nets (GNs) for the modeling the annulment proceedings before the Supreme Court of Cassation. The model enables the proceedings before the Supreme Court of Cassation to be analyzed in more detail and shortcomings to be discovered.

Part IV, "Medical and health care applications", is concerned with one of the main application areas of the methods considered which is strongly impacted by its economic and social importance.

Dariusz Mikołajewski and Emilia Mikołajewska ("Selected artificial intelligence technologies in the practice of the clinician and researcher in physiotherapy") consider the determination of the scope of application of adaptive computational intelligence systems based on fuzzy logic to analyze the results of physiotherapy. The paper confirms the potential of artificial intelligence, including fuzzy logic, to improve various areas of physiotherapy. Fuzzy logic as a tool for describing uncertainty and linguistically described parameters can contribute particularly here.

Martin Lubich, Elenko Popov, Radostina Georgieva, Dmitrii Dmitrenko. Borislav Bojkov, Chavdar Slavov, Ludmila Todorova, Vassia Atanassova, Peter Vassilev, and Krassimir Atanassov ("A generalized net model of some nephrological diseases") present a generalized net model of the process of diagnosing of the basic nephrological diseases and conditions: AKI, oliguria, proteinuria, and hematuria and shows an implementation in everyday clinical practice for education, differential diagnosis, and multidisciplinary collaboration.

Diana Petkova and Krassimir Atanassov ("A generalized net model of acute respiratory distress syndrome") show a Generalized Net (GN) model of acute respiratory distress syndrome which can be used for the investigation of blood parameters after HCL administration, after inhalation of liposomes for improvement of lung functions as well as for training medical students to take decisions in case of ARDS pathology after HCl—inhalation or other pathological models.

Simeon Ribagin and Gergana Angelova-Popova ("Generalized net model of rehabilitation algorithm for patients with shoulder impingement syndrome") present an example of a Generalized Nets (GNs) application in orthopedics and traumatology rehabilitation. The model describes a possible algorithm protocol for rehabilitation treatment of the patients with shoulder impingement syndrome and can be implemented in the decision-making support systems, tele-rehabilitation platforms, and optimization of the physiotherapy protocols for shoulder impingement syndrome and better rehabilitation strategies.

Evdokia Sotirova, Valentin Stoyanov, Sotir Sotirov, Zlatina Mirincheva, Hristo Bozov, and Todor Kostadinov ("Application of the InterCriteria Analysis Approach to a burnout syndrome data") present a study of statistical data related to burnout syndrome among the medical employees collected through a survey among the staff in five medical centers in Bulgaria (two university general hospitals for Active Treatment, one general hospital for active treatment (municipal), and one specialized rehabilitation hospital). The dependencies between the various parameters describing the studied objects are studied by InterCriteria approach (ICA).

Part V, "Decision making, optimization and problem solving", is concerned with an analysis of some basic aspects, tools and techniques, algorithms, and systems to deal with various broadly perceived decision making, including optimization, and decision analysis-related problems.

Jan W. Owsiński ("On the use of 'ideal structures' in opinion profile identification") is concerned with the author's concept of "ideal structures" and its use, notably in multimodal aggregation of observations, generated through some mechanism, which offers by itself little insight into the potential structure of the resulting dataset. The problem is exemplified by opinion analysis, based, e.g., on some survey, involving several response dimensions. The purpose of the study is to gain knowledge of the structure of the set of opinions with an emphasis on the potential variety of coherent opinions. Connections with cluster analysis are indicated, and the distinction from this domain is also shown.

Leszek Klukowski ("Estimation of the strict preference relation on the basis of pairwise comparisons for moderate and large size sets") presents two approaches for determining estimates of the preference relation on the basis of multiple binary pairwise comparisons with random errors. This requires to obtain an optimal solution of a discrete programming problem which minimizes the sum of differences between the relation form and comparisons which is NP hard and can be solved with the use of exact algorithms for moderate size of sets only, i.e., below 100 elements. For larger sizes, heuristic algorithms are employed. The paper presents how to find an optimal or suboptimal results at an acceptable computational cost using some statistical procedure.

Hubert Zarzycki ("Improved CSO algorithm in practical applications") presents the cat swarm optimization (CSO) nature-inspired algorithm and its application to the problem of creating a diet, with a comprehensive analysis of results and specifics of the CSO algorithm.

Stepan A. Rogonov, Ilia S. Soldatenko, and Alexander V. Yazenin ("On the Quasi-Efficient Frontier of the Set of Optimal Portfolios Under Hybrid Uncertainty with Short Sales Allowed") describe methods for constructing a quasi-efficient frontier of minimum risk portfolio under conditions of hybrid uncertainty to make short sales possible. An acceptable level of expected return for an investor is defined in crisp and fuzzy forms and illustrated on an example. The dependance of the quasi-efficient frontier on the value of α-level is investigated.

Part VI, "Applications", is a wide presentation of various applications of the tools and techniques considered, ranging from socioeconomic systems through business, to industry.

Łukasz Apiecionek, Dawid Ewald, Jacek M. Czerniak, and Jan Baumgart ("Fuzzy mechanism for data transmission adaptation") present an application of artificial intelligence and the protocol of parallel data transmission TCP for increasing the stability of information transfer in the 5G networks. The main characteristics of the Protocol MPTCP are presented as well as the use of artificial intelligence mechanisms to adapt the choice of data transmission path depending on the current error rate.

Barbara Mażbic-Kulma, Jan W. Owsiński, Jarosław Stańczak, Aleksy Barski, and Krzyszof Sęp ("Using the hub & spoke structure to improve the organization of public transport in Warsaw") propose the use of the hub and spoke concept for designing public transport connections. All fast means of urban transport (metro, fast urban rail, suburban rail, high-speed trams) are used as a kind of backbone, a structure of connections of high-speed and capacity communication hubs to which commuters are brought, and slower means of transport (buses, ordinary trams) are used to a large extent to transport people to these communication hubs from other system components. The location of such main changeover stations using some aspects of the hub and spoke method is proposed by applying a specialized evolutionary algorithm to analyze the actual city transportation system on the basis of publicly available timetable of Warsaw Public Transport vehicles.

Ewa Rak and Jaromir Sarzyński ("Evaluation Metrics for a Hybrid Classification System based on the Distributivity Equation and the UNSW-NB15 cyberattack dataset") consider the usefulness of the hybrid classifier to detect network intrusion threats, more specifically for ensemble classifiers involving the distributivity law which aggregates classifiers accordingly. Results of experiments on the UNSW-NB 15 dataset are considered using a fivefold cross-validation for the performance evaluation of classifiers on the dataset using the scikit-learn tool. Then authors' hybrid algorithm based on selected classification algorithms (Multilayer Perceptron Network, k-Nearest Neighbors and Naive Bayes) is used for comparison.

Sotir Sotirov and Simeon Ribagin ("Hybrid sensor system for robot control with nonlinear autoregressive network with exogenous inputs") present a nonlinear autoregressive exogenous input (NARX) neural network and its use for modeling a sensor-based hybrid system for possible control of robotic devices. The signals for the model are obtained from the MyoWare surface muscle sensors. The hybrid sensor system proposed represents a sensor-based control for various task execution and direct human–robot interaction.

Simeon Ribagin, Iasen Hristozov, and Krassimir Atanassov ("Generalized net model of a collective work of a DOBOT Magician robot system with a centralized control") propose an application of the theory of Generalized Nets (GNs) for developing a model of the DOBOT Magician robot system with centralized control. The GN model describes the collective work of five robots (DOBOT Magician serial manipulators) and two conveyor belts the joint operation of which is controlled by a centralized system. The model can be extended to a more complex GN model to describe a collective work of a group of self-controlled system of robots.

Hubert Zarzycki, Łukasz Apiecionek, Dawid Ewald, and Jacek Czerniak ("Optimizing the operation of a wireless sensor network using a swarm intelligence algorithm") present the results obtained after testing and comparing the whale algorithm (WOA) and the artificial bee colony (ABC) algorithm to support the operation of wireless sensor networks, notably for studying the number of active sensors and energy consumption.

We hope that a comprehensive and wide coverage of the volume and its constituting parts and papers, which present both the state-of-the-art and original contributions, will be of much interest and use for a wide research community and will inspire and trigger new research efforts.

We wish to express our deep gratitude to the contributors for their great works, as well as to other participants of the BOS/SOR-2022 and IWIFSGN-2022 conferences, whose presence and active participation have helped make the conferences a success. Special thanks are due to anonymous peer reviewers who, through deep and constructive remarks and suggestions, have greatly helped to improve the quality and clarity of the contributions.

And last but not least, we wish to thank Dr. Tom Ditzinger, Dr. Leontina di Cecco, and Ms. Zainab Liagat for their dedication and help to implement and finish this important publication project on time, while maintaining the highest publication standards.

Krassimir T. Atanassov
Vassia Atanassova
Janusz Kacprzyk
Andrzej Kałuszko
Maciej Krawczak
Jan W. Owsiński
Sotir S. Sotirov
Evdokia Sotirova
Eulalia Szmidt
Sławomir Zadrożny

Contents

Fuzzy Sets, Intuitionistic Fuzzy Sets, Rough Sets

On Intuitionistic Fuzzy Extended Modal Topological Structures

Krassimir Atanassov[1,2](✉)

[1] Department of Bioinformatics and Mathematical Modelling, Institute of Biophysics and Biomedical Engineering, Bulgarian Academy of Sciences, Acad. G. Bonchev Str., Block 105, 1113 Sofia, Bulgaria
krat@bas.bg

[2] Intelligent Systems Laboratory, Prof. Dr. Assen Zlatarov University, 1 "Prof. Yakimov" Blvd., 8010 Burgas, Bulgaria

Abstract. The aim of the current investigation is to present intuitionistic fuzzy evaluation of the usability of facts in a database. The access to the facts by the users is evaluated. Depending on the time, the most used facts are determined according to the frequency of the users access to them. In each step of the fact selection its intuitionistic fuzzy evaluation is aggregated. An example using relational database with user calls is presented. The proposed investigation can be executed in Big Data systems having the support for the relational operations. Nowadays, many operations in the field of relational databases and NoSQL databases are implemented in Big Data systems.

Keywords: Database · Intuitionistic fuzzy set · NewSQL

1 Introduction

On the basis of [8,10,12], the concept of an Intuitionistic Fuzzy Modal Topological Structure (IFMTS) is introduced by the author in [4]. It is proved that the two standard intuitionistic fuzzy topological operators $\mathcal{C}$ and $\mathcal{I}$ (see, e.g., [1,3]), and the two standard intuitionistic fuzzy modal operators $\square$ and $\diamondsuit$ (see, e.g., [1,3]) generate two different IFMTSs. There, some basic properties of these structures are discussed.

The idea for the new object is developed by the author in [5] and [6]. In the first paper, the concepts of a first and a second type of Intuitionistic Fuzzy Feeble Topological Structures (IFFTS1 and IFFTS2), and of a first and a second type of Intuitionistic Fuzzy Modal Feeble Topological Structures (IFMFTS1 and IFMFTS2) are introduced. In the second paper, the concepts of a third type of Intuitionistic Fuzzy Feeble Topological Structures (IFFTS3), and of a third type of Intuitionistic Fuzzy Modal Feeble Topological Structures (IFMFTS3) are introduced. In [6], there are two (dual) topological operators, that are extensions of the standard intuitionistic fuzzy topological operators.

K. T. Atanassov et al. (Eds.): BOS/SOR/IWIFSGN 2022, LNNS 793, pp. 3–14, 2023.
https://doi.org/10.1007/978-3-031-45069-3_1

In the present paper, the extended intuitionistic fuzzy topological operators, introduced for the first time in [6] will be used, while the standard modal operators in the structures will be changed with extended ones.

2 Preliminaries

First, following [1,3], we mention that each IFS A^* has the form

$$A^* = \{\langle x, \mu_A(x), \nu_A(x)\rangle | x \in E\},$$

where E is a fixed universe, $A \subseteq E$ and for each $x \in E$ both functions satisfy the condition:

$$\mu_A(x) + \nu_A(x) \leq 1.$$

As usually, instead of A^* for brevity below we use A.

The IFSs have more than 10 different geometrical interpretations, but the most important one is shown on Fig. 1.

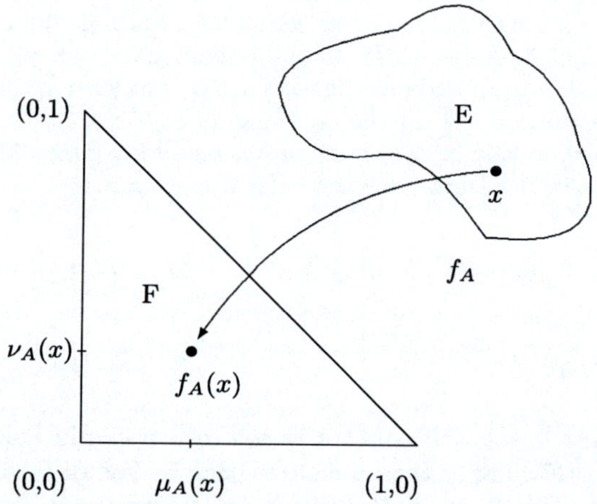

Fig. 1. A geometrical interpretation of an IFS element

Over the IFSs a lot of operations, relations and operators are defined. The most used among them, which we will need below are the following (see [1,3]):

$$\begin{array}{l}
A \subset B \; iff \; (\forall x \in E)(\mu_A(x) \leq \mu_B(x) \;\&\; \nu_A(x) \geq \nu_B(x));\\
A \supset B \; iff \; B \subset A;\\
A = B \; iff \; (\forall x \in E)(\mu_A(x) = \mu_B(x) \;\&\; \nu_A(x) = \nu_B(x));
\end{array}$$

$$\begin{array}{lcl}
\neg A & = & \{\langle x, \nu_A(x), \mu_A(x)\rangle | x \in E\};\\
A \cap B & = & \{\langle x, \min(\mu_A(x), \mu_B(x)), \max(\nu_A(x), \nu_B(x))\rangle | x \in E\};\\
A \cup B & = & \{\langle x, \max(\mu_A(x), \mu_B(x)), \min(\nu_A(x), \nu_B(x))\rangle | x \in E\};
\end{array}$$

$$\Box A = \{\langle x, \mu_A(x), 1 - \mu_A(x)\rangle | x \in E\};$$
$$\Diamond A = \{\langle x, 1 - \nu_A(x), \nu_A(x)\rangle | x \in E\}.$$

The last two operators are the intuitionistic fuzzy versions of the standard modal logic operators (see, e.g., [7,9,11]).

The following operations "negation", "union" and "intersection" are introduced, too (see [2]):

$$\neg^{\varepsilon,\eta} A = \{\langle x, \min(1, \nu_A(x) + \varepsilon), \max(0, \mu_A(x) - \eta)\rangle | x \in E\},$$

$$A \cup^{\varepsilon,\eta} B = \{\langle x, \min(1, \max(\mu_A(x), \mu_B(x)) + \varepsilon), \max(0, \min(\nu_A(x), \nu_B(x)) - \eta)\rangle | x \in E\},$$

$$A \cap^{\varepsilon,\eta} B = \{\langle x, \max(0, \min(\mu_A(x), \mu_B(x)) - \eta), \min(1, \max(\nu_A(x), \nu_B(x)) + \varepsilon)\rangle | x \in E\},$$

where $0 \le \varepsilon \le \eta \le 1$.

The first topological operators over IFSs were introduced by the author in 1985 and their properties were described sequentially (see [1,3]). We can see directly that these operators (analogues of the topological operators "closure" and "interior") which have the forms

$$\mathcal{C}(A) = \{\langle x, \sup_{y \in E} \mu_A(y), \inf_{y \in E} \nu_A(y)\rangle | x \in E\},$$

$$\mathcal{I}(A) = \{\langle x, \inf_{y \in E} \mu_A(y), \sup_{y \in E} \nu_A(y)\rangle | x \in E\}$$

are direct modifications of the intuitionistic fuzzy operations "union" ($\cup$) and "intersection" ($\cap$).

Their geometrical interpretations are shown on Figs. 2 and 3.

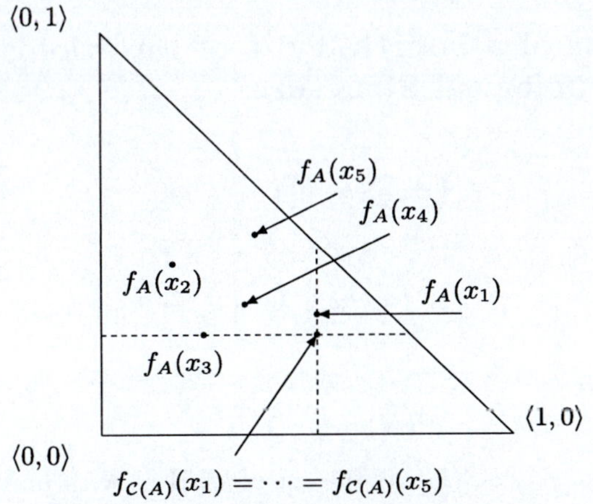

Fig. 2. A geometrical interpretation of operator $\mathcal{C}$

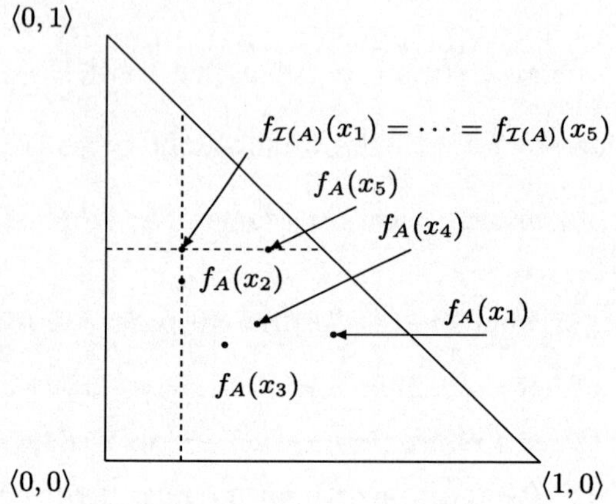

Fig. 3. A geometrical interpretation of operator $\mathcal{I}$

For every fixed ε and η, such that $0 \le \varepsilon \le \eta \le 1$, by analogy with the intuitionistic fuzzy operations $\cup^{\varepsilon,\eta}$ and $\cap^{\varepsilon,\eta}$, the two intuitionistic fuzzy topological operators $\mathcal{C}^{\varepsilon,\eta}$ and $\mathcal{I}^{\varepsilon,\eta}$ are introduced in [6] as follows:

$$\mathcal{C}^{\varepsilon,\eta}(A) = \{\langle x, \min(1, \sup_{y\in E} \mu_A(y) + \varepsilon), \max(0, \inf_{y\in E} \nu_A(y) - \eta)\rangle | x \in E\},$$

$$\mathcal{I}^{\varepsilon,\eta}(A) = \{\langle x, \max(0, \inf_{y\in E} \mu_A(y) - \eta), \min(1, \sup_{y\in E} \nu_A(y) + \varepsilon)\rangle | x \in E\}.$$

Obviously, the intuitionistic fuzzy operations $\cup^{\varepsilon,\eta}$ and $\cap^{\varepsilon,\eta}$ are extensions of the intuitionistic fuzzy operations $\cup$ and $\cap$, that are obtained from the first ones when $\varepsilon = \eta = 0$. By the same reason, the intuitionistic fuzzy topological operators $\mathcal{C}^{\varepsilon,\eta}$ and $\mathcal{I}^{\varepsilon,\eta}$ are extensions of the intuitionistic fuzzy operations $\mathcal{C}$ and $\mathcal{I}$.

3 Definitions of a Fourth Type of Intuitionistic Fuzzy Feeble Topological Structures

Let

$$O^* = \{\langle x, 0, 1\rangle | x \in E\},$$
$$U^* = \{\langle x, 0, 0\rangle | x \in E\},$$
$$E^* = \{\langle x, 1, 0\rangle | x \in E\}.$$

Let for each set X

$$\mathcal{P}(X) = \{Y | Y \subseteq X\}.$$

Let for each set E, $IFS(E)$ be the set of all IFSs with universe E. Then we see that

$$\mathcal{P}(E^*) = \{A | A \subseteq E^*\},$$

where A is an IFS over the universe E.

As has been mentioned in the Introduction, three different types of intuitionistic fuzzy modal feeble topological structures are described. In [5], for the first time, an extended modal operator (D_α) was added in a topological structure and its properties were discussed, but it was used together with the self-dual topological "weight-center" operator $\mathcal{W}$.

Here, for a first time we will include in a topological structure the extended modal operators $G_{\alpha,\beta}$, $H_{\alpha,\beta}$ and $G_{\alpha,\beta}$ in their partial forms, where $\alpha, \beta \in [0, 1]$ (see [3]) and will introduce two fourth types of Intuitionistic Fuzzy Extended Modal Feeble Topological Structures (IFEMFTS4). These operators are

$$G_{\alpha,\beta}(A) = \{\langle x, \alpha.\mu_A(x), \beta.\nu_A(x)\rangle | x \in E\},$$

$$H_{\alpha,\beta}(A) = \{\langle x, \alpha.\mu_A(x), \nu_A(x) + \beta.\pi_A(x)\rangle | x \in E\},$$

$$J_{\alpha,\beta}(A) = \{\langle x, \mu_A(x) + \alpha.\pi_A(x), \beta.\nu_A(x)\rangle | x \in E\}$$

and they have the geometrical interpretations on Figs. 4, 5 and 6, respectively.

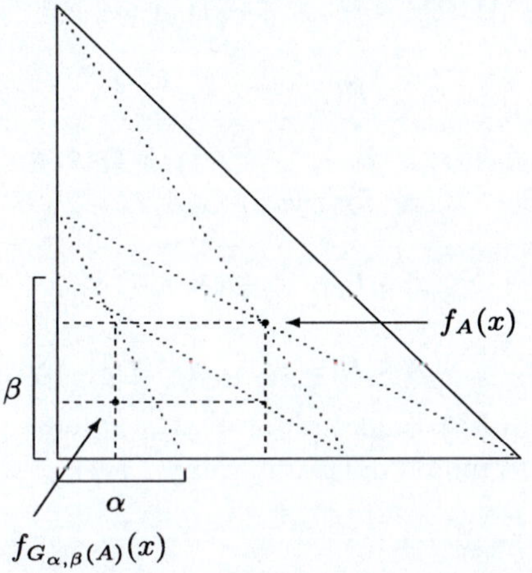

Fig. 4. A geometrical interpretation of operator $G_{\alpha,\beta}$

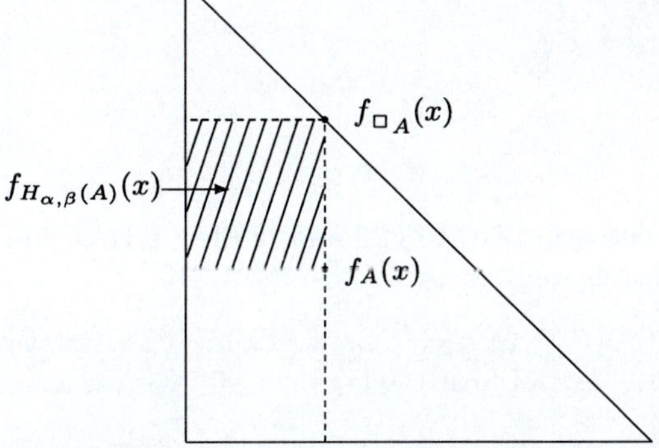

Fig. 5. A geometrical interpretation of operator $H_{\alpha,\beta}$

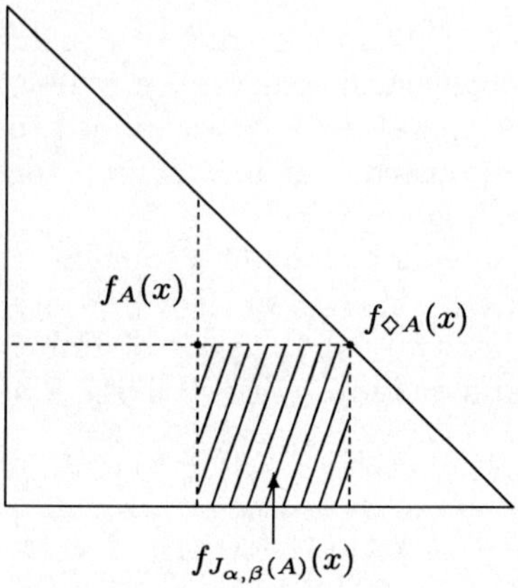

Fig. 6. A geometrical interpretation of operator $J_{\alpha,\beta}$

The first of the IFEMFTS4 is the object that we will denote as *cl*-IFEMFTS4 with the form

$$\langle \mathcal{P}(E^*), \mathcal{O}, \bullet, \circ \rangle,$$

where E is a fixed universe, $\bullet, \blacksquare \ : IFS(E^*) \times IFS(E^*) \to IFS(E^*)$ are the operations defined in Sect. 2 and for two IFSs $A, B \in \mathcal{P}(E^*)$ satisfying equalities

$$A \bullet B = \neg(\neg A \blacksquare \neg B),$$

$$A \blacksquare B = \neg(\neg A \bullet \neg B),$$

$\mathcal{O} : IFS(E^*) \to IFS(E^*)$ is an operator of a closure type, $\circ : IFS(E^*) \to IFS(E^*)$ is an extended modal operator, and for every two IFSs $A, B \in \mathcal{P}(E^*)$:

C1 $\mathcal{O}(A \bullet B) = \mathcal{O}(A) \bullet \mathcal{O}(B)$,
C2 $A \subseteq \mathcal{O}(A)$,
C3 $\mathcal{O}(A) \subseteq \mathcal{O}(\mathcal{O}(A))$,
C4 $O^* \subseteq \mathcal{O}(O^*)$,
C5 $\circ\mathcal{O}(A) \subseteq \mathcal{O}(\circ A)$,
C6 $\circ(A \blacksquare B) \subseteq \circ A \blacksquare \circ B$,
C7 $\circ \circ A = \circ A$,
C8 $\circ A \subseteq A$,
C9 $\circ E^* = E^*$.

The difference between IFEMFTS3 and IFEMFTS4 is that here the relation "$\subseteq$" in C6 in the first structure is "$=$".

Theorem 1. $\langle \mathcal{P}(E^*), \mathcal{C}^{\epsilon,\eta}, \cup, H_{\alpha,0} \rangle$ is a *cl*-IFEMFTS4 with the extended modal operator $H_{\alpha,0}$ (in its partial form), where $\alpha \in [0, 1]$ is an arbitrary number.

Proof. The validity of conditions C2–C4 are checked in [6]. So, we check the rest six conditions as follows.

$$\begin{aligned}
\mathcal{C}^{\varepsilon,\eta}(A \cup B) &= \mathcal{C}^{\varepsilon,\eta}\{\langle x, \max(\mu_A(x), \mu_B(x)), \min(\nu_A(x), \nu_B(x))\rangle | x \in E\} \\
&= \{\langle x, \min(1, \sup_{y \in E} \max(\mu_A(y), \mu_B(y)) + \varepsilon), \\
&\quad \max(0, \inf_{y \in E} \min(\nu_A(y), \nu_B(y)) - \eta)\rangle | x \in E\} \\
&= \{\langle x, \min(1, \max(\sup_{y \in E} \mu_A(y), \sup_{y \in E} \mu_B(y)) + \varepsilon), \\
&\quad \max(0, \min(\inf_{y \in E} \nu_A(y), \inf_{y \in E} \nu_B(y)) - \eta)\rangle | x \in E\} \\
&= \{\langle x, \min(1, \max(\sup_{y \in E} \mu_A(y) + \varepsilon, \sup_{y \in E} \mu_B(y) + \varepsilon)), \\
&\quad \max(0, \min(\inf_{y \in E} \nu_A(y) - \eta, \inf_{y \in E} \nu_B(y) - \eta))\rangle | x \in E\} \\
&= \{\langle x, \max(\min(1, \sup_{y \in E} \mu_A(y) + \varepsilon), \\
&\quad \min(1, \sup_{y \in E} \mu_B(y) + \varepsilon)), \\
&= \min(\max(0, \inf_{y \in E} \nu_A(y) - \eta), \max(0, \inf_{y \in E} \nu_B(y) - \eta))\rangle | x \in E\} \\
&= \{\langle x, \min(1, \sup_{y \in E} \mu_A(y) + \varepsilon), \max(0, \inf_{y \in E} \nu_A(y) - \eta)\rangle | x \in E\} \\
&\cup \{\langle x, \min(1, \sup_{y \in E} \mu_B(y) + \varepsilon), \max(0, \inf_{y \in E} \nu_B(y) - \eta)\rangle | x \in E\} \\
&= \mathcal{C}^{\varepsilon,\eta}(A) \cup \mathcal{C}^{\varepsilon,\eta}(B),
\end{aligned}$$

i.e., condition C1 is valid. The checks of conditions C5–C9 are the following. The geometrical interpretation of $H_{\alpha,0}$ is shown on Fig. 7.

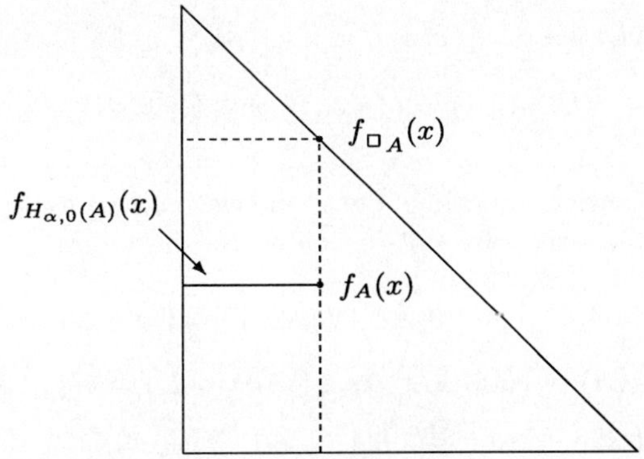

Fig. 7. A geometrical interpretation of operator $H_{\alpha,0}$

$$H_{\alpha,0}(\mathcal{C}^{\varepsilon,\eta}(A)) = H_{\alpha,0}(\{\langle x, \min(1, \sup_{y\in E} \mu_A(y) + \varepsilon),$$
$$\max(0, \inf_{y\in E} \nu_A(y) - \eta))\rangle | x \in E\})$$
$$= \{\langle x, \alpha \min(1, \sup_{y\in E} \mu_A(y) + \varepsilon), \max(0, \inf_{y\in E} \nu_A(y) - \eta))\rangle | x \in E\}$$
$$= \{\langle x, \min(\alpha, \sup_{y\in E} \alpha\mu_A(y) + \alpha\varepsilon), \max(0, \inf_{y\in E} \nu_A(y) - \eta))\rangle | x \in E\}$$
$$\subseteq \{\langle x, \min(1, \sup_{y\in E} \alpha\mu_A(y) + \varepsilon), \max(0, \inf_{y\in E} \nu_A(y) - \eta))\rangle | x \in E\}$$
$$= \mathcal{C}^{\varepsilon,\eta}(\{\langle x, \alpha\mu_A(x), \nu_A(x)\rangle | x \in E\}$$
$$= \mathcal{C}^{\varepsilon,\eta}(H_{\alpha,0}(A));$$

$$H_{\alpha,0}(A \cap B) = H_{\alpha,0}(\{\langle x, \min(\mu_A(x), \mu_B(x)),$$
$$\max(\nu_A(x), \nu_B(x)))\rangle | x \in E\})$$
$$= \{\langle x, \alpha \min(\mu_A(x), \mu_B(x)), \max(\nu_A(x), \nu_B(x)))\rangle | x \in E\}$$
$$= \{\langle x, \min(\alpha\mu_A(x), \alpha\mu_B(x)), \max(\nu_A(x), \nu_B(x)))\rangle | x \in E\}$$
$$= \{\langle x, \alpha\mu_A(x), \nu_A(x)\rangle | x \in E\} \cap \{\langle x, \alpha\mu_B(x), \nu_B(x)\rangle | x \in E\}$$
$$= H_{\alpha,0}(A) \cap H_{\alpha,0}(B);$$

$$H_{\alpha,0}(H_{\beta,0}(A)) = H_{\alpha,0}(\{\langle x, \beta\mu_A(x), \nu_A(x)\rangle | x \in E\}$$
$$= \{\langle x, \alpha\beta\mu_A(x), \nu_A(x)\rangle | x \in E\}$$
$$H_{\alpha\beta,0}(A);$$

$$H_{\alpha,0}(A) = \{\langle x, \alpha\mu_A(x), \nu_A(x)\rangle | x \in E\} \subseteq A;$$
$$H_{\alpha,0}(E^*) = \{\langle x, \alpha, 0\rangle | x \in E\} \subseteq E^*. \qquad \square$$

Now, we will mention equations between two pairs of extended intuitionistic fuzzy modal operators that are not discussed by the moment. For each IFS A:

$$H_{\alpha,0}(A) = \{\langle x, \alpha\mu_A(x), \nu_A(x)\rangle | x \in E\} = G_{\alpha,1}(A),$$
$$J_{0,\alpha}(A) = \{\langle x, \mu_A(x), \alpha\nu_A(x)\rangle | x \in E\} = G_{1,\alpha}(A),$$
$$\neg H_{\alpha,0}(\neg A) = \neg H_{\alpha,0}(\{\langle x, \nu_A(x), \mu_A(x)\rangle | x \in E\}$$
$$= \neg\{\langle x, \alpha\nu_A(x), \mu_A(x)\rangle | x \in E\}$$

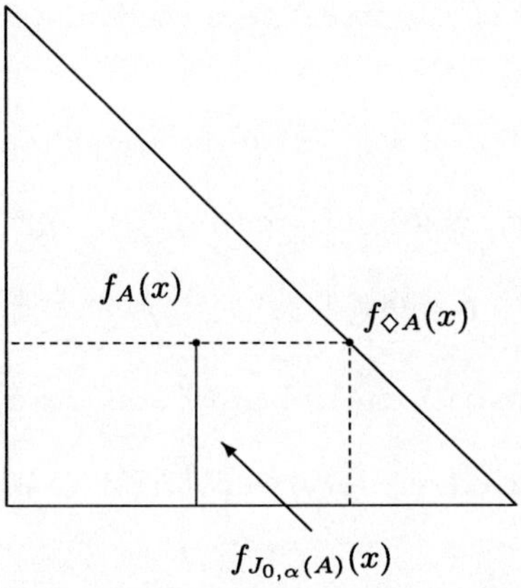

Fig. 8. A geometrical interpretation of operator $J_{0,\alpha}$

$$= \{\langle x, \mu_A(x), \alpha\nu_A(x)\rangle | x \in E\} = J_{0,\alpha}(A).$$

We mention that the geometrical interpretation of operator $J_{0,\alpha}$ is shown on Fig. 8.

Therefore, we can formulate and prove by the above manner

Theorem 2. $\langle \mathcal{P}(E^*), \mathcal{C}^{\varepsilon,\eta}, \cup, G_{\alpha,1}\rangle$ is a *cl*-IFEMFTS4 with the extended modal operator $G_{1,\alpha}$ (in its partial form), where $\alpha \in [0,1]$ is an arbitrary number.

By analogy with the above, we call the object $\langle \mathcal{P}(E^*), \mathcal{O}, \blacksquare, *\rangle$, an *in*-IFMFTS4, where E is a fixed universe, $\bullet, \blacksquare : IFS(E^*) \times IFS(E^*) \to IFS(E^*)$ are the operations defined above, $\mathcal{O} : IFS(E^*) \to IFS(E^*)$ is an operator from an interior type, and $* : IFS(E^*) \to IFS(E^*)$ is an extended modal operator, and for every two IFSs $A, B \in \mathcal{P}(E^*)$:

$$*A = \neg \circ \neg A,$$

and the following conditions are valid:

I1 $\mathcal{O}(A \blacksquare B) = \mathcal{O}(A) \blacksquare \mathcal{O}(B)$,
I2 $\mathcal{O}(A) \subseteq A$,
I3 $\mathcal{O}(\mathcal{O}(A)) \subseteq \mathcal{O}(A)$,
I4 $\mathcal{O}(E^*) \subseteq E^*$,
I5 $*\mathcal{O}(A) \supseteq \mathcal{O}(*A)$,
I6 $*(A \blacksquare B) = *A \blacksquare *B$,
I7 $**A = *A$,
I8 $A \subseteq *A$,
I9 $*O^* = O^*$.

Theorem 3. $\langle \mathcal{P}(E^*), \mathcal{I}^{\varepsilon,\eta}, \cap, J_{0,\alpha}\rangle$ is an *in*-IFEFMTS4.

Proof. As in the proof of Theorem 1, we check the validity of conditions I1, I5–I9, as follows.

$$\begin{aligned}
\mathcal{I}^{\varepsilon,\eta}(A\cap B) &= \mathcal{I}^{\varepsilon,\eta}\{\langle x, \min(\mu_A(x),\mu_B(x)), \max(\nu_A(x),\nu_B(x))\rangle | x\in E\}\\
&= \{\langle x, \max(0, \inf_{y\in E}\min(\mu_A(y),\mu_B(y))-\eta),\\
&\quad \min(1, \sup_{y\in E}\max(\nu_A(y),\nu_B(y))+\varepsilon)\rangle | x\in E\}\\
&= \{\langle x, \max(0, \min(\inf_{y\in E}\mu_A(y), \inf_{y\in E}\mu_B(y))-\eta),\\
&\quad \min(1, \max(\sup_{y\in E}\nu_A(y), \sup_{y\in E}\nu_B(y))+\varepsilon)\rangle | x\in E\}\\
&= \{\langle x, \max(0, \min(\inf_{y\in E}\mu_A(y)-\eta, \inf_{y\in E}\mu_B(y)-\eta)),\\
&\quad \min(1, \max(\sup_{y\in E}\nu_A(y)+\varepsilon, \sup_{y\in E}\nu_B(y)+\varepsilon))\rangle | x\in E\}\\
&= \{\langle x, \min(\max(0, \inf_{y\in E}\mu_A(y)-\eta), \max(0, \inf_{y\in E}\mu_B(y)-\eta)),\\
&\quad \max(\min(1, \sup_{y\in E}\nu_A(y)+\varepsilon), \min(1, \sup_{y\in E}\nu_B(y)+\varepsilon))\rangle | x\in E\}\\
&= \{\langle x, \max(0, \inf_{y\in E}\mu_A(y)-\eta), \min(1, \sup_{y\in E}\nu_A(y)+\varepsilon)\rangle | x\in E\}\\
&\quad \cap\{\langle x, \max(0, \inf_{y\in E}\mu_B(y)-\eta), \min(1, \sup_{y\in E}\nu_B(y)+\varepsilon)\rangle | x\in E\}\\
&= \mathcal{I}^{\varepsilon,\eta}(A)\cap\mathcal{I}^{\varepsilon,\eta}(B).
\end{aligned}$$

Therefore, condition I1 is valid. Below, sequentially, we check conditions I5–I9 as follows.

$$\begin{aligned}
J_{0,\alpha}(\mathcal{I}^{\varepsilon,\eta}(A)) &= J_{0,\alpha}(\{\langle x, \min(1, \sup_{y\in E}\mu_A(y)+\varepsilon),\\
&\quad \max(0, \inf_{y\in E}\nu_A(y)-\eta)\rangle | x\in E\})\\
&= \{\langle x, \min(1, \sup_{y\in E}\mu_A(y)+\varepsilon), \alpha\max(0, \inf_{y\in E}\nu_A(y)-\eta)\rangle | x\in E\}\\
&= \{\langle x, \min(1, \sup_{y\in E}\mu_A(y)+\varepsilon), \max(0, \inf_{y\in E}\alpha\nu_A(y)-\alpha\eta)\rangle | x\in E\}\\
&\supseteq \{\langle x, \min(1, \sup_{y\in E}\mu_A(y)+\varepsilon), \max(0, \inf_{y\in E}\alpha\nu_A(y)-\eta)\rangle | x\in E\})\\
&= \mathcal{I}^{\varepsilon,\eta}(\{\langle x, \mu_A(x), \alpha\nu_A(x)\rangle | x\in E\})\\
&= \mathcal{I}^{\varepsilon,\eta}(J_{0,\alpha}(A));
\end{aligned}$$

$$\begin{aligned}
J_{0,\alpha}(A \cup B) &= J_{0,\alpha}(\{\langle x, \max(\mu_A(x), \mu_B(x)),\\
&\qquad \min(\nu_A(x), \nu_B(x))\rangle | x \in E\})\\
&= \{\langle x, \max(\mu_A(x), \mu_B(x)), \alpha \min(\nu_A(x), \nu_B(x))\rangle | x \in E\}\\
&= \{\langle x, \max(\mu_A(x), \mu_B(x)), \min(\alpha\nu_A(x), \alpha\nu_B(x))\rangle | x \in E\}\\
&= \{\langle x, \mu_A(x), \alpha\nu_A(x)\rangle | x \in E\} \cup \{\langle x, \mu_B(x), \alpha\nu_B(x)\rangle | x \in E\}\\
&= J_{0,\alpha}(A) \cup J_{0,\alpha}(B);
\end{aligned}$$

$$\begin{aligned}
J_{0,\alpha}(J_{0,\beta}(A)) &= J_{0,\alpha}(\{\langle x, \mu_A(x), \beta\nu_A(x)\rangle | x \in E\}\\
&= \{\langle x, \mu_A(x), \alpha\beta\nu_A(x)\rangle | x \in E\}\\
&\quad J_{0,\alpha\beta}(A);
\end{aligned}$$

$$\begin{aligned}
A &\subseteq \{\langle x, \mu_A(x), \alpha\nu_A(x)\rangle | x \in E\} = J_{0,\alpha}(A);\\
O^* &\subseteq \{\langle x, 0, \alpha\rangle | x \in E\} + J_{0,\alpha}(O^*).
\end{aligned}$$ □

Theorem 4. $\langle \mathcal{P}(E^*), \mathcal{I}^{\varepsilon,\eta}, \cup, G_{1,\alpha}\rangle$ is an *in*-IFEMFTS4 with the extended modal operator $J_{0,\alpha}$ (in its partial form), where $\alpha \in [0, 1]$ is an arbitrary number.

4 Conclusion

In a next research, we will study new intuitionistic fuzzy topological structures in which operations "union" and "intersection" have other forms. They will generate new intuitionistic fuzzy topological operators.

A classification of the different types of intuitionistic fuzzy topological structures will be prepared in near future, too.

Acknowledgement. This research was funded by the Bulgarian National Science Fund, grant number KP-06-N22/1/2018 "Theoretical research and applications of InterCriteria Analysis".

References

1. Atanassov, K.: Intuitionistic Fuzzy Sets. Springer, Heidelberg (1999)
2. Atanassov, K.: Intuitionistic fuzzy implication $\rightarrow^{\varepsilon,\eta}$ and intuitionistic fuzzy negation $\neg^{\varepsilon,\eta}$. Dev. Fuzzy Sets, Intuitionistic Fuzzy Sets, Generalized Nets Relat. Top. **1**, 1–10 (2008)
3. Atanassov, K.: On Intuitionistic Fuzzy Sets Theory. Springer, Berlin (2012). https://doi.org/10.1007/978-3-642-29127-2
4. Atanassov, K.: Intuitionistic fuzzy modal topological structure. Mathematics **10**, 3313 (2022). https://doi.org/10.3390/math10183313
5. Atanassov, K.: On the intuitionistic fuzzy modal feeble topological structures. Notes Intuitionistic Fuzzy Sets **28**(3), 211–222 (2022)
6. Atanassov, K.: On four intuitionistic fuzzy feeble topological structures. In: Proceedings of the IEEE 11th International Conference on "Intelligent Systems", Warsaw, 12 - 14 Oct 2022. (in press)
7. Blackburn, P., J. van Bentham, F.: Wolter Modal Logic. North Holland, Amsterdam (2006)
8. Bourbaki, N. Éléments De Mathématique, Livre III: Topologie Générale, Chapitre 1: Structures Topologiques, Chapitre 2: Structures Uniformes. Herman, Paris (Third Edition, in French) (1960)
9. Feys, R.: Modal Logics. Gauthier, Paris (1965)
10. Kuratowski, K.: Topology, vol. 1. Academic Press, New York (1966)
11. Mints, G.: A Short Introduction to Modal Logic. University of Chicago Press, Chicago (1992)
12. Munkres, J.: Topology. Prentice Hall Inc., New Jersey (2000)

Intuitionistic Fuzzy Evaluation of User Requests Frequency

Veselina Bureva[1(✉)], Petar Petrov[1], Velin Andonov[2], and Krassimir Atanassov[3]

[1] Intelligent Systems Laboratory, Prof. Dr. Assen Zlatarov University, 1 "Prof. Yakimov" Blvd., 8010 Burgas, Bulgaria
vbureva@btu.bg

[2] Institute of Mathematics and Informatics, Bulgarian Academy of Sciences, Acad. G. Bonchev Str., Block 8, 1113 Sofia, Bulgaria
velin_andonov@math.bas.bg

[3] Institute of Biophysics and Biomedical Engineering, Bulgarian Academy of Sciences, Acad. G. Bonchev Str., Block 105, 1113 Sofia, Bulgaria
krat@bas.bg

Abstract. The aim of the current investigation is to present intuitionistic fuzzy evaluation of the usability of facts in a database. The access to the facts by the users is evaluated. Depending on the time, the most used facts are determined according to the frequency of the users access to them. In each step of the fact selection its intuitionistic fuzzy evaluation is aggregated. An example using relational database with user calls is presented. The proposed investigation can be executed in Big Data systems having the support for the relational operations. Nowadays, many operations in the field of relational databases and NoSQL databases are implemented in Big Data systems.

Keywords: CDatabase · Intuitionistic fuzzy set · NewSQL

1 Introduction

In the present investigation, an approach to frequency estimation of the user access to certain facts is described. The frequency is determined using the notation of Intuitionistic Fuzzy Sets (IFSs, see, e.g., [2–4]).

In the field of IFSs, the values of functions μ_A and ν_A determine the degree of membership ($\mu_A(x)$) and the degree of non-membership ($\nu_A(x)$) of an element $x \in E$ to a set $A \subseteq E$, evaluated in the interval $[0, 1]$, where E is a fixed universe. The degree of uncertainty is defined as

$$\pi_A(x) = 1 - \mu_A(x) - \nu_A(x).$$

The work of Veselina Bureva, Velin Andonov and Krassimir Atanassov is supported by the research project "Perspective Methods for Quality Prediction in the Next Generation Smart Informational Service Networks" (KP-06-N52/2) financed by the Bulgarian National Science Fund.

K. T. Atanassov et al. (Eds.): BOS/SOR/IWIFSGN 2022, LNNS 793, pp. 15–21, 2023.
https://doi.org/10.1007/978-3-031-45069-3_2

The IFS has the form

$$A = \{\langle x, \mu_A(x), \nu_A(x)\rangle | x \in E\}.$$

The Intuitionistic Fuzzy Pair (IFP, see [5]) is an object $\langle a, b\rangle$, where $a, b \in [0, 1]$ and $a + b \leq 1$. The components of the IFP (a and b) are presented as degrees of membership (validity, etc.) and non-membership (non-validity, etc.).

The IFPs are used as evaluations of some objects or processes. Elements of any two IFSs are compared by their respective elements' degrees of membership and non-membership. In the current investigation, the Intuitionistic Fuzzy Evaluations (IFEs) are defined to evaluate the facts usage.

Here, we will discuss the possibility to add to each fact in a DataBase some IFE that has to be changed in time with respect to the frequence of the use of this fact.

An example of IFEs of facts usage in a relational database is constructed in Sect. 3. The application can be extended in the field of Big Data systems having support for relational operations. The huge amount of facts can be monitored and their IFEs can be calculated.

2 Main Idea

Let us have a DataBase and let to each of its fact an IFE be juxtaposed. These IFEs will present the most used facts and the most unused facts.

Let initially the facts have IFE $\langle 0, 1\rangle$. If the appropriate fact is selected again, its IFE will be increased to $\langle c, 1 - c\rangle$ where c is a fixed constant in the interval $[0, 1]$.

Let in some time-moment the fact F have the IFE $\langle a, b\rangle$. Then, after its next use, the evaluation will be $\langle a + c - ac, b - bc\rangle$.

If the fact is not used for a long time period (fixed in advance), then its IF evaluation will obtain the value $\langle a - c + ac, b + c - ac\rangle$.

We can see immediately that all evaluations are IFPs, because

$$a + c - ac,\ \ b - bc,\ \ a - c + ac,\ \ b + c - ac \in [0, 1]$$

and

$$\begin{aligned} 0 \leq b + c - bc &\leq a + c - ac + b - bc \\ &= a + b - c(a + b) + c \\ &= (a + b)(1 - c) + c \\ &\leq 1 - c + c = 1, \end{aligned}$$

$$\begin{aligned} 0 \leq a + b &= a - c + ac + b + c - ac \\ &= a + b - c + ac + c(1 - a) \\ &\leq a + b - c + ac + bc \end{aligned}$$

$$= a + b - c + (a + b)c$$
$$\leq a + b \leq 1.$$

Nowadays, large amounts of data in different formats are collected from many data sources. The performance of data requests using huge datasets is improved to allow good parallel or distributed query execution. The variety of information needs different storages for storing and processing data. Insensibly, the NoSQL databases are introduced to satisfy the needs of additional data processing. Thereafter, NewSQL databases are introduced as combination of the two types of databases: NoSQL and relational databases. The functions from NoSQL databases can be added to relational databases. NewSQL databases can be considered as "the bridge" between SQL and NoSQL databases. NewSQL databases provide functionalities as highly scalable, online transaction processing (OLTP) combining the ACID guarantees of SQL based relational database engines with the horizontal scalability of NoSQL systems [1,6]. Examples of NewSQL databases are VoltDB, NuoDB, Postgres-XC, Drizzle, H-Store, MySQL cluster. There are small parts of big data systems allowing triggers execution. These fields can be extended in the next years. The introduced IFEs in the current investigation can be implemented in the area of Big Data systems allowing the support of NewSQL databases. The same tables and queries have to be written. If the big data system does not support triggers, the intuitionistic fuzzy estimations can be calculated using SQL query or procedure. The realization of the IFEs in the area of DataBases in discussed in Sect. 3.

3 An Example

The process of phone calls between phone subscribers in a phone call exchange is described. The phone subscribers call other phone subscribers. A monitoring procedure of investigating the successful and unsuccessful calls is developed. The frequency of user requests for calls is observed. The active telephone lines and inactive telephone lines are determined.

An example of the discussed process is presented. Database phoneCalls is constructed. It contains tables **users**(*phoneNumber, FirstName, LastName*) and **userCalls**(*CallDay, CallTime, SuccessOrNot, Duration, Outgoing, Incoming, DegreeMembership, DegreeNonMembership*). Additional table **userCallsLogs** (*ID, Outgoing, Incoming, DegreeMembership, DegreeNonMembership*) for recording calculated values after triggers execution. Two triggers attached to the table **userCalls** are created, namely: **IFE_INSERT** and **IFE_UPDATE**. They activate after inserting or updating data in table userCalls and calculating the intuitionistic fuzzy evaluations of the user calls (see Fig. 1).

The table userCallsLogs contains the fields ID, Outcome, Income, DegreeMembership and DegreeNonMembership. It is used for the triggers activated after insert and update statements. After the triggers activation new values for the intuitionistic fuzzy evaluations are written to the table (Fig. 2).

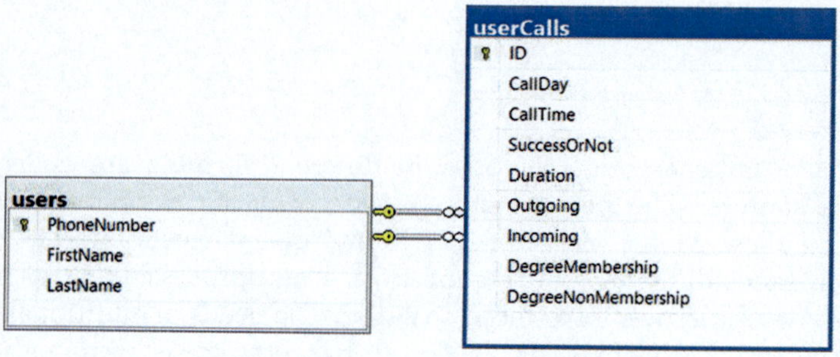

Fig. 1. Database diagram userCalls.

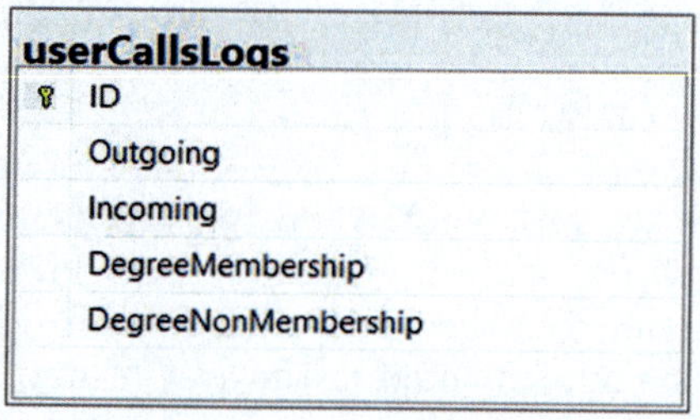

Fig. 2. Table userCallsLogs.

DESKTOP-GR24H6D\S...allsT - dbo.users

PhoneNumber	FirstName	LastName
0887834234	Radostina	Qneva
0888978235	Maya	Hristova
0889765432	George	Dimitrov
0895678678	Dimitrina	Prodanova
0899123987	Dimitar	Ivanov

Fig. 3. The data of table users. An example.

ID	CallDay	CallTime	SuccessOr...	Duration	Outgoing	Incoming	DegreeMe...	DegreeNo...
3	2022-09-14	05:11:01	True	3,34	0889765432	0888978235	0,300	0,600
4	2022-09-15	15:00:01	True	6,34	0889765432	0888978235	0,200	0,500
5	2022-09-16	22:00:33	True	17,04	0889765432	0888978235	0,280	0,450
6	2022-09-17	22:00:33	False	0,00	0889765432	0888978235	0,352	0,405

Fig. 4. The data of table *userCalls*. An example.

The sample data for tables users and userCalls is presented in Fig. 3 and Fig. 4.

Different queries can be executed over the database to monitor the userCalls. An example of creating views with information for successful and unsuccessful calls is given below.

The table of successful calls is created. It can be updated when new data is inserted. The table is shown in Fig. 5.

ID	CallDay	CallTime	SuccessOrNot	Duration	Outgoing	Incoming	DegreeMembership	DegreeNonMembership
3	2022-09-14	05:11:01.0000000	1	3.34	0889765432	0888978235	0.300	0.600
4	2022-09-15	15:00:01.0000000	1	6.34	0889765432	0888978235	0.200	0.500
5	2022-09-16	22:00:33.0000000	1	17.04	0889765432	0888978235	0.280	0.450

Fig. 5. Successful user calls.

In the same way, the table of unsuccessful calls is created. It can be also updated when new data is inserted. The view is presented in Fig. 6.

ID	CallDay	CallTime	SuccessOrNot	Duration	Outgoing	Incoming	DegreeMembership	DegreeNonMembership
6	2022-09-17	22:00:33.0000000	0	0.00	0889765432	0888978235	0.352	0.405

Fig. 6. Unsuccessful user calls.

The table userCalls contains the initial values for the degree of membership and the degree of non-membership of each call record $\langle 0, 1 \rangle$. Depending on the frequency of the user calls the intuitionistic fuzzy evaluations are modified. The trigger IFE_INSERT expects inserting new records in table userCalls to activate its actions. It contains conditional statement which checks if the outgoing phone number and the incoming phone number have had a conversation or not. If the users have not phoned each other until the current time moment, the trigger subtracts 0.1 from the degree of non-membership and assigns 0.1 to the degree of membership. If the users have phoned each other until the current time moment, the trigger calculates the degree of membership by the formula Membership+0.1-Membership*0.1 and the degree of non-membership by the formula NonMembership*(1-0.1). The value 0.1 can be different and must be determined previously (or be provided as variable with assigned value from the list/column of numbers). The new values of the IFEs are written into table userCallsLogs.

The SQL trigger is tested using INSERT statement. The modified values are written to the table userCallsLogs. The next SQL query selects the user calls before appropriate date and after it (i.e., the SELECT statements are executed). The results are intersected to receive the phone calls that do not exist in the two parts of data (Fig. 7). These calls are not frequent in time and their IFEs have to be modified.

The IFE_Update trigger is created to check for update statements over the table userCalls and to copy the new values for the degree of membership and the degree of non-membership into table userCallsLogs.

The IFE_UPDATE trigger is tested by UPDATE statement. The table userCallsLogs contains the actual information after the insert and update statements (Fig. 8).

```sql
SELECT * FROM userCalls WHERE CallDay<='2022-09-20'
INTERSECT
SELECT * FROM userCalls WHERE CallDay>='2022-09-15'
```

Results | Messages

ID	CallDay	CallTime	SuccessOrNot	Duration	Outgoing	Incoming	DegreeMembership	DegreeNonMembership
4	2022-09-15	15:00:01.0000000	1	6.34	0889765432	0888978235	0.200	0.500
5	2022-09-16	22:00:33.0000000	1	17.04	0889765432	0888978235	0.280	0.450
6	2022-09-17	22:00:33.0000000	0	0.00	0889765432	0888978235	0.352	0.405

Fig. 7. The call that is not repeated in time.

ID	Outgoing	Incoming	DegreeMembership	DegreeNonMembership
1	0889765432	0888978235	0,200	0,500
2	0889765432	0888978235	0,280	0,450
3	0889765432	0888978235	0,352	0,405
4	0889765432	0888978235	0,417	0,365

Fig. 8. Table userCallsLogs containing the actual intuitionistic fuzzy estimations after activation of the IFE_INSERT and IFE_UPDATE triggers.

In the end the SELECT query formatting the values for the degree of membership and the degree of non-membership is executed. In the table, the values can be calculated with many numbers after the decimal point (Fig. 9).

```sql
SELECT Outgoing,
       Incoming,
       CAST(DegreeMembership AS DECIMAL(4,2)) AS DegreeMembership,
       CAST(DegreeNonMembership AS DECIMAL(4,2)) AS DegreeNonMembership
FROM userCallsLogs
```

200 % | Results | Messages

	Outgoing	Incoming	DegreeMembership	DegreeNonMembership
1	0889765432	0888978235	0.20	0.50
2	0889765432	0888978235	0.28	0.45
3	0889765432	0888978235	0.35	0.41
4	0889765432	0888978235	0.42	0.37

Fig. 9. Displaying the IFEs with converted data type length.

The same query is modified to add concatenation of the columns for the degree of membership and the degree of non-membership. The new column with intuitionistic fuzzy evaluations (pairs) is visualized (Fig. 10).

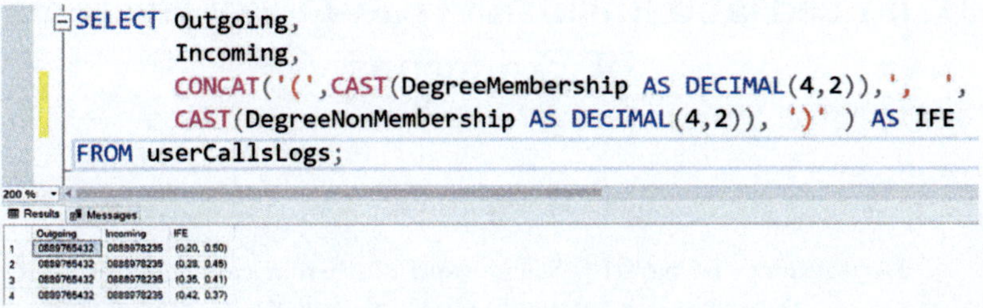

Fig. 10. Concatenated IFEs.

4 Conclusion

In the current investigation an approach for frequency estimation of the user access to certain facts is presented. The formulas for intuitionistic fuzzy estimations are defined. An example of the relational database for phone center is constructed. SQL statements for intuitionistic fuzzy evaluations are executed. The IFEs present the most used and the most unused facts.

The proposed approach can be used in the QoS modeling in an overall telecommunication system.

References

1. Almassabi, A., Bawazeer, O., Adam, S.: Top NewSQL databases and features classification. Int. J. Database Manage. Syst. (IJDMS) **10**(2), 1–31 (2018)
2. Atanassov, K.: Intuitionistic Fuzzy Sets: Theory and Applications. Springer, Heidelberg (1999). https://doi.org/10.1007/978-3-7908-1870-3
3. Atanassov, K.: On Intuitionistic Fuzzy Sets Theory. Springer, Berlin (2012). https://doi.org/10.1007/978-3-642-29127-2
4. Atanassov, K.: Generalized Nets and Intuitionistic Fuziness in Data Mining. Professor Marin Drinov Publishing House of the Bulgarian Academy of Sciences, Sofia, Bulgaria (2020)
5. Atanassov, K., Szmidt, E., Kacprzyk, J.: On intuitionistic fuzzy pairs. Notes Intuitionistic Fuzzy Sets **19**(3), 1–13 (2013)
6. Harrison, G.: Next Generation Databases: NoSQL, NewSQL, and Big Data. Apress, New York (2015)

Balanced and Intuitionistic Fuzzy Systems of Equations

Zofia Matusiewicz(✉)

Department of Cognitive Science and Mathematical Modeling,
University of Information Technology and Management, Al. Sucharskiego 2,
35-225 Rzeszow, Poland
zmatusiewicz@wsiz.edu.pl
https://wsiz.edu.pl

Abstract. Human decision-making is an essential eld of research. Fuzzy sets were created for a mathematical concept to support modeling human behavior, reasoning, and decision-making. However, sociological, psychological, and financial investigations have shown that omitting negative information does not entirely interpret human behavior. Balanced and intuitionistic sets and relations are positive and negative data analysis tools. This work presents the differences and similarities in using sets, relations, and systems of balanced and intuitionistic equations. In this paper, we compare the results of the application of intuitionistic systems of equations with a product ($\max - \min, \min - \max$) and balanced systems of equations with balanced MAX and MIN operations. Situations are pointed out, where applying the relationships discussed is more beneficial. In addition, we will discuss possible problems that hinder their application. Examples are introduced to understand the differences between the approaches discussed.

Keywords: balanced fuzzy sets · balanced relations · balanced systems of equations · intuitionistic fuzzy sets · intuitionistic fuzzy relations · intuitionistic systems of equations

1 Introduction

Fuzzy set and relation-based techniques have become an important component of information technology development. In this field, fuzzy sets and relations have had a significant role to date in the collection and processing of imprecise linguistic information. They have also contributed to the development of intelligent decision support systems, expert systems.

Systems of $\max - \min$ equations were introduced by E. Sanchez in 1979 (see [19]) in connection with fuzzy models in medical diagnosis. He studied the problem of the existence of solutions and presented formulae for finding the greatest solution (the solution most desirable in terms of optimizing the diagnosis).

The work is supported by the grant *Analysis of the problem of collecting, processing and model building from real data*, University of Information Technology and Management, Poland.

K. T. Atanassov et al. (Eds.): BOS/SOR/IWIFSGN 2022, LNNS 793, pp. 22–31, 2023.
https://doi.org/10.1007/978-3-031-45069-3_3

Authors of later works investigated the family of all solutions to systems of equations of the max − min type (see [5,7,10,16]). At the same time, they started to consider systems with the product of max $-T$ (see [6]) and systems of type $S-T$ (see [17]), where T denotes a triangular norm and S a triangular conorm. In the last three decades, the most attention has been paid to relational systems of sup − inf and sup $-prod$ equations (see monographs [6,8,18]). The most recent results concern max $-T$ and max $-*$ systems with additional assumptions on the triangular norm T and the operation $*$ (see [9,13,20,21]).

Since their conception, fuzzy sets and their extensions have become popular. The essence is that the idea of fuzzy sets nullifies bivalent judgements (black-or-white, all-or-nothing) and allows models of reality to be created. The genesis of introducing extensions to fuzzy sets was the observation that classical fuzzy set theory, which concentrates all negative information in the 0-value of the membership function, does not allow negative information to be taken into account.

The main objective of this article is to present differences and similarities in the concepts and constructions of extensions of fuzzy sets and relations. The concept of intuitionistic fuzzy set was introduced in 1983 and decried in [1] and [2]. Intuitionistic fuzzy sets are viewed as a proper mathematical tool for representing hesitancy concerning both membership and non-membership of an element in a set. Later, Burillo and Bustince, in their articles [3,4] introduced and examined many properties of intuitionistic fuzzy relations (IFRs). The IFR is also explored and extended in numerous articles dealing, inter alia, with systems of intuitionistic fuzzy equations (e.g. [14]).

Further practical applications of fuzzy sets contributed to the introduction of balanced sets. Balanced sets were introduced by Homenda in 2006 (see [11]), and balanced relations (BR) and systems of equations in 2022 by Matusiewicz and Homenda ([15]). The main motivation for the study of balanced sets, relations and relational systems is the potentiality of considering the positive and negative stimuli that condition human decisions in the decision-making process. This extension allows us to describe data and situations we can characterize as gain or loss (analogously levels of pessimism/optimism, scarcity/fulfillment, aversion/interest, etc.). Such concepts are applicable in many fields, e.g. finance, marketing, social, and engineering.

Further on, we will show that balanced and intuitionistic fuzzy sets, relations and systems of equations are not in competition with each other, but complementary in a wide range of applications.

2 Mathematical Background

2.1 Extensions of Fuzzy Sets

Intuitionistic fuzzy sets [1] and balanced fuzzy sets [11] are the extensions of the concept of fuzzy sets [23]. Let $X \neq \emptyset$ be ordinary finite non-empty set. By intuitionistic fuzzy sets, we understand

$$A = \{(x, \mu_A(x), \nu_A(x)) : x \in X\} \text{ and } 0 \leq \mu_A(x) + \nu_A(x) \leq 1, \qquad (1)$$

where $\mu_A : X \to [0,1]$ is the degree of membership and $\nu_A : X \to [0,1]$ is the degree of non-membership of the element x in the set A. Moreover, we would like to recall that $\Pi(x) = 1 - \mu_A(x) - \nu_A(x)$ is called a hesitancy degree of x to A. So, if $1-(\nu_A(x)+\mu_A(x)) = 0$, the set A is a fuzzy set. In addition, we would like to introduce a balancing factor θ and selecting factor τ as follows:

$$\theta(x) = \mu_A(x) - \nu_A(x), \quad \tau(x) = \begin{cases} \mu_A(x),\ \nu_A(x) < \mu_A(x) \\ -\nu_A(x),\ \mu_A(x) < \nu_A(x) \ . \\ 0,\ \mu_A(x) = \mu_A(x) \end{cases} \tag{2}$$

Balanced fuzzy sets disperse positive information over the interval $[0,1]$ and negative information over the interval $[-1,0]$. Therefore, unipolar scales, with a unit interval of $[0,1]$ are replaced by bipolar scales, with an interval of $[-1,1]$. Thus, a 'degree of membership' is given when an element belongs to a set and a degree of 'non-membership' when an element is outside the set.

2.2 Balanced Sets and Relations

In this work, by balanced operation we mean an operation $\star : [-1,1]^2 \to [-1,1]$. Firstly, we recall the definition of balanced extended triangular norms [12]. The notion of balanced triangular norms was introduced and described in the works ([11]). The concept of extension consists in spreading the negative information centered at point 0 over the interval $[-1,0]$ and preserving symmetry.

Definition 1 ([11], Definition 5). *The balanced t-norm and t-conorm are mappings $P : [-1,1]^2 \to [-1,1]$, where P stands for both the balanced t-norm and balanced t-conorm, where they satisfy the following properties*

1. $P(a, P(b,c)) = P(P(a,b),c)$ *- associativity,*
2. $P(a,b) = P(b,a)$, *commutativity,*
3. $P(a,b) \le P(c,d)$ *for* $a \le c$ *and* $b \le d$ *- monotonicity,*
4. $S(0,a) = a$ *and* $T(1,a) = a$ *- for* $a \in [0,1]$ *- boundary condition:*
5. $P(x,y) = N(P(N(x), P(N(y))))$ *- symmetry.*

In this paper, classical maximum and minimum are denoted by min, max, and balanced MAX, MIN, based on the Definition 1, are defined as MIN and MAX operations (see [15]):

$$MIN(a,b) = \begin{cases} \min(a,b) & a,b \in [0,1], \\ -\min(-a,-b) & a,b \in [-1,0], \ , \\ 0 & a \cdot b < 0. \end{cases} \tag{3}$$

$$MAX(a,b) = \begin{cases} \max(a,b) & a,b \in [0,1] \\ -\max(-a,-b) & a,b \in [-1,0] \\ b & |a| < |b|,\ a \cdot b \le 0 \\ a & |b| < |a|,\ a \cdot b \le 0 \\ |a| & |b| = |a|,\ a \cdot b < 0, \end{cases} \tag{4}$$

In our study we introduce (after Zadeh [23]) the product $\rho - \delta$ and in particular, we consider the $MAX - MIN$ product.

Table 1. The MIN and MAX operations.

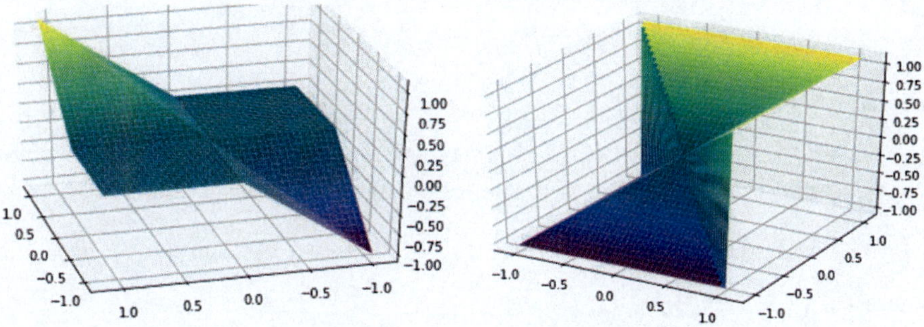

Definition 2. *By the* $\rho - \delta$ *product of balanced relations matrices* $A \in [-1,1]^{mn}$ *and* $B \in [-1,1]^{np}$ *we denote* $C = A \bullet B$*, in the following form:*

$$C = \begin{bmatrix} a_{11} & \dots & a_{1n} \\ \dots & \dots & \dots \\ a_{1n} & \dots & a_{mn} \end{bmatrix} \bullet \begin{bmatrix} b_{11} & \dots & b_{1p} \\ \dots & \dots & \dots \\ b_{n1} & \dots & b_{np} \end{bmatrix} \tag{5}$$

$c_{ij} = \rho(\dots\rho(\rho(\delta(a_{i1}, b_{1j}), \delta(a_{i2}, b_{2j})), \delta(a_{i3}, b_{3j})), \dots, \delta(a_{in}, b_{nj}))$*, where* ρ, δ *are balanced norms and* $i \in \{1,\dots,m\}$, $j \in \{1,\dots,n\}$.
This will be written as follows $c_{ij} = \rho_{k=1}^{n}\delta(a_{ik}, b_{kj})$.

Let $L = [-1,1]$ or $L = [0,1]$. For the matrices $A \in L^{m\times n}, B \in L^{m\times n}$ and vectors $c, d \in L^n$ of fuzzy and balanced relations the following order is used:

$$(A \le B) \Leftrightarrow (a_{ij} \le b_{ij}),\ (c \le d), (c_j \le d_j) \tag{6}$$

where $i \in \{1,\dots,m\}$, $j \in \{1,\dots,n\}$.

2.3 Intuitionistic Fuzzy Relations

Let $X, Y \neq \emptyset$ be ordinary finite non-empty sets. The set of all intuitionistic fuzzy relations is denoted by $IFS(X \times Y)$ in $X \times Y$.

Definition 3 *([3], Definition 1)***.** *An intuitionistic fuzzy relation* R *of* $X \times Y$

$$R = \{((x,y), \mu_R(x,y), \nu_R(x,y)) : (x,y) \in X \times Y\}, \tag{7}$$

where $\mu_R : X \times Y \to [0,1], \nu_R : X \times Y \to [0,1]$ *with the condition* $0 \le \mu_R(x,y) + \nu_R(x,y) \le 1$ *for all* $(x,y) \in X \times Y$.

The matrix representation of the intuitionistic fuzzy relation $R \in IFS(X\times Y)$ *is called an intuitionistic fuzzy membership matrix* A

$$(a_{ij}^{\mu}, a_{ij}^{\nu}) = R(x_i, y_j) = (\mu_R(x_i, y_j), \nu_R(x_i, y_j)), \tag{8}$$

where $M = 1,\dots,m$ *and* $N = 1,\dots,n$ *and*

$$a_{ij}^{\mu} + a_{ij}^{\nu} \le 1. \tag{9}$$

Corollary 1. *Each intuitionistic fuzzy membership matrix A can be decomposed into two matrices A^{μ} and A^{ν}, having the form*

$$a_{ij}^{\mu} = \mu_R(x_i, y_j),\ a_{ij}^{\nu} = \nu_R(x_i, y_j),\ i \in \{1, \ldots, m\},\ j \in \{1, \ldots, n\}, \tag{10}$$

where $(A^{\mu} + A^{\nu})_{ij} \leq 1$.

Let $L2 = \{(x, y) \in [0, 1]^2 : x + y \leq 1\}$. For the matrices $A \in L2^{m \times n}, B \in L2^{m \times n}$ and vectors $c, d \in L2^n$ of the IFR the following order is used [14]

$$(A \preceq B) \Leftrightarrow (a_{ij}^{\mu} \leq b_{ij}^{\mu} \text{ and } a_{ij}^{\nu} \geq b_{ij}^{\nu}),\ \ (c \preceq d) \Leftrightarrow (c_j^{\mu} \leq d_j^{\mu} \text{ and } c_j^{\nu} \geq d_j^{\nu}), \tag{11}$$

where $i \in \{1, \ldots, m\}$, $j \in \{1, \ldots, n\}$.

Corollary 2. *Let $A \in L2^{m \times n}, B \in L2^{m \times n}$ and $c, d \in L2^n$ we have*

$$(A \preceq B) \Leftrightarrow ((A^{\mu} \leq B^{\mu}) \text{ and } (B^{\nu} \leq A^{\nu})), \tag{12}$$

$$(c \preceq d) \Leftrightarrow ((c^{\mu} \leq d^{\mu}) \text{ and } (d^{\nu} \leq c^{\nu})),$$

Based on Definition 3 from [4] we have got:

Definition 4. *Let $*$, $\star$ be some operations and α, γ the associative operations and intuitionistic fuzzy membership matrices $A \in L2^{m \times p}$ and $C \in L2^{p \times n}$ of the relations $IFR(X \times Y)$ and $P \in IFR(Y \times Z)$. The product of the matrices A and C is defined by*

$$(A \circ_{\gamma,\star}^{\alpha,*} C)_{ij} = (\alpha_{k=1}^{p}(a_{ik}^{\mu} * c_{kj}^{\mu}), \beta_{k=1}^{p}(a_{ik}^{\nu} \star c_{kj}^{\nu})) \tag{13}$$

where $(A \circ_{\gamma,\star}^{\alpha,*} C)_{ij}^{\mu} + (A \circ_{\gamma,\star}^{\alpha,*} C)_{ij}^{\nu} \leq 1$ and $i \in \{1, \ldots, m\}$, $j \in \{1, \ldots, n\}$.

Corollary 3. *Let $A^{\mu}, A^{\nu} \in [0, 1]^{m \times p}, B^{\mu}, B^{\nu} \in [0, 1]^{p \times n}$ and α, γ be associative operations*

$$(A \circ_{\gamma,\star}^{\alpha,*} C) = (A^{\mu} \circ^{\alpha,*} C^{\mu}, A^{\nu} \circ^{\gamma,\star} C^{\nu}), \tag{14}$$

and $\circ^{\alpha,*}$, $\circ^{\gamma,\star}$ are $\alpha - *$ and $\gamma - \star$ fuzzy matrix products:

$$(A^{\mu} \circ^{\alpha,*} C^{\mu})_{ij} = \alpha_{1 \leq k \leq p}\, (a_{ip}^{\mu} * c_{pj}^{\mu}),$$

$$(A^{\nu} \circ^{\gamma,\star} C^{\nu})_{ij} = \gamma_{1 \leq k \leq p}\, (a_{ip}^{\nu} \star c_{pj}^{\nu}),$$

where $(A^{\mu} + A^{\nu})_{ij} \leq 1$ and $(B^{\mu} + B^{\nu})_{ij} \leq 1$ for $i \in \{1, \ldots, m\}$, $j \in \{1, \ldots, n\}$.

Theorem 1 (cf. [4], Proposition 1). *Let α, β, $*$, $\star$ be increasing, associative, binary operations in $[0, 1]$. Moreover, if (γ, γ') and $(\star, \star')$ are the dual operations satisfying the conditions $\alpha \leq \gamma'$ and $* \leq \star'$, then*

$$(A \circ_{\gamma,\star}^{\alpha,*} C)_{ij}^{\mu} + (A \circ_{\gamma,\star}^{\alpha,*} C)_{ij}^{\nu} \leq 1 \tag{15}$$

is true for all $i \in \{1, \ldots, m\}$, $j \in \{1, \ldots, n\}$.

Thus, a particular case of the product of fuzzy matrices of intuitionistic relations is $\circ_{\min,\max}^{\max,\min}$. For short, $\circ^{\max,\min}$ will be denoted by $\circ$.

3 Connections Between Balanced and Intuitionistic Fuzzy Systems of Equations

Since we are familiar with some methods for solving balanced and intuitionistic equations (see [14,15]), we would like to present connections between these systems of equations.

Based on (4) and (3) we obtain:

Proposition 1. *Let us restrict the operations of* max, min *and* MAX *and* MIN *to the interval* $[0, 1]$. *Then the products satisfy the property:* $MAX - MIN|_{[0,1]} = \max - \min|_{[0,1]}$.

Proposition 2. *Let us restrict the operations of* max, min *and* MAX *and* MIN *to the interval* $[-1, 0]$. *Then the products satisfy the property:* $MAX - MIN|_{[-1,0]} = \min - \max|_{[-1,0]}$.

Proposition 3. *Let be given matrices* $A \in [-1, 0]^{mn}$, $B \in [-1, 0]^{np}$, $\bullet = MAX - MIN$. *Let us restrict the operations of* max, min *and* MAX *and* MIN *to the interval* $[-1, 0]$. *Then, the products satisfy the property:* $A \bullet B = -((-A) \circ (-B))$.

Proof. For any matrices $A \in [-1, 0]^{mn}$, $B \in [-1, 0]^{np}$, we determine the product $C = A \bullet B$ using (4) and (3):

$$c_{ij} = MAX_{k=1}^{n}(MIN(a_{ik}, b_{kj})) = MAX_{k=1}^{n}(-\min(-a_{ik}, -b_{kj})) =$$

$= -\max_{k=1}^{n}(\min(-a_{ik}, -b_{kj}))$. So $(A \bullet B)_{ij} = -((-A) \circ (-B))_{ij}$ for any $i \in \{1, \ldots, m\}$ and $j \in \{1, \ldots, n\}$.

To convert intuitionistic fuzzy systems $A \bullet x = b$ into balanced ones, we use functions θ and τ on all of IFR's elements of the matrix A and vector b. The matrices and vectors obtained are called balancing and selecting matrices and vectors, respectively.

Proposition 4. *Let* $\bullet = MAX - MIN$, $A \in L2^{mn}$ *and* $b \in L2^{m}$. *Both (with balancing and selecting matrices and vectors) balanced systems of equations obtained from (2)* $D \circ y = c$ *give the following properties:*

1. *If* $A^{\nu} \le A^{\mu}$ *and* $b^{\nu} \le b^{\mu}$, *then we get classical relational equation* $D \circ y = c$.
2. *If* $A^{\mu} \le A^{\nu}$ *and* $b^{\mu} \le b^{\nu}$, *then we obtain equation* $(-D) \circ (-y) = (-c)$.

Proposition 4 comes from the facts that in 1. D and c have got elements greater than 0, and in 2. D and c - lower than 0.

Proposition 5. *Let there be a given balanced system* $D \bullet y = c$, *such that in each column of the matrix* D, *there are only non-positive or non-negative elements. Furthermore, in the balanced system in the corresponding rows of a given matrix and vector, we have only non-negative or non-positive elements. In this case, such a system can be decomposed into two systems and solved separately.*

Converting the balanced system $D \bullet y = c$ into an intuitionistic fuzzy system determines the family of equations $A \circ_{\min,\max}^{\max,\min} x = b$. The corresponding elements of the intuitionistic matrices and vectors can be determined under the assumption that the balanced matrix and vector are of balancing type:

$$a_{ij} = \begin{cases} a_{ij}^{\mu} \leq \frac{1+d_{ij}}{2} \\ a_{ij}^{\nu} \leq \frac{1-d_{ij}}{2} \end{cases} \tag{16}$$

or selecting type:

$$a_{ij} = \begin{cases} a_{ij}^{\mu} = d_{ij} \quad and\ a_{ij}^{\nu} \in [0, 1-d_{ij}]\ d_{ij} \geq 0 \\ a_{ij}^{\nu} = -d_{ij}\ and\ a_{ij}^{\mu} \in [0, 1+d_{ij}]\ d_{ij} \leq 0 \end{cases} \tag{17}$$

However, we wish to emphasize that the transformation of one type of a system of equations into another does not transfer the family of solutions. The usefulness of these systems sometimes varies, so their transformation can be utilized for studying other properties of the phenomenon described by these systems.

4 Examples of Applications of Systems of Equations

The study of relationships in datasets containing data describing problems of gain and loss, levels of pessimism and optimism, scarcity and fullness, aversion and interest requires the introduction and analysis of balanced systems of equations.

Example 1. The morphological examination is one of the basic tests performed almost at the beginning of any medical treatment. The coefficients (see Table 2) have specific levels of correctness. However, the listed coefficients may be in excess or deficiency. Suppose we take 0 as the absence of a coefficient abnormality level. Let the interval $(0, 1]$ denote the membership of the set we will call surplus, while the values in the interval $[-1, 0)$ mean not only that the data does not occur in excess, but we note its deficiency. Consider also that the normalization of data will not only proceed differently for each coefficient but also for data referred to as deficiency and for supernormal data. Therefore, for example, let us see the distribution of low HGB levels. With a value of not more than $10dL$, we talk about mild anemia, moderate anemia occurs when the hemoglobin level is between $8 - 9.9\,g/dL$, and severe anemia when the hemoglobin level is between $6.5 - 7.9\,g/dL$. The last stage is life-threatening anemia - hemoglobin levels below $6.5\,g/dL$. So, in this case, the normalization of the data (here: negative) is based on medical knowledge.

By further studying the relationship between the deficiency and excess of various coefficients, we can look for relationships between the various changes that occur in the human body. As a further consequence, this will enable us to plan an outline of diagnostic steps.

However, there are areas where - depending on the data we have - we can model the problem using both types of fuzzy relations: intuitionistic and balanced.

Table 2. Normal morphology in adults

Parameter	Standard for women	Standard for men
HGB	12.0 – 16.0 g/dL	13.5 – 18 g/dL
HCT	33 - 51%	37 - 53%
RBC	$4.0 - 5.2 \times 10^6/\mu l$	$4.5 - 5.9 \times 10^6/\mu l$
MCV	$80 - 100fl$	$80 - 100fl$
MCH	$26 - 34pg$	$26 - 34pg$
MCHC	32 – 36%	32 - 36%
RDW	11.5 – 13.1%	11.5 – 13.1%
WBC	$4.5 - 11.0 \times 10^3/\mu l$	$4.5 - 11.0 \times 10^3/\mu l$
EOS	0 – 3%	0 – 3%
BASO	0 – 1%	0 – 1%
LYMPH	24 – 44%	24 – 44%
MONO	4 – 10%	4 – 10%
PLT	$150 - 450 \times 10^3/\mu l$	$150 - 450 \times 10^3/\mu l$
MPV	$6.5 - 10.0fl$	$6.5 - 10.0fl$
PDW	$9.8 - 12.5fl$	$9.8 - 12.5fl$
PCT	0.2 – 0.4%	0.2 – 0.4%
P-LCR	19.1 – 46.6%	19.1 – 46.6%

Example 2. Let us consider the topic of examining a patient's mental state. In this case, we will focus on questions related to clinical depression. The characteristics, symptoms, or features of depression can range from very mild to severe. During the medical interview/questionnaire, the severity of specific symptoms can be determined. Some of the patient's symptoms are present, and some are not. For example, one symptom being, "Constant feelings of sadness" which is scalable from $[-1, 1]$ (i.e. $[-1, 0)$ - level of sadness, 0 - state considered average, $(0, 1]$ - state ranging from elevated contentment to euphoria).

However, such questions are used in clinical trials containing terms such as: "continuous", "persistent", or referring to a specific time interval. For example, "low energy levels", "persistent concentration problems", etc., often cause problems for respondents. It is then easier for them to define their state using the statement, "at what level true and at what level false". Thus, intuitionistic and balanced sets and relations also seem valuable mathematical apparatus in this case.

Other applications of fuzzy intuitionistic sets are also reported in the literature. Some of the first considerations appeared in [22]. The data collected takes different forms. It can describe our hesitation about belonging to a set. In a situation where we have a collection of data that contains certain degrees of belonging and not belonging to a given set, we can, according to Corollary 1, decompose it into two separate systems of equations. However, when we have a sequence of mixed positive and negative data, such an operation is unfeasible.

Example 3. This example illustrates that by splitting a balanced system into two sub-systems: with non-negative and non-positive coefficients. We do not obtain the exact solutions. Let us start by presenting the solution to the original system of equations (with $\bullet = MAX - MIN$ product):

$$A = \begin{bmatrix} 0.3 & -0.4 & 0.7 \\ -0.5 & -0.8 & 1 \\ 0.8 & 0.9 & 0.3 \end{bmatrix}, b = \begin{bmatrix} 0.7 \\ -0.8 \\ 0.5 \end{bmatrix}, x = \begin{bmatrix} x_1 \\ x_2 \\ x_3 \end{bmatrix},$$

where $x_1 = 0.5$, $x_2 \in [-1, -0.8]$, $x_3 \in [0.7, 0.8)$. Now solve (with $\bullet|_{[0,1]}$ and $\bullet|_{[-1,0]}$ products, respectively),

$$A^+ = \begin{bmatrix} 0.3 & 0 & 0.7 \\ 0 & 0 & 1 \\ 0.8 & 0.9 & 0.3 \end{bmatrix}, b^+ = \begin{bmatrix} 0.7 \\ 0 \\ 0.5 \end{bmatrix}, x^+ = \begin{bmatrix} x_1^+ \\ x_2^+ \\ x_3^+ \end{bmatrix},$$

$$A^- = \begin{bmatrix} 0 & -0.4 & 0 \\ -0.5 & -0.8 & 0 \\ 0 & 0 & 0 \end{bmatrix}, b^- = \begin{bmatrix} 0 \\ -0.8 \\ 0 \end{bmatrix}, x^- = \begin{bmatrix} x_1^- \\ x_2^- \\ x_3^- \end{bmatrix},$$

the above systems of equations have no solution.

5 Summary

This work considers applications of balanced and intuitionistic sets, relations, and equations. The similarities and differences of these theories are practically illustrated. Since fuzzy intuitionistic and balanced sets carry with them the information of belonging and/or not belonging to a set, they are often contrasted. This approach is wrong, and it should be emphasized that the effectiveness of many models depends on the correct choice of concepts. Regarding the type of a problem and the scope of data, two kinds of relations and relational equations can be used in a model - to achieve comparative effects and improve model quality.

References

1. Atanassov, K.T.: Intuitionistic fuzzy sets. Fuzzy Sets Syst. **20**(1), 87–96 (1986)
2. Atanassov, K. T.: Intuitionistic Fuzzy Sets: Theory and Applications. Springer, Heidelberg (1999). https://doi.org/10.1007/978-3-7908-1870-3
3. Burillo, P., Bustince, H.: Intuitionistic fuzzy relations (Part I). Mathware Soft Comput. **2**(1), 5–38 (1995)
4. Bustince, H., Burillo, P.: Structures on intuitionistic fuzzy relations. Fuzzy Sets Syst. **78**(3), 293–303 (1996)

5. Czogała, E., Drewniak, J., Pedrycz, W.: Fuzzy relation equations on a finite set. Fuzzy Sets Syst. **7**, 89–101 (1982)
6. Di Nola, A., Sessa, S., Pedrycz, W., Sanchez, E.: Fuzzy Relation Equations and their Applications to Knowledge Engineering. Kluwer Academic Publishing, Dordrecht (1989)
7. Drewniak, J.: Fuzzy relation equations and inequalities. Fuzzy Sets Syst. **14**, 237–247 (1984)
8. Drewniak, J.: Fuzzy Relation Calculus. Silesian University, Katowice (1989)
9. Drewniak, J., Matusiewicz, Z.: Properties of max-* fuzzy relation equations. Soft Computut. **14**(10), 1037–1041 (2010)
10. Higashi, M., Klir, G.J.: Resolution of finite fuzzy relation equations. Fuzzy Sets Syst. **13**, 65–82 (1984)
11. Homenda, W.: Balanced fuzzy sets. Inf. Sci. **176**, 2467–2506 (2006)
12. Klement, P., Mesiar, R., Pap, E.: Triangular Norms. Springer, Dordrecht (2000)
13. Matusiewicz, Z., Drewniak, J.: Increasing continuous operations in fuzzy max $-*$ equations and inequalities. Fuzzy Sets Syst. **232**, 120–133 (2013)
14. Matusiewicz, Z.: Systems of intuitionistic fuzzy equations, New developments in fuzzy sets, intuitionistic fuzzy sets, generalized nets and related topics. Volume I: Foundations: Instytut Badań Systemowych. Polska Akademia Nauk, Warszawa (2012)
15. Matusiewicz, Z., Homenda, W.: Balanced relations equations. In: 2022 IEEE International Conference on Fuzzy Systems (FUZZ-IEEE), pp. 1-8 (2022)
16. Miyakoshi, M., Shimbo, M.: Sets of solution-set equivalent coefficient matrices of fuzzy relation equations. Fuzzy Sets Syst. **35**, 357–387 (1990)
17. Pedrycz, W.: $S-T$ fuzzy relational equations. Fuzzy Sets Syst. **59**, 189–195 (1993)
18. Peeva, K., Kyosev, Y.: Fuzzy Relational Calculus: Theory, Applications and Software. Advances in Fuzzy Systems-Applications and Theory. World Scientific, Singapore (2004)
19. Sanchez, E.: Resolution of composite fuzzy relation equations. Inf. Control **30**, 38–48 (1976)
20. Shieh, B.-S.: Solutions of fuzzy relation equations based on continuous t-norms. Inf. Sci. **177**(19), 4208–4215 (2007)
21. Stamou, G.B., Tzafestas, S.G.: Resolution of composite fuzzy relation equations based on Archimedean triangular norms. Fuzzy Sets Syst. **120**(3), 395–407 (2001)
22. Szmidt, E., Kacprzyk, J.: Remarks on some applications of intuitionistic fuzzy sets in decision making. Notes IFS **2**, 22–31 (1996)
23. Zadeh, L.A.: Similarity relations and fuzzy orderings. Inf. Sci. **3**(2), 177–200 (1971)

On the Solutions of Some Equations with Intuitionistic Fuzzy Index Matrices

Stela Todorova(✉)

Prof. Dr. Assen Zlatarov University, 1 "Prof. Yakimov" Blvd., 8010 Burgas, Bulgaria
stodorovabtu@gmail.com

Abstract. The solutions of three equations with intuitionistic fuzzy index matrices are described. The conditions for their existing are given.

Keywords: Index matrix · Intuitionistic fuzzy index matrix

1 Introduction

The concept of the Index Matrix (IM) is developed sequentially in [3,4,6]. It is an object with the form

$$[K, L, a_{k_i,l_j}] = \begin{array}{c|cccc} & l_1 & l_2 & \dots & l_n \\ \hline k_1 & a_{11} & a_{12} & \dots & a_{1n} \\ k_2 & a_{21} & a_{22} & \dots & a_{2n} \\ \vdots & \vdots & \vdots & \dots & \vdots \\ k_m & a_{m1} & a_{m2} & \dots & a_{mn} \end{array}$$

where K $= \{k_1, k_2, ..., k_m\}$ and L $= \{l_1, l_2, ..., l_n\}$ amd for $1 \leq$ i $\leq$ m, $1 \leq$ j $\leq$ n: $a_{k_i,l_j} \in \mathcal{R}$, where $\mathcal{R}$ is some fixed set.

Over two IMs, in [3,4,6] a lot of operations and operators are defined.

Here, we will use the following three of them: $\oplus_{(\circ)}$, $\otimes_{(\circ)}$, $\ominus_{(\circ)}$, where $\circ : \mathcal{R} \times \mathcal{R} \to \mathcal{R}$.

Addition

$$A \oplus_{(\circ)} B = [K \cup P, L \cup Q, \{c_{t_u,v_w}\}],$$

where

$$c_{t_u,v_w} = \begin{cases} a_{k_i,l_j}, & \text{if } t_u = k_i \in K \text{ and } v_w = l_j \in L - Q \\ & \text{or } t_u = k_i \in K - P \text{ and } v_w = l_j \in L; \\ b_{p_r,q_s}, & \text{if } t_u = p_r \in P \text{ and } v_w = q_s \in Q - L \\ & \text{or } t_u = p_r \in P - K \text{ and } v_w = q_s \in Q; \\ a_{k_i,l_j} \circ b_{p_r,q_s}, & \text{if } t_u = k_i = p_r \in K \cap P \\ & \text{and } v_w = l_j = q_s \in L \cap Q \\ e_\circ, & \text{otherwise} \end{cases},$$

K. T. Atanassov et al. (Eds.): BOS/SOR/IWIFSGN 2022, LNNS 793, pp. 32–38, 2023.
https://doi.org/10.1007/978-3-031-45069-3_4

where $e_\circ$ is the unit element of $\mathcal{R}$ related to operation $\circ$. For example, if $\mathcal{R}$ is the set of real numbers and $\circ \in \{+, -\}$, then $e_\circ$ is "0"; while if $\circ \in \{\times, :\}$, then $e_\circ$ is "1".

Termwise Multiplication

$$A \otimes_{(\circ)} B = [K \cap P, L \cap Q, \{c_{t_u, v_w}\}],$$

where

$$c_{t_u, v_w} = a_{k_i, l_j} \circ b_{p_r, q_s},$$

for $t_u = k_i = p_r \in K \cap P$ and $v_w = l_j = q_s \in L \cap Q$.

Structural Subtraction

$$A \ominus B = [K - P, L - Q, \{c_{t_u, v_w}\}],$$

where "–" is the set–theoretic difference operation and

$$c_{t_u, v_w} = a_{k_i, l_j}, \text{ for } t_u = k_i \in K - P \text{ and } v_w = l_j \in L - Q.$$

Let us assume that everywhere the IMs $A = [K, L, \{a_{k_i, l_j}\}]$ and $B = [P, Q, \{b_{p_r, q_s}\}]$ be given. We will determine the condition for solving the equations

$$A \oplus_{(\circ)} X = B, \tag{1}$$

$$A \otimes_{(\circ)} X = B, \tag{2}$$

$$A \ominus X = B, \tag{3}$$

and will construct the IMs X with minimal and maximal number of rows and columns for the case, when the elements of the matrices are Intuitionistic Fuzzy Pairs (IFPs, see [7,8]). Each IFP has the form $\langle a, b\rangle$, where $a, b \in [0, 1]$ and $a + b \le 1$. For two IFPs, different relations and operations are described in [7], but here we will use only the following:

$$\langle a, b\rangle \le \langle c, d\rangle \text{ iff } a \le c \text{ and } b \ge d,$$

$$\langle a, b\rangle = \langle c, d\rangle \text{ iff } a = c \text{ and } b = d,$$

$$\langle a, b\rangle \wedge \langle c, d\rangle = \langle \min(a, c), \max(b, d)\rangle,$$

$$\langle a, b\rangle \vee \langle c, d\rangle = \langle \max(a, c), \min(b, d)\rangle.$$

When $\mathcal{R}$ is a set of IFPs, the IM is called *an Intuitionistic Fuzzy IM* (IFIM, see [5,6]). In the present paper, we will discuss the solutions of (1)–(3) when A and B are IFIMs.

2 The Equation with Operation $\oplus_\circ$

Let us suppose that

$$X = [Y, Z, \{y_{u_v,w_t}, z_{u_v,w_t}\}].$$

Let

$$a_{k_i,l_j} = \langle \alpha_{k_i,l_j}, \beta_{k_i,l_j} \rangle,$$

$$b_{p_r,q_s} = \langle \gamma_{p_r,q_s}, \delta_{p_r,q_s} \rangle.$$

Let everywhere below

$$e_\vee = \langle 0, 1 \rangle$$

and

$$e_\wedge = \langle 1, 0 \rangle.$$

When we must solve Eq. (1), we see that if $K - P \neq \emptyset$, i.e., when there is $k \in K - P$, then $k \in (K \cup Y) - P$ that is impossible. Therefore, one of the conditions for existing of solutions of (1) is $K \subseteq P$. By analogy, a second condition will be: $L \subseteq Q$.

2.1 The Equation with Operation $\oplus_\wedge$

When operation $\circ$ is operation $\wedge$, then the third condition is the following

$$(\forall k_i = p_r \in K)(\forall l_j = q_s \in L)(\langle \alpha_{k_i,l_j}, \beta_{k_i,l_j} \rangle) \geq \langle \gamma_{p_r,q_s}, \delta_{p_r,q_s} \rangle. \tag{4}$$

Let us assume below that these three conditions are valid. Now, we will search the minimal and maximal solutions of (1), for the case when operation $\circ$ is operation $\wedge$.

For the maximal solution we have

$$X = [P, Q, \{\langle y_{p_r,q_s}, z_{p_r,q_s} \rangle\}],$$

where

$$(\forall p_r \in P)(\forall q_s \in Q)(\langle y_{p_r,q_s}, z_{p_r,q_s} \rangle = \langle \gamma_{p_r,q_s}, \delta_{p_r,q_s} \rangle. \tag{5}$$

The check of the validity of this solution is the following:

$$A \oplus_\wedge X = [K \cup Y, L \cup Z, \{\langle \varepsilon_{\eta,\theta}, \zeta_{\eta,\theta} \rangle\}]$$

(from $K \cup Y = K \cup P = P$ and $L \cup Z = L \cup Q = Q$)

$$A \oplus_\wedge X = [P, Q, \{\langle \varepsilon_{\eta,\theta}, \zeta_{\eta,\theta} \rangle\}],$$

where from (5):

$$\langle \varepsilon_{\eta,\theta}, \zeta_{\eta,\theta} \rangle = \begin{cases} \langle \alpha_{k_i,l_j}, \beta_{k_i,l_j} \rangle, & \text{if } \ u_v = k_i \in K \\ & \text{and } w_t = l_j \in L - Q \\ & \text{or } \eta_\theta = k_i \in K - Y \\ & \text{and } \eta_\theta = l_j \in L \\ \langle \gamma_{p_r,q_s}, \delta_{p_r,q_s} \rangle, & \text{if } \eta_\theta = p_r \in K \\ & \text{and } \eta_\theta = q_s \in Z - L \\ & \text{or } \eta_\theta = p_r \in Y - K \\ & \text{and } \eta_\theta = q_s \in Z \\ \langle \alpha_{k_i,l_j}, \beta_{k_i,l_j} \rangle & \text{if } \eta_\theta = k_i = p_r \in K \cap Y \\ \quad \wedge \langle \gamma_{p_r,q_s}, \delta_{p_r,q_s} \rangle, & \text{and } \eta_\theta = l_j = q_s \in L \cap Z \\ \langle 1, 0 \rangle, & \text{otherwise} \end{cases},$$

(because $K \subseteq Y$ and $L \subseteq Z$)

$$\langle \varepsilon_{\eta,\theta}, \zeta_{\eta,\theta} \rangle = \begin{cases} \langle \gamma_{p_r,q_s}, \delta_{p_r,q_s} \rangle, & \text{if } \eta_\theta = p_r \in K \\ & \text{and } \eta_\theta = q_s \in Z - L \\ & \text{or } \eta_\theta = p_r \in Y - K \\ & \text{and } \eta_\theta = q_s \in Z \\ \langle \alpha_{k_i,l_j}, \beta_{k_i,l_j} \rangle & \text{if } \eta_\theta = k_i = p_r \in K \cap Y \\ \quad \wedge \langle \gamma_{p_r,q_s}, \delta_{p_r,q_s} \rangle, & \text{and } \eta_\theta = l_j = q_s \in L \cap Z \end{cases},$$

(from (4))

$$\langle \varepsilon_{\eta,\theta}, \zeta_{\eta,\theta} \rangle = \langle \gamma_{p_r,q_s}, \delta_{p_r,q_s} \rangle.$$

Therefore,

$$A \oplus_\circ X = [P, Q, \{\langle \gamma_{p_r,q_s}, \delta_{p_r,q_s} \rangle\}] = B.$$

For the minimal solution of (1) there are three cases.

Case 1: if $L = Q$ and if

$$(\exists C : \emptyset \neq C \subseteq K)(\forall k_i \in C)(\forall l_j \in L)(\langle \alpha_{k_i,l_j}, \beta_{k_i,l_j} \rangle = \langle \gamma_{p_r,q_s}, \delta_{p_r,q_s} \rangle).$$

Then

$$X = [P - C, Q, \{\langle y_{p_r,q_s}, z_{p_r,q_s} \rangle\}],$$

where as above

$$(\forall p_r \in P)(\forall q_s \in Q)(\langle y_{p_r,q_s}, z_{p_r,q_s} \rangle = \langle \gamma_{p_r,q_s}, \delta_{p_r,q_s} \rangle.$$

Case 2: if $K = P$ and if

$$(\exists D : \emptyset \neq D \subseteq L)(\forall l_j \in D)(\forall k_i \in K)(\langle \alpha_{k_i,l_j}, \beta_{k_i,l_j} \rangle = \langle \gamma_{p_r,q_s}, \delta_{p_r,q_s} \rangle).$$

Then

$$X = [P, Q - D, \{\langle \gamma_{p_r,q_s}, \delta_{p_r,q_s} \rangle\}].$$

Case 3: if $K = P$ and $L = Q$ and if

$$(\exists C : \emptyset \neq C \subseteq K)(\forall k_i \in C)(\forall l_j \in L)(\langle \alpha_{k_i,l_j}, \beta_{k_i,l_j} \rangle = \langle \gamma_{p_r,q_s}, \delta_{p_r,q_s} \rangle)$$

$$\& \ (\exists D : \emptyset \neq D \subseteq L)(\forall l_j \in D)(\forall k_i \in K)(\langle \alpha_{k_i,l_j}, \beta_{k_i,l_j} \rangle = \langle \gamma_{p_r,q_s}, \delta_{p_r,q_s} \rangle).$$

Then

$$X = [P - C, Q - D, \{\langle \gamma_{p_r,q_s}, \delta_{p_r,q_s} \rangle\}].$$

We will check of the validity of the third case, because the checks of the first two are similar.

$$A \oplus_\wedge X = [P - C, Q - D, \{\langle \varepsilon_{\eta,\theta}, \zeta_{\eta,\theta} \rangle\}],$$

where from (5):

$$\langle \varepsilon_{\eta,\theta}, \zeta_{\eta,\theta} \rangle = \begin{cases} \langle \alpha_{k_i,l_j}, \beta_{k_i,l_j} \rangle, & \text{if } u_v = k_i \in K \\ & \text{and } w_t = l_j \in L - (Q - D) \\ & \text{or } \eta_\theta = k_i \in K - (P - C) \\ & \text{and } \eta_\theta = l_j \in L \\ \langle \gamma_{p_r,q_s}, \delta_{p_r,q_s} \rangle, & \text{if } \eta_\theta = p_r \in P - C \\ & \text{and } \eta_\theta = q_s \in (Q - D) - L \\ & \text{or } \eta_\theta = p_r \in (P - C) - K \\ & \text{and } \eta_\theta = q_s \in Q - D \\ \langle \alpha_{k_i,l_j}, \beta_{k_i,l_j} \rangle & \text{if } \eta_\theta = k_i = p_r \in K \cap (P - C) \\ \quad \wedge \langle \gamma_{p_r,q_s}, \delta_{p_r,q_s} \rangle, & \text{and } \eta_\theta = l_j = q_s \in L \cap (Q - D) \\ e_\circ, & \text{otherwise} \end{cases},$$

(because $K - (P - C) = \emptyset$ and $L - (Q - D) = \emptyset$)

$$\langle \varepsilon_{\eta,\theta}, \zeta_{\eta,\theta} \rangle = \begin{cases} \langle \gamma_{p_r,q_s}, \delta_{p_r,q_s} \rangle, & \text{if } \eta_\theta = p_r \in P - C \\ & \text{and } \eta_\theta = q_s \in (Q - D) - L \\ & \text{or } \eta_\theta = p_r \in (P - C) - K \\ & \text{and } \eta_\theta = q_s \in Q - D \\ \langle \alpha_{k_i,l_j}, \beta_{k_i,l_j} \rangle & \text{if } \eta_\theta = k_i = p_r \in K \cap (P - C) \\ \quad \wedge \langle \gamma_{p_r,q_s}, \delta_{p_r,q_s} \rangle, & \text{and } \eta_\theta = l_j = q_s \in L \cap (Q - D) \end{cases},$$

(from (4))

$$\langle \varepsilon_{\eta,\theta}, \zeta_{\eta,\theta} \rangle = \langle \gamma_{p_r,q_s}, \delta_{p_r,q_s} \rangle.$$

Therefore,

$$A \oplus_\circ X = [P - C, Q - D, \{\langle \gamma_{p_r,q_s}, \delta_{p_r,q_s} \rangle\}] = B.$$

2.2 The Equation with Operation $\oplus_\vee$

When operation $\circ$ is operation $\vee$, then the third condition is the following

$$(\forall k_i = p_r \in K)(\forall l_j = q_s \in L)(\langle \alpha_{k_i,l_j}, \beta_{k_i,l_j} \rangle) \leq \langle \gamma_{p_r,q_s}, \delta_{p_r,q_s} \rangle. \tag{6}$$

If we assume that these three conditions are valid, then we will see that the maximal solution have the same form, as above:

$$X = [P, Q, \{\langle y_{p_r,q_s}, z_{p_r,q_s} \rangle\}],$$

where

$$(\forall p_r \in P)(\forall q_s \in Q)(\langle y_{p_r,q_s}, z_{p_r,q_s}\rangle = \langle \gamma_{p_r,q_s}, \delta_{p_r,q_s}\rangle.$$

The same is valid for the minimal solutions of (1), as well as the checks.

3 The Equation with Operation $\otimes_\circ$

When we must solve Eq. (2), we see that if $P - K \neq \emptyset$, i.e., when there is $k \in P-K$, then $k \notin (K \cap Y) = P$ that is impossible. Therefore, one of the conditions for existing of solutions of (2) is $P \subseteq K$. By analogy, a second condition will be: $Q \subseteq L$.

3.1 The Equation with Operation $\otimes_\wedge$

When operation $\circ$ is operation $\wedge$, then the third condition is: the inequalities from (4) to be valid. Similarly to the above, we see that the solution is

$$X = [P, Q, \{\langle y_{p_r,q_s}, z_{p_r,q_s}\rangle\}]$$

and its elements satisfy (5).

It is important that in this case is that IFIM X is the unique (minimal) solution and for every two sets C and D, so that $C \cap K = \emptyset$, $D \cap L = \emptyset$, the IFIM

$$X = [P \cup C, Q \cup D, \{\langle y_{p_r,q_s}, z_{p_r,q_s}\rangle\}]$$

is a solution of (2) when its elements with indices from sets p and Q satisfy (5) and the other indices are arbitrary IFPs.

3.2 The Equation with Operation $\otimes_\vee$

By analogy with Sects. 2.2 and 3.1, in the present case the (minimal) solution is

$$X = [P, Q, \{\langle y_{p_r,q_s}, z_{p_r,q_s}\rangle\}]$$

when its elements satisfy (6).

4 The Equation with Operation $\ominus$

In this case, we see again that IFIMs A and B must satisfy the conditions $P \subseteq K, Q \subseteq L$. Now, the (minimal) solution is

$$X = [K - P, L - Q, \{\langle y_{p_r,q_s}, z_{p_r,q_s}\rangle\}],$$

where the elements of IFIM X are arbitrary IFPs.

5 Conclusion

In the present research, we studied the solutions of equations (1)–(3), when operation "$\circ$" is "$\wedge$" or "$\vee$". In future, the research can have some directions.

First, it can be checked these equations with different forms of operation "$\circ$", e.g., using the operations from [1,2].

Second, these equations can use operation $\odot_{(\circ,*)}$.

Third, $\mathcal{R}$ can be the set of real numbers and in this case, the solutions will have essentially different forms.

References

1. Angelova, N., Stoenchev, M.: Intuitionistic fuzzy conjunctions and disjunctions from first type. Ann. "Inform." Section, Union Scientists Bulgaria, **8**, 1–17 (2015)
2. Angelova, N., Stoenchev, M.: Intuitionistic fuzzy conjunctions and disjunctions from third type. Notes Intuitionistic Fuzzy Sets **23**(5), 29–41 (2017)
3. Atanassov, K.: Generalized index matrices. Comptes rendus de l'Academie Bulgare des Sci. **40**(11), 15–18 (1987)
4. Atanassov, K.: On index matrices, part 1: standard cases. Adv. Stud. Contemp. Math. **20**(2), 291–302 (2010)
5. Atanassov, K.: On index matrices, part 2: intuitionistic fuzzy case. Proc. Jangjeon Math. Soc. **13**(2), 121–126 (2010)
6. Atanassov, K.: Index Matrices: Towards an Augmented Matrix Calculus. Springer, Cham (2014)
7. Atanassov, K.: Intuitionistic Fuzzy Logics. Springer, Cham (2017)
8. Atanassov, K., Szmidt, E., Kacprzyk, J.: On intuitionistic fuzzy pairs. Notes Intuitionistic Fuzzy Sets **19**(3), 1–13 (2013)

Convergence of Functions of Several Intuitionistic Fuzzy Observables

Katarína Čunderlíková(✉)

Mathematical Institute, Slovak Academy of Sciences, Štefánikova 49, 814 73 Bratislava, Slovakia
cunderlikova.lendelova@gmail.com

Abstract. The aim of this paper is to show the connection between the random variables in classical probability space and the intuitionistic fuzzy observables. We study a convergence in distribution, a convergence in measure and an almost everywhere convergence of function of several intuitionistic fuzzy observables induced by a Borel measurable function.

Keywords: Kolmogorov probability space · Intuitionistic fuzzy observable · Intuitionistic fuzzy state · Borel function · Function of several intuitionistic fuzzy observables · Convergence in distribution · Convergence in measure · Almost everywhere convergence

1 Introduction

In the Kolmogorov theory a random variable is a measurable function $\xi : \Omega \to R$ defined on a space $(\Omega, \mathcal{S})$, i.e. such a function that the preimage $\xi^{-1}(I)$ of any interval $I \subset R$ belongs to the given σ-algebra $\mathcal{S}$ of subsets of Ω. It induces a mapping $I \to \xi^{-1}(I) \in \mathcal{S}$ from the family $\mathcal{J}$ of all intervals to a σ-algebra $\mathcal{S}$. On the other hand in the intuitionistic fuzzy probability theory instead of the random variable $\xi : \Omega \to R$ an intuitionistic fuzzy observable $x : \mathcal{B}(R) \to \mathcal{F}$ is considered.

In paper [7] we presented the Kolmogorov construction and we showed the way how we can construct a sequence $(\xi_n)_1^\infty$ of a random variables to each sequence of intuitionistic fuzzy observables $(x_n)_1^\infty$. In this paper we work with the functions of several intuitionistic fuzzy observables $g_n(x_1, \ldots, x_n) : \mathcal{B}(R) \to \mathcal{F}$ given by the formula

$$g_n(x_1, \ldots, x_n)(A) = h_n\big(g_n^{-1}(A)\big)$$

for each $A \in \mathcal{B}(R)$, where h_n is a joint intuitionistic fuzzy observable of intuitionistic fuzzy observables $x_1, \ldots, x_n : \mathcal{B}(R) \to \mathcal{F}$ and $g_n : R^n \to R$ is a Borel measurable function.

Our main idea is a construction a sequence $(\eta_n)_1^\infty$ of a random variables to each sequence of intuitionistic fuzzy observables $\big(g_n(x_1, \ldots, x_n)\big)_1^\infty$. We study a convergence in distribution, a convergence in measure and an almost everywhere

K. T. Atanassov et al. (Eds.): BOS/SOR/IWIFSGN 2022, LNNS 793, pp. 39–48, 2023.
https://doi.org/10.1007/978-3-031-45069-3_5

convergence of function of several intuitionistic fuzzy observables induced by a Borel measurable function in this paper. We show a connection with the different types of convergences in classical probability space, too.

Remark that in a whole text we use a notation "IF" in short as the term "intuitionistic fuzzy".

2 IF-Events, IF-States and IF-Observables

Recall that the notion of intuitionistic fuzzy sets was introduced by K. T. Atanassov in 1983 as a generalization of the fuzzy sets introduced by L. Zadeh (see [1–3,14,15]). In this section we explane the basic notions and defintions.

Definition 1. *Let Ω be a nonempty set. An IF-set* $\mathbf{A}$ *on Ω is a pair (μ_A, ν_A) of mappings $\mu_A, \nu_A : \Omega \to [0,1]$ such that $\mu_A + \nu_A \leq 1_\Omega$.*

Definition 2. *Start with a measurable space $(\Omega, \mathcal{S})$. Hence $\mathcal{S}$ is a σ-algebra of subsets of Ω. An IF-event is called an IF-set $\mathbf{A} = (\mu_A, \nu_A)$ such that $\mu_A, \nu_A : \Omega \to [0,1]$ are $\mathcal{S}$-measurable.*

The family of all IF-events on $(\Omega, \mathcal{S})$ will be denoted by $\mathcal{F}$, $\mu_A : \Omega \longrightarrow [0,1]$ will be called the membership function, $\nu_A : \Omega \longrightarrow [0,1]$ will be called the non-membership function.

If $\mathbf{A} = (\mu_A, \nu_A) \in \mathcal{F}$, $\mathbf{B} = (\mu_B, \nu_B) \in \mathcal{F}$, then we define the Łukasiewicz binary operations $\oplus, \odot$ on $\mathcal{F}$ by:

$$\mathbf{A} \oplus \mathbf{B} = ((\mu_A + \mu_B) \wedge 1_\Omega, (\nu_A + \nu_B - 1_\Omega) \vee 0_\Omega)),$$
$$\mathbf{A} \odot \mathbf{B} = ((\mu_A + \mu_B - 1_\Omega) \vee 0_\Omega, (\nu_A + \nu_B) \wedge 1_\Omega))$$

and the partial ordering is then given by

$$\mathbf{A} \leq \mathbf{B} \Longleftrightarrow \mu_A \leq \mu_B, \nu_A \geq \nu_B.$$

In paper we use max-min connectives defined by:

$$\mathbf{A} \vee \mathbf{B} = (\mu_A \vee \mu_B, \nu_A \wedge \nu_B),$$
$$\mathbf{A} \wedge \mathbf{B} = (\mu_A \wedge \mu_B, \nu_A \vee \nu_B)$$

and the De Morgan rules: $(a \vee b)^* = a^* \wedge b^*$, $(a \wedge b)^* = a^* \vee b^*$, where $a^* = 1 - a$.

In the IF-probability theory ([11,13]) instead of the notion of probability we use the notion of state.

Definition 3. *Let $\mathcal{F}$ be the family of all IF-events in Ω. A mapping $\mathbf{m} : \mathcal{F} \to [0,1]$ is called an IF-state, if the following conditions are satisfied:*

(i) $\mathbf{m}((1_\Omega, 0_\Omega)) = 1, \mathbf{m}((0_\Omega, 1_\Omega)) = 0$;
(ii) *if* $\mathbf{A} \odot \mathbf{B} = (0_\Omega, 1_\Omega)$ *and* $\mathbf{A}, \mathbf{B} \in \mathcal{F}$, *then* $\mathbf{m}(\mathbf{A} \oplus \mathbf{B}) = \mathbf{m}(\mathbf{A}) + \mathbf{m}(\mathbf{B})$;
(iii) *if* $\mathbf{A}_n \nearrow \mathbf{A}$ *(i.e.* $\mu_{A_n} \nearrow \mu_A$, $\nu_{A_n} \searrow \nu_A$*), then* $\mathbf{m}(\mathbf{A}_n) \nearrow \mathbf{m}(\mathbf{A})$.

Probably the most useful result in the IF-state theory is the following representation theorem ([11]):

Theorem 1. *To each IF-state* $\mathbf{m} : \mathcal{F} \to [0,1]$ *there exists exactly one probability measure* $P : \mathcal{S} \to [0,1]$ *and exactly one* $\alpha \in [0,1]$ *such that*

$$\mathbf{m}(\mathbf{A}) = (1-\alpha)\int_{\Omega} \mu_A dP + \alpha\Big(1 - \int_{\Omega} \nu_A dP\Big)$$

for each $\mathbf{A} = (\mu_A, \nu_A) \in \mathcal{F}$.

The third basic notion in the probability theory is the notion of an observable. Let $\mathcal{J}$ be the family of all intervals in R of the form

$$[a,b) = \{x \in R : a \leq x < b\}.$$

Then the σ-algebra $\sigma(\mathcal{J})$ is denoted $\mathcal{B}(R)$ and it is called the σ-algebra of Borel sets, its elements are called Borel sets.

Definition 4. *By an IF-observable on* $\mathcal{F}$ *we understand each mapping* $x : \mathcal{B}(R) \to \mathcal{F}$ *satisfying the following conditions:*

(i) $x(R) = (1_\Omega, 0_\Omega)$, $x(\emptyset) = (0_\Omega, 1_\Omega)$;
(ii) if $A \cap B = \emptyset$, *then* $x(A) \odot x(B) = (0_\Omega, 1_\Omega)$ *and* $x(A \cup B) = x(A) \oplus x(B)$;
(iii) if $A_n \nearrow A$, *then* $x(A_n) \nearrow x(A)$.

If we denote $x(A) = \big(x^\flat(A), 1_\Omega - x^\sharp(A)\big)$ *for each* $A \in \mathcal{B}(R)$, *then* $x^\flat, x^\sharp : \mathcal{B}(R) \to \mathcal{T}$ *are observables, where* $\mathcal{T} = \{f : \Omega \to [0,1]; f \text{ is } \mathcal{S} - \text{measurable}\}$.

Remark 1. Sometimes we need to work with n-dimensional IF-observable $x : \mathcal{B}(R^n) \to \mathcal{F}$ defined as a mapping with the following conditions:

(i) $x(R^n) = (1_\Omega, 0_\Omega)$, $x(\emptyset) = (0_\Omega, 1_\Omega)$;
(ii) if $A \cap B = \emptyset$, $A, B \in \mathcal{B}(R^n)$, then $x(A) \odot x(B) = (0_\Omega, 1_\Omega)$ and $x(A \cup B) = x(A) \oplus x(B)$;
(iii) if $A_n \nearrow A$, then $x(A_n) \nearrow x(A)$ for each $A, A_n \in \mathcal{B}(R^n)$.

If $n = 1$ we simply say that x is an IF-observable.

If $x : \mathcal{B}(R) \longrightarrow \mathcal{F}$ is an IF-observable, and $\mathbf{m} : \mathcal{F} \longrightarrow [0,1]$ is an IF-state, then the **IF-distribution function** of x is the function $\mathbf{F} : R \longrightarrow [0,1]$ defined by the formula

$$\mathbf{F}(t) = \mathbf{m}\big(x((-\infty, t))\big)$$

for each $t \in R$.

Similarly as in the classical case the following two theorems can be proved ([13]).

Theorem 2. *Let* $\mathbf{F} : R \longrightarrow [0,1]$ *be the IF-distribution function of an IF-observable* $x : \mathcal{B}(R) \longrightarrow \mathcal{F}$. *Then* $\mathbf{F}$ *is non-decreasing on* R, *left continuous in each point* $t \in R$ *and*

$$\lim_{n\to-\infty} \mathbf{F}(t) = 0, \ \lim_{n\to\infty} \mathbf{F}(t) = 1.$$

Theorem 3. *Let* $x : \mathcal{B}(R) \longrightarrow \mathcal{F}$ *be an IF-observable,* $\mathbf{m} : \mathcal{F} \longrightarrow [0,1]$ *be an IF-state. Define the mapping* $\mathbf{m}_x : \mathcal{B}(R) \longrightarrow [0,1]$ *by the formula*

$$\mathbf{m}_x(C) = \mathbf{m}(x(C)).$$

Then $\mathbf{m}_x : \mathcal{B}(R) \longrightarrow [0,1]$ *is a probability measure.*

3 Product Operation, Joint IF-Observable and Independence

In [10] we introduced the notion of product operation on the family of IF-events $\mathcal{F}$ and showed an example of this operation.

Definition 5. *We say that a binary operation* $\cdot$ *on* $\mathcal{F}$ *is product if it satisfying the following conditions:*

(i) $(1_\Omega, 0_\Omega) \cdot (a_1, a_2) = (a_1, a_2)$ *for each* $(a_1, a_2) \in \mathcal{F}$;
(ii) the operation $\cdot$ *is commutative and associative;*
(iii) if $(a_1, a_2) \odot (b_1, b_2) = (0_\Omega, 1_\Omega)$ *and* $(a_1, a_2), (b_1, b_2) \in \mathcal{F}$, *then*

$$(c_1, c_2) \cdot ((a_1, a_2) \oplus (b_1, b_2)) = ((c_1, c_2) \cdot (a_1, a_2)) \oplus ((c_1, c_2) \cdot (b_1, b_2))$$

and

$$((c_1, c_2) \cdot (a_1, a_2)) \odot ((c_1, c_2) \cdot (b_1, b_2)) = (0_\Omega, 1_\Omega)$$

for each $(c_1, c_2) \in \mathcal{F}$;
(iv) if $(a_{1n}, a_{2n}) \searrow (0_\Omega, 1_\Omega)$, $(b_{1n}, b_{2n}) \searrow (0_\Omega, 1_\Omega)$ *and* $(a_{1n}, a_{2n}), (b_{1n}, b_{2n}) \in \mathcal{F}$, *then* $(a_{1n}, a_{2n}) \cdot (b_{1n}, b_{2n}) \searrow (0_\Omega, 1_\Omega)$.

The following theorem provides an example of product operation for IF-events (see [10, Theorem 1]).

Theorem 4. *The operation* $\cdot$ *defined by*

$$(x_1, y_1) \cdot (x_2, y_2) = (x_1 \cdot x_2, y_1 + y_2 - y_1 \cdot y_2)$$

for each $(x_1, y_1), (x_2, y_2) \in \mathcal{F}$ *is product operation on* $\mathcal{F}$.

In [12] B. Riečan defined the notion of a joint IF-observable and proved its existence (see [12, Theorem 3.3]).

Definition 6. *Let* $x, y : \mathcal{B}(R) \to \mathcal{F}$ *be two IF-observables. The joint IF-observable of the IF-observables* x, y *is a mapping* $h : \mathcal{B}(R^2) \to \mathcal{F}$ *satisfying the following conditions:*

(i) $h(R^2) = (1_\Omega, 0_\Omega)$, $h(\emptyset) = (0_\Omega, 1_\Omega)$;
(ii) if $A, B \in \mathcal{B}(R^2)$ *and* $A \cap B = \emptyset$, *then* $h(A \cup B) = h(A) \oplus h(B)$ *and* $h(A) \odot h(B) = (0_\Omega, 1_\Omega)$;
(iii) if $A, A_n \in \mathcal{B}(R^2)$ *and* $A_n \nearrow A$, *then* $h(A_n) \nearrow h(A)$;
(iv) $h(C \times D) = x(C) \cdot y(D)$ *for each* $C, D \in \mathcal{B}(R)$.

Theorem 5. *For each two IF-observables* $x, y : \mathcal{B}(R) \to \mathcal{F}$ *there exists their joint IF-observable.*

Remark 2. The joint IF-observable of IF-observables x, y from Definition 6 is a twodimensional IF-observable.

4 Kolmogorov Construction

In this section we explain a notation and a construction of Kolmogorov probability space introduced in [7]. Consider the space R^N and the family of cylinders with finitely dimensional basis B. Such **a cylinder** has the form

$$A = \{(t_n)_1^\infty \in R^N : (t_1, \dots, t_n) \in B\}.$$

If we introduce the notion of **a projection (n-th coordinate random vector)** $\pi_J : R^N \to R^n$,

$$\pi_J\big((t_n)_1^\infty\big) = (t_1, \dots, t_n),$$

then the cylinder A can be expressed in the form

$$A = \pi_J^{-1}(B).$$

The volume of the cylinder A is product of the content $P_J(B)$ of its basis B and its height 1, hence

$$P(A) = P_J(B).$$

The following two Propositions say about the connections between Kolmogorov probability space with probability measure P and IF-space with IF-state $\mathbf{m}$, see [7, Proposition 4.7, Proposition 4.8].

Proposition 1. *Let $\mathcal{C}$ be the family of all cylinders in R^N, i.e.*

$$\mathcal{C} = \{\pi_J^{-1}(A) \mid \emptyset \neq J \subset N, J \ finite, A \in \mathcal{B}(R^{|J|})\}.$$

Then there exists exactly one probability measure $P : \sigma(\mathcal{C}) \to [0,1]$ such that

$$P(\pi_J^{-1}(A)) = P_J(A)$$

for each cylinders $\pi_J^{-1}(A)$. Particularly

$$P(\{(t_n)_1^\infty : t_i \in A_i, i = 1, 2, \dots, n\}) = \mathbf{m}(h_n(A_1 \times \ldots \times A_n)) = \mathbf{m}(x_1(A_1) \cdot \ldots \cdot x_n(A_n)).$$

Remark 3. >From Proposition 1 we have that

$$P \circ \pi_J^{-1} = \mathbf{m} \circ h_n,$$

where h_n is a joint IF-observable of IF-observables $x_1, \dots, x_n$.

Proposition 2. *Define the coordinate function $\xi_n : R^N \to R$ by the formula*

$$\xi_n((t_i)_1^\infty) = t_n.$$

Then ξ_n is a random variable with respect to $\sigma(\mathcal{C})$ such that

$$P_{\xi_n} = \mathbf{m} \circ x_n = \mathbf{m}_{x_n}.$$

Remark 4. By the preceding procedure to each sequence $(x_n)_1^\infty$ we can construct a sequence $(\xi_n)_1^\infty$ of a random variables.

5 Function of Several IF-Observables

In this paper we study a convergence of functions of several IF-observables. If we have several IF-observables and a Borel measurable function, we can define the IF-observable, which is a function of several IF-observables. In this regard, we provide the following definition, see [6].

Definition 7. *Let $x_1, \ldots, x_n : \mathcal{B}(R) \to \mathcal{F}$ be IF-observables, h_n their joint IF-observable and $g_n : R^n \to R$ a Borel measurable function. Then we define the IF-observable $g_n(x_1, \ldots, x_n) : \mathcal{B}(R) \to \mathcal{F}$ by the formula*

$$g_n(x_1, \ldots, x_n)(A) = h_n\big(g_n^{-1}(A)\big).$$

for each $A \in \mathcal{B}(R)$.

Example 1. Let $x_1, \ldots, x_n : \mathcal{B}(R) \to \mathcal{F}$ be the IF-observables and $h_n : \mathcal{B}(R^n) \to \mathcal{F}$ be their joint IF-observable. Then

1. the IF-observable $y_n = g_n(x_1, \ldots, x_n) = \frac{\sqrt{n}}{\sigma}\Big(\frac{1}{n}\sum_{i=1}^{n} x_i - a\Big)$ is defined by the equality $y_n = h_n \circ g_n^{-1}$, where $g_n(t_1, \ldots, t_n) = \frac{\sqrt{n}}{\sigma}\Big(\frac{1}{n}\sum_{i=1}^{n} t_i - a\Big)$;
2. the IF-observable $y_n = g_n(x_1, \ldots, x_n) = \frac{1}{n}\sum_{i=1}^{n} x_i$ is defined by the equality $y_n = h_n \circ g_n^{-1}$, where $g_n(t_1, \ldots, t_n) = \frac{1}{n}\sum_{i=1}^{n} t_i$;
3. the IF-observable $y_n = g_n(x_1, \ldots, x_n) = \frac{1}{n}\sum_{i=1}^{n} (x_i - \mathbf{E}(x_i))$ is defined by the equality $y_n = h_n \circ g_n^{-1}$, where $g_n(t_1, \ldots, t_n) = \frac{1}{n}\sum_{i=1}^{n} (t_i - \mathbf{E}(x_i))$;
4. the IF-observable $y_n = g_n(x_1, \ldots, x_n) = \frac{1}{a_n}\big(\max(x_1, \ldots, x_n) - b_n\big)$ is defined by the equality $y_n = h_n \circ g_n^{-1}$, where $g_n(t_1, \ldots, t_n) = \frac{1}{a_n}\big(\max(t_1, \ldots, t_n) - b_n\big)$,

for all real numbers $t_1, \ldots, t_n$.

6 Convergence of Functions of Several IF-Observables

In this section we study the convergence in measure, the convergence in distribution and the almost everywhere convergence of functions of several IF-observables. First we recall the definitions of these types of convergence for random variables.

Definition 8. *Let $(\eta_n)_1^\infty$ be a sequence of random variables in a probability space $(\Omega, \mathcal{S}, P)$. We say that*

(i) the sequence $(\eta_n)_1^\infty$ converges in distribution to a function $F : R \longrightarrow [0, 1]$, if for each $t \in R$

$$\lim_{n\to\infty} P\big(\eta_n^{-1}((-\infty, t))\big) = F(t);$$

(ii) the sequence $(\eta_n)_1^\infty$ *converges in measure* P *to* 0*, if for each* $\varepsilon > 0$, $\varepsilon \in R$

$$\lim_{n\to\infty} P\big(\eta_n^{-1}((-\varepsilon,\varepsilon))\big) = 1;$$

(iii) the sequence $(\eta_n)_1^\infty$ *converges* P*-almost everywhere to 0, if*

$$P\left(\bigcap_{p=1}^{\infty}\bigcup_{k=1}^{\infty}\bigcap_{n=k}^{\infty}\eta_n^{-1}\left(\left(-\frac{1}{p},\frac{1}{p}\right)\right)\right) = 1,$$

i.e.

$$\lim_{p\to\infty}\lim_{k\to\infty}\lim_{i\to\infty} P\left(\bigcap_{n=k}^{k+i}\eta_n^{-1}\left(\left(-\frac{1}{p},\frac{1}{p}\right)\right)\right) = 1.$$

The above classical definitions have the following intuitionistic fuzzy modifications for IF-observables, see [4–6,8,9].

Definition 9. *Let* $(y_n)_1^\infty$ *be a sequence of IF-observables in the IF-space* $(\mathcal{F},\mathbf{m})$*, where* $\mathbf{m}$ *be an IF-state. We say that*

(i) the sequence $(y_n)_1^\infty$ *converges in distribution to a function* $\Psi : R \longrightarrow [0,1]$*, if for each* $t \in R$

$$\lim_{n\to\infty} \mathbf{m}\big(y_n((-\infty,t))\big) = \Psi(t);$$

(ii) the sequence $(y_n)_1^\infty$ *converges in measure* $\mathbf{m}$ *to* 0*, if for each* $\varepsilon > 0$, $\varepsilon \in R$

$$\lim_{n\to\infty} \mathbf{m}\big(y_n((-\varepsilon,\varepsilon))\big) = 1;$$

(iii) the sequence $(y_n)_1^\infty$ *converges* $\mathbf{m}$*-almost everywhere to 0, if*

$$\lim_{p\to\infty}\lim_{k\to\infty}\lim_{i\to\infty} \mathbf{m}\left(\bigwedge_{n=k}^{k+i} y_n\left(\left(-\frac{1}{p},\frac{1}{p}\right)\right)\right) = 1.$$

In the following proposition we introduce the connection between a convergence in the classical probability space and in the IF-space. We use the functions of several IF-observables.

Proposition 3. *Let* $(x_i)_1^\infty$ *be a sequence of IF-observables in the IF-space* $(\mathcal{F},\mathbf{m})$*,* $h_n : \mathcal{B}(R^n) \to \mathcal{F}$ *be the joint IF-observable of* $x_1,\ldots,x_n$ *and* $g_n : R^n \to R$ *be a Borel measurable function. Let IF-observable* $y_n = g_n(x_1,\ldots,x_n) : \mathcal{B}(R) \to \mathcal{F}$ *be given by* $y_n = h_n \circ g_n^{-1}$ *and random variable* $\eta_n = g_n(t_1,\ldots,t_n) : R^N \to R$ *be defined by* $\eta_n = g_n \circ \pi_n$*, where* $\pi_n : R^N \to R^n$ *is the n-th coordinate random vector defined by* $\pi_n\big((t_n)_1^\infty\big) = (t_1,\ldots,t_n)$*. It follows that*

$$P_{\eta_n} = P\circ\eta_n^{-1} = \mathbf{m}\circ y_n = \mathbf{m}_{y_n} \text{ and}$$

(i) the sequence $(y_n)_1^\infty$ *converges in distribution to a function* F *if and only if so does the sequence* $(\eta_n)_1^\infty$*;*

(ii) the sequence $(y_n)_1^\infty$ converges in measure $\mathbf{m}$ *to 0 if and only if $(\eta_n)_1^\infty$ converges in measure P to 0;*

(iii) if the sequence $(\eta_n)_1^\infty$ converges P-almost everywhere to 0, then the sequence $(y_n)_1^\infty$ converges $\mathbf{m}$*-almost everywhere to 0.*

Proof. By Proposition 1 we have that $P \circ \pi_n^{-1} = \mathbf{m} \circ h_n$. Therefore

$$\begin{aligned}\mathbf{m}_{y_n} &= \mathbf{m} \circ y_n = \mathbf{m}\big(h_n \circ g_n^{-1}\big) = \mathbf{m} \circ h_n \circ g_n^{-1} = P \circ \pi_n^{-1} \circ g_n^{-1} \\ &= P\big(\pi_n^{-1} \circ g_n^{-1}\big) = P\big((g_n \circ \pi_n)^{-1}\big) = P(\eta_n^{-1}) = P_{\eta_n}.\end{aligned}$$

(i) "$\Rightarrow$" Let the sequence $(y_n)_1^\infty$ converges in distribution to a function F. From Definition 9 we have that $\lim_{n\to\infty} \mathbf{m}\big(y_n((-\infty, t))\big) = F(t)$ for each $t \in R$. Since $P \circ \eta_n^{-1} = \mathbf{m} \circ y_n$, then

$$\lim_{n\to\infty} P\big(\eta_n^{-1}((-\infty, t))\big) = \lim_{n\to\infty} \mathbf{m}\big(y_n((-\infty, t))\big) = F(t).$$

Hence $(\eta_n)_1^\infty$ converges in distribution to a function F.
"$\Leftarrow$" It is analogy to proof "$\Rightarrow$".

(ii) "$\Rightarrow$" Let $(y_n)_1^\infty$ converges in measure $\mathbf{m}$ to 0. Then $\lim_{n\to\infty} \mathbf{m}\big(y_n((-\varepsilon, \varepsilon))\big) = 1$ for each $0 < \varepsilon$, $\varepsilon \in R$. Hence for each $0 < \varepsilon$, $\varepsilon \in R$

$$\lim_{n\to\infty} P\big(\eta_n^{-1}((-\varepsilon, \varepsilon))\big) = \lim_{n\to\infty} \mathbf{m}\big(y_n((-\varepsilon, \varepsilon))\big) = 1.$$

"$\Leftarrow$" It is analogy to proof "$\Rightarrow$".

(iii) Let $(\eta_n)_1^\infty$ converges P-almost everywhere to 0. Then

$$\lim_{p\to\infty} \lim_{k\to\infty} \lim_{i\to\infty} P\Big(\bigcap_{n=k}^{k+i} \eta_n^{-1}\Big(\Big(-\frac{1}{p}, \frac{1}{p}\Big)\Big)\Big) = 1. \tag{1}$$

But from $h_n(A \cap B) \le h_n(A) \wedge h_n(B)$ and $P \circ \pi_n^{-1} = \mathbf{m} \circ h_n$ we have

$$\begin{aligned}P\Big(\bigcap_{n=k}^{k+i} \eta_n^{-1}\Big(\Big(-\frac{1}{p}, \frac{1}{p}\Big)\Big)\Big) &= P\Big(\bigcap_{n=k}^{k+i} (g_n \circ \pi_n)^{-1}\Big(\Big(-\frac{1}{p}, \frac{1}{p}\Big)\Big)\Big) \\ &= P\Big(\bigcap_{n=k}^{k+i} \pi_n^{-1}\Big(g_n^{-1}\Big(\Big(-\frac{1}{p}, \frac{1}{p}\Big)\Big)\Big)\Big) P\Big(\pi_{k+i}^{-1}\Big(\bigcap_{n=k}^{k+i} g_n^{-1}\Big(\Big(-\frac{1}{p}, \frac{1}{p}\Big)\Big)\Big)\Big) \\ &= \mathbf{m}\Big(h_{k+i}\Big(\bigcap_{n=k}^{k+i} g_n^{-1}\Big(\Big(-\frac{1}{p}, \frac{1}{p}\Big)\Big)\Big)\Big) \le \mathbf{m}\Big(\bigwedge_{n=k}^{k+i} h_{k+i}\Big(g_n^{-1}\Big(\Big(-\frac{1}{p}, \frac{1}{p}\Big)\Big)\Big)\Big) = \\ &= \mathbf{m}\Big(\bigwedge_{n=k}^{k+i} (h_n \circ g_n^{-1})\Big(\Big(-\frac{1}{p}, \frac{1}{p}\Big)\Big)\Big) = \mathbf{m}\Big(\bigwedge_{n=k}^{k+i} y_n\Big(-\frac{1}{p}, \frac{1}{p}\Big)\Big).\end{aligned}$$

Therefore

$$P\Big(\bigcap_{n=k}^{k+i} \eta_n^{-1}\Big(\Big(-\frac{1}{p}, \frac{1}{p}\Big)\Big)\Big) \le \mathbf{m}\Big(\bigwedge_{n=k}^{k+i} y_n\Big(-\frac{1}{p}, \frac{1}{p}\Big)\Big). \tag{2}$$

By (1) and (2) we obtain that

$$\lim_{p\to\infty}\lim_{k\to\infty}\lim_{i\to\infty}\mathbf{m}\Big(\bigwedge_{n=k}^{k+i} y_n\Big(-\frac{1}{p},\frac{1}{p}\Big)\Big)=1.$$

7 Conclusion

In this paper we showed the connection between the convergences of random variables in Kolmogorov probability space and the convergences of IF-observables in IF-space $(\mathcal{F}, \mathbf{m})$. We studied the convergence in distribution, the convergence in measure and the almost everywhere convergence. This result can be applied for proving the modifications of the Strong law of large numbers, the Weak law of large numbers, the Central limit theorem and the Fisher-Tippett-Gnedenko theorem for function of several IF-observables. Using Proposition 3 we obtain shorter proof of above mentioned theorems.

Acknowledgements. This publication was supported by the Joint Polish - Slovak project under the agreement on scientific cooperation between the Polish Academy of Sciences and the Slovak Academy of Sciences, reg. num.15 and by the Operational Programme Integrated Infrastructure (OPII) for the project 313011BWH2: InoCHF - Research and development in the field of innovative technologies in the management of patients with CHF, co-financed by the European Regional Development Fund.

References

1. Atanassov, K.T.: Intuitionistic fuzzy sets. VII ITKR Session, Sofia, 20-23 June 1983 (Deposed in Centr. Sci.-Techn. Library of the Bulg. Acad. of Sci., 1697/84) (in Bulgarian). Reprinted: Int. J. Bioautomation 2016, **20**(S1), S1–S6 (1983)
2. Atanassov, K.T.: Intuitionistic Fuzzy Sets: Theory and Applications. Physica Verlag, New York (1999)
3. Atanassov, K.T.: On Intuitionistic Fuzzy Sets. Springer, Berlin (2012). https://doi.org/10.1007/978-3-642-29127-2
4. Bartková, R., Čunderlíková, K.: About fisher-tippett-gnedenko theorem for intuitionistic fuzzy events. In: Kacprzyk, J., Szmidt, E., Zadrożny, S., Atanassov, K.T., Krawczak, M. (eds.) IWIFSGN/EUSFLAT -2017. AISC, vol. 641, pp. 125–135. Springer, Cham (2018). https://doi.org/10.1007/978-3-319-66830-7_12
5. Čunderlíková, K.: Upper and lower limits and m-almost everywhere convergence of intuitionistic fuzzy observables. Notes Intuitionistic Fuzzy Sets **24**(4), 40–49 (2018)
6. Čunderlíková, K.: m-almost everywhere convergence of intuitionistic fuzzy observables induced by Borel measurable function. Notes Intuitionistic Fuzzy Sets **25**(2), 29–40 (2019)
7. Čunderlíková, K.: The individual ergodic theorem for intuitionistic fuzzy events using intuitionistic fuzzy state. Iran. J. Fuzzy Syst. **17**(5), 13–22 (2020)
8. Čunderlíková, K., Babicová, D.: Convergence in measure of intuitionistic fuzzy observables. Notes Intuitionistic Fuzzy Sets **28**(3), 228–237 (2022)
9. Čunderlíková, K., Riečan, B.: Convergence of intuitionistic fuzzy observables. In: Atanassov, K.T., et al. (eds.) IWIFSGN 2018. AISC, vol. 1081, pp. 29–39. Springer, Cham (2021). https://doi.org/10.1007/978-3-030-47024-1_4

10. Lendelová, K.: Conditional IF-probability. In: Lawry, J., et al. (eds.) Soft Methods for Integrated Uncertainty Modelling. Advances in Soft Computing, vol. 37, pp. 275–283. Springer, Heidelberg (2006). https://doi.org/10.1007/3-540-34777-1_33
11. Riečan, B.: On a problem of Radko Mesiar: general form of IF-probabilities. Fuzzy Sets Syst. **152**, 1485–1490 (2006)
12. Riečan, B.: On the probability and random variables on IF events. In: Ruan, D., et al. (eds.) Applied Artifical Intelligence, FLINS 2006, pp. 138–145. World Scientific (2006). https://doi.org/10.1142/6049
13. Riečan, B.: Analysis of fuzzy logic models. In: Koleshko, V. (eds.) Intelligent Systems, pp. 219–244. IntechOpen (2012). https://doi.org/10.5772/36172
14. Zadeh, L.A.: Fuzzy sets. Inf. Control **8**, 338–358 (1965)
15. Zadeh, L.A.: Probability measures on fuzzy sets. J. Math. Anal. Appl. **23**(2), 421–427 (1968)

Review of Fuzzification Functionals Dedicated to OFN

Dawid Ewald[1(✉)], Jacek M. Czerniak[1], Jan Baumgart[1], and Huber Zarzycki[2]

[1] Kazimierz Wielki University,Bydgoszcz, Kujawsko-Pomorskie, Poland
dawidewald@ukw.edu.pl

[2] General Tadeusz Kościuszko Military Academy of Land Forces in Wrocław, Wrocław, Poland

Abstract. Ordered fuzzy numbers are a very useful idea when we try to describe the real world in a natural human way. Terms such as "far or "near" are naturally understood by people. Properties of ordered fuzzy numbers make it possible to more accurately represent the natural human understanding of the world and convert this into a useful code understandable to computers. This article describes methods for fuzzification of real numbers to the model of ordered fuzzy numbers. The professional literature provides many order-sensitive functions for defuzzification of OFN numbers, but there are no fuzzification functions dedicated to OFN. The article puts forward theoretical backgrounds of ordered fuzzy numbers arithmetic and describes fuzzification functionals along with examples.

Keywords: fuzzy number · fuzzy logic · ordered fuzzy number · OFN

1 Ordered Fuzzy Numbers

The well-known problem of increasing the fuzziness of a number with each successive arithmetic operations on LR numbers can be avoided by applying ordered fuzzy numbers proposed by V. Kosiński, P. Propokowich and D. Ślęzak. This chapter discusses the idea and arithmetic of these numbers [1,4,5,19–21,23–25].

1.1 OFN Arithmetic

An important aspect of ordered fuzzy numbers, which was used in the proposed hybrid optimization method, is their arithmetic. Basic arithmetic operations are presented further in this chapter. The definition of arithmetic of ordered fuzzy numbers can be found in the literature [3,6–11,13,15,17]:

– the sum – let there be given directed fuzzy numbers $A = (f_A, g_A), B = (f_B, g_B)$ and $C = (f_C, g_C)$. One can say that C is the sum noted as $C = A+B$, if:

$$\forall_{y \in]0,1]} [f_A(y) + f_B(y) = f_C(y) \wedge g_A(y) + g_B(y) = g_C(y)] \tag{1}$$

K. T. Atanassov et al. (Eds.): BOS/SOR/IWIFSGN 2022, LNNS 793, pp. 49–66, 2023.
https://doi.org/10.1007/978-3-031-45069-3_6

- difference – let there be ordered fuzzy numbers $A = (f_A, g_A), B = (f_B, g_B)$ and $C = (f_C, g_C)$. One can say that Cis the result of subtracting B from A noted as $C = A - B$, if:

$$\forall_{y \in]0,1]} [f_A(y) - f_B(y) = f_C(y) \wedge g_A(y) - g_B(y) = g_C(y)] \tag{2}$$

Subtraction consists in the addition of the additive inverse of the given number.

$$A - B = A + (-B) \tag{3}$$

- multiplication by a scalar – let there be ordered fuzzy numbers $A = (f_A, g_A)$ and $C = (f_C, g_C)$. One can say that C is the product of A multiplied by the scalar $(C = A * r)$, if:

$$\forall_{y \in [0,1]} [r \cdot f_A(y) = f_C(y) \text{ i } r \cdot g_A(y) = g_C(y)] \tag{4}$$

- division by a scalar – let there be ordered fuzzy numbers $A = (f_A, g_A)$ and $C = (f_C, g_C)$. One can say that C is the quotient of A divided by the scalar $(C = A/r)$, if:

$$\forall_{y \in [0,1]} [f_A(y)/r = f_C(y) \text{ i } g_A(y)/r = g_C(y)] \tag{5}$$

- OFN product – let there be given ordered fuzzy numberse $A = (f_A, g_A), B = (f_B, g_B)$ and $C = (f_C, g_C)$. One can say that e C is the product of the numbers A and B $(C = A * B)$, if:

$$\forall_{y \in [0,1]} [f_A(y) \cdot f_B(y) = f_C(y) \wedge g_A(y) \cdot g_B(y) = g_C(y)] \tag{6}$$

- OFN quotient – let there be given directed fuzzy numbers $A = (f_A, g_A), B = (f_B, g_B)$ and $C = (f_C, g_C)$. One can say that C is the result of A divided by B noted as $C = A/B$, if:

$$\forall_{y \in [0,1]} [f_A(y)/f_B(y) = f_C(y) \wedge g_A(y)/g_B(y) = g_C(y)] \tag{7}$$

- the multiplicative inverse to a given directed fuzzy number $A = (f_A, g_A)$ is a number noted as follows:

$$A^{-1} = (\frac{1}{f_A}, \frac{1}{g_A}) \tag{8}$$

- number A raised to the power of B – let there be given directed fuzzy numbers [18] $A = (f_A, g_A), B = (f_B, g_B)$ and $C = A^B$:

$$f_C(x) = f_A(x)^{f_B(x)} \wedge g_C(x) = g_A(x)^{g_B(x)}, \forall x \in [0,1] \tag{9}$$

- absolute value of OFN – let $A = (f_A, g_A)$ then $\mid A \mid = C = (f_C, g_C)$

$$f_C(x) = \mid f_A(x) \mid \wedge g_C(x) = \mid g_A(x) \mid, \forall x \in [0,1] \tag{10}$$

2 Fuzzifiers

Fuzzifiers are functions used to transform a real number to OFN notation while taking into account its characteristic feature, i.e. the order. The proposals presented herein enable to easily implement the solution and easily use OFN algorithms

2.1 Inverse Golden Ratio Method

The inverse Golden Ratio method is based on the Golden Ratio concept, derived from the Fibonacci sequence. The proposed solution is inspired by the defuzzifier presented by W. Dobrosielski which uses the same relationship [2]. The following terms can be found in earlier literature: harmonic division, golden ratio, divine proportion.

Many intellectuals have been fascinated by this issue for more than 2000 years. As stated by Mario Livia in his paper:

"Some of the greatest mathematical minds of all ages, from Pythagoras and Euclid in ancient Greece, through the medieval Italian mathematician Leonardo of Pisa and the Renaissance astronomer Johannes Kepler, to present-day scientific figures such as Oxford physicist Roger Penrose, have spent endless hours over this simple ratio and its properties. But the fascination with the Golden Ratio is not confined just to mathematicians. Biologists, artists, musicians, historians, architects, psychologists, and even mystics have pondered and debated the basis of its ubiquity and appeal. In fact, it is probably fair to say that the Golden Ratio has inspired thinkers of all disciplines like no other number in the history of mathematics" [12,14,16,22].

At present, the best-known interpretation of the golden ratio is the division of a line segment into two parts, so that the ratio of the length of the longer part to the shorter part is the same as the ratio of the length of the entire segment to the longer part, as shown in Fig. 1. In other words: the length of the longer part is to be the geometric mean of the length of the shorter part and the entire segment.The golden ratio, expressed algebraically, is as follows:

$$\frac{a+b}{a} = \frac{a}{b} = \Phi \tag{11}$$

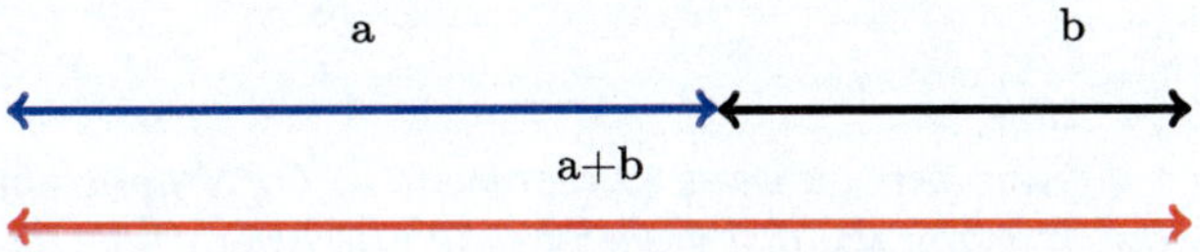

Fig. 1. Golden ratio of line segments

To find the values of Φ, mthe left-hand side of the equation No. 11 can be transformed to:

$$\frac{a+b}{a} = 1 + \frac{b}{a} = 1 + \frac{1}{\Phi},\ gdzie\ \frac{b}{a} = \frac{1}{\Phi} \tag{12}$$

then:

$$\Phi^2 - \Phi - 1 = 0 \tag{13}$$

which results in:

$$\Phi_1 = \frac{1+\sqrt{5}}{2}\ lub\ \Phi_2 = \frac{1-\sqrt{5}}{2} \tag{14}$$

Bearing in mind that the domain of Φ is limited to positive numbers, we obtain the following solution of the equation:

$$\Phi = \Phi_1 = \frac{1+\sqrt{5}}{2} = 1,618033998875\ldots \tag{15}$$

Thus, the quotient of two factors a and b is called the "golden ratio" if its value is equal to $\Phi = 1,61803398875\ldots$. The fuzzifier is defined as follows:

Theorem 1.
Number A represented in OFN notation using GRf operator :

- *The first and the last value of the number with positive order:*

$$GRf = \begin{cases} R1 = A - 2.4721, \\ R4 = A + 1.5278. \end{cases} \tag{16}$$

where:
GRf is the fuzzifier,
A stands for the number we want to represent in OFN notation,
R1, R4 are the first and the last values in the OFN notation [R1, R2, R3, R4],
R2, R3 are calculated according to the OFN shape.

- *The first and the last value of the number with negative order:*

$$GRf = \begin{cases} R1 = A + 2.4721 \\ R4 = A - 1.5278 \end{cases} \tag{17}$$

where:
GRf is the fuzzifier,
A stands for the number we want to represent in OFN notation,
R1, R4 are the first and the last values in the OFN notation [R1, R2, R3, R4].

- *Number A is represented in OFN notation using GRf operator of triangular shape:*

- *positive order (Fig. 2):*

$$GRf = \begin{cases} R1, \\ R1+1, \\ R1+1, \\ R4 \end{cases} \tag{18}$$

where:
GRf is the fuzzifier,
Values of R1, R4 are according to formula 16.

- *negative order (Fig. 4):*

$$: GRf = \begin{cases} R1, \\ R1-1, \\ R1-1, \\ R4 \end{cases} \tag{19}$$

where:
GRf is the fuzzifiera,
Values of R1, R4are according to formula 17.

– *Number A expressed in OFN notation using the GRf operator in the shape of a trapezoid:*

- *positive order (Fig. 3):*

$$GRf = \begin{cases} R1, \\ R1+1, \\ R4-1, \\ R4 \end{cases} \tag{20}$$

where:
GRf is the fuzzifier,
of R1, R4 are according to formula 16.

- *negative order (Fig. 5):*

$$GRf = \{ R1, -1, +1, \tag{21}$$

where:
GRf is the fuzzifier,
of R1, R4 are according to formula 17.

Example:

Let's fuzzify number 7 to the form of OFN with positive order:
$[R1, R2, R3, R4] \rightarrow [4.5279, 5.5279, 5.5279, 8.5278]$ – triangle (Fig. 2)
$[R1, R2, R3, R4] \rightarrow [4.5279, 5.5279, 7.5278, 8.5278]$ – trapezoid (Fig. 3)

Let's fuzzify number 7 to the form of OFN with negative order:
$[R1, R2, R3, R4] \rightarrow [9.4721, 8.4721, 8.4721, 5.4722]$ – triangle (Fig. 4)
$[R1, R2, R3, R4] \rightarrow [9.4721, 8.4721, 6.4722, 5.4722]$ – trapezoid (Fig. 5)

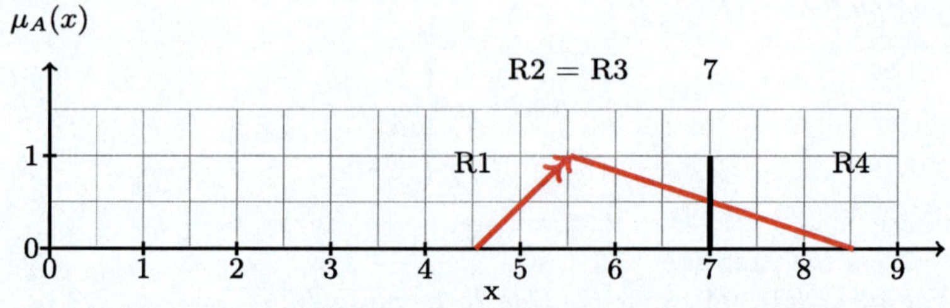

Fig. 2. Fuzzy number 7 - positive order, triangle

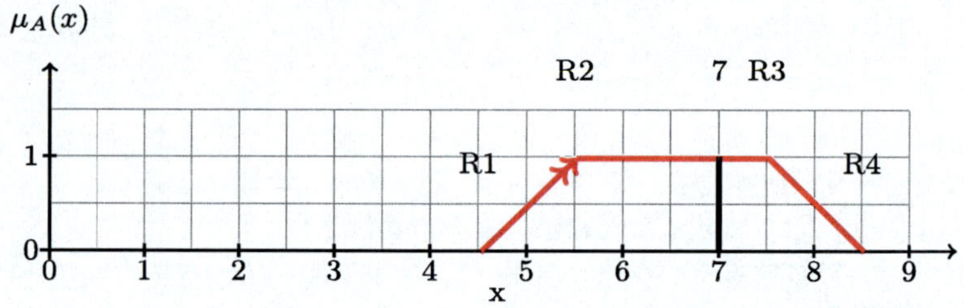

Fig. 3. Fuzzy number 7 - positive order, trapezoid

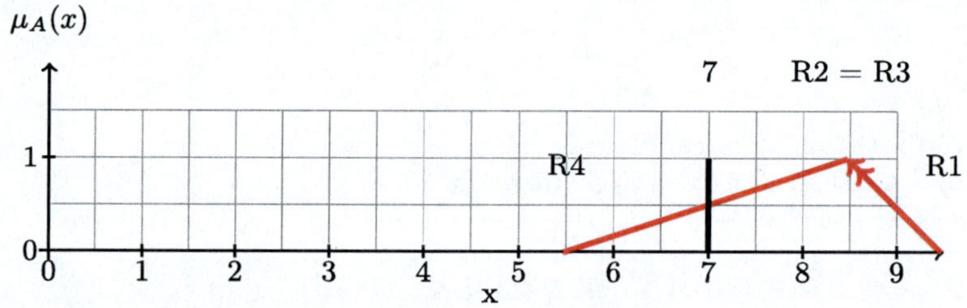

Fig. 4. Fuzzy number 7 - negative order, triangle

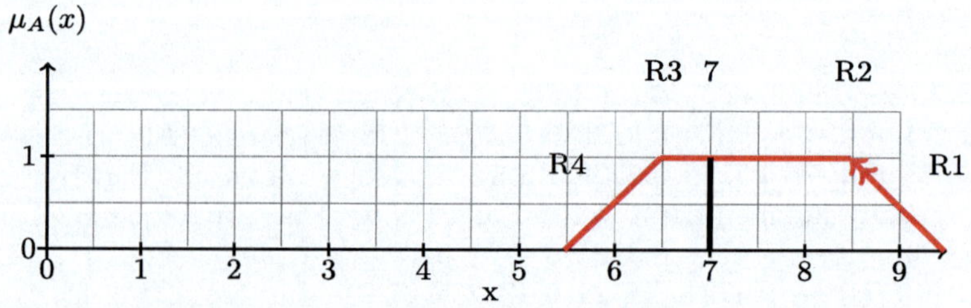

Fig. 5. Fuzzy number 7 - negative order, trapezoid

2.2 BeeT Fuzzifier

The fuzzifier inspired by the golden ratio idea was described in the previous subsection. More fuzzification functionals are presented in the next two subsections. Those functionals are divided so as beeT functionals can take any shape. BeeT functionals are a group of order-sensitive fuzzifiers that can be classified according to their shape. These functions are necessary to represent real numbers in OFN notation. The numbers assume classic shapes commonly known from the literature. The fuzzification methods were developed in the course of the experiment and were used in the study.

BeeT Fuzzification Functionals – Trapezoid Shape. he beeT fuzzy number of a trapezoidal shape is created according to the following principle:

$$beeTf\alpha\beta\gamma\delta = \begin{cases} R1 = number, \\ R2 = number + \mid number * \beta \mid, \\ R3 = number + \mid number * \gamma \mid, \\ R4 = number + \mid number * \delta \mid, \end{cases} \tag{22}$$

where:
$R1, R2, R3, R4$ – subsequent values of the ordered fuzzy number,

α – subsequent values of the ordered fuzzy number,:

- p – number with positive order – $[R1, R2, R3, R4]$ (Fig. 6),
- n – number with negative order – $[R4, R3, R2, R1]$ (Fig. 7).

liczba – means a number we want to fuzzify into the form of the ordered fuzzy number acc. to beeT type,
β, γ, δ – values specifying the level of fuzziness.

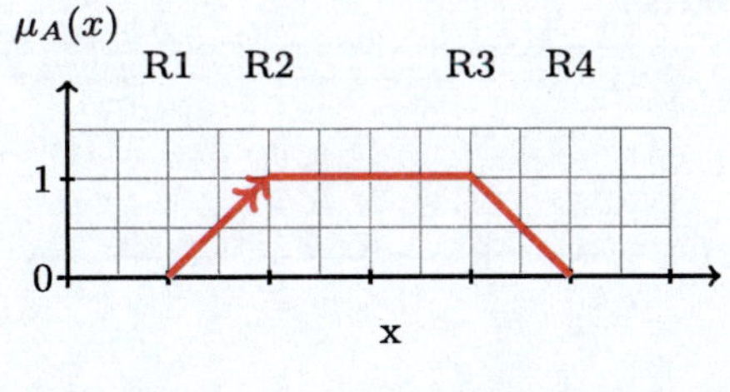

Fig. 6. Graphical interpretation of the beeT fuzzifier - positively ordered trapezoidal number

Example:
Let's fuzzify number 3 using the $beeTfp010207$ functional to the form of a positively ordered OFN with the shape of a trapezoid (Fig. 8):

$$[R1, R2, R3, R4] \rightarrow [3, 3.3, 3.6, 5.1]$$

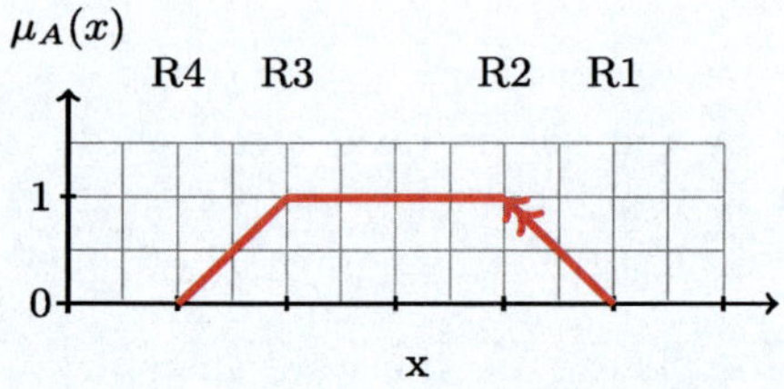

Fig. 7. Graphical interpretation of the beeT fuzzifier - negatively ordered trapezoidal number

Let's fuzzify number 3 using the $beeTfn010207$ functional to the form of a negatively ordered OFN with the shape of a trapezoid (Fig. 9):

$$[R4, R3, R2, R1] \rightarrow [5.1, 3.6, 3.3, 3]$$

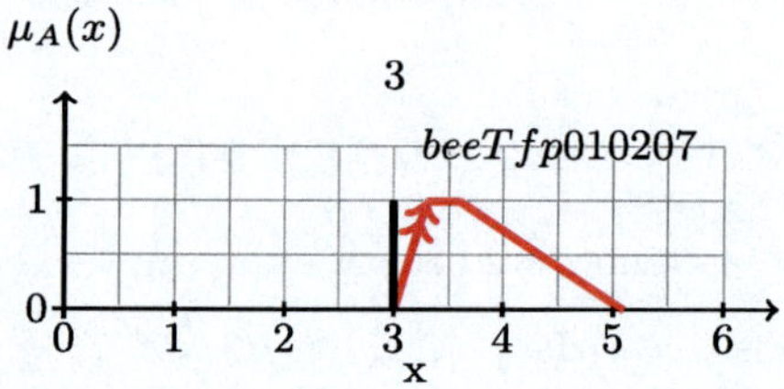

Fig. 8. BeeT type fuzzy number 3 - positive order, trapezoid

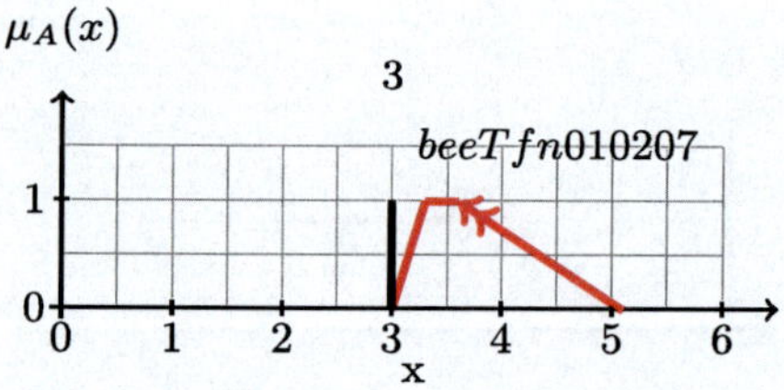

Fig. 9. BeeT type fuzzy number 3 - negative order, trapezoid

BeeT Fuzzification Functionals – Right-Angled Triangle Shapeo. The beeT fuzzy number of a right-angled triangle shape is created according to the following principle:

$$beeTf\alpha\beta = \begin{cases} R1 = number, \\ R2 = number, \\ R3 = number, \\ R4 = number + \mid number * \beta \mid, \end{cases} \tag{23}$$

where:
$R1, R2, R3, R4$ – subsequent values of the ordered fuzzy number,
α – defines the order of the number and assumes the following values:

- p – number with positive order – $[R1, R2, R3, R4]$ (Fig. 10),
- n – number with negative order – $[R4, R3, R2, R1]$ (Fig. 11).

liczba – means a number we want to fuzzify into the form of the ordered fuzzy number acc. to beeT type,
β – values specifying the level of fuzziness.

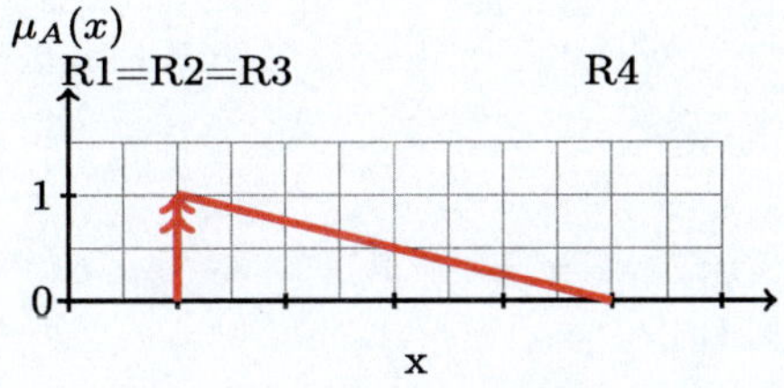

Fig. 10. Graphical interpretation of the beeT fuzzifier - positively ordered number of a right-angled triangle shape

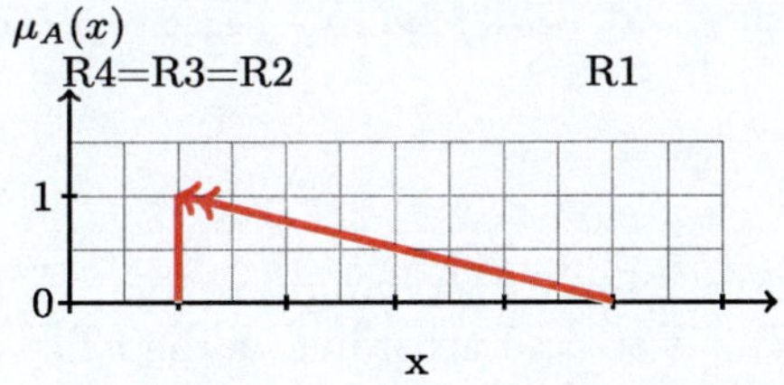

Fig. 11. Graphical interpretation of the beeT fuzzifier - negatively ordered number of a right-angled triangle shape

Example:
Let's fuzzify number 3 using $beeTrfp05$ functional to the form of a positively ordered OFN with the shape of a right-angled triangle (Fig. 12):

$$[R1, R2, R3, R4] \rightarrow [3, 3, 3, 4.5]$$

Let's fuzzify number 3 using the $beeTrfn05$ functional to the form of a negatively ordered OFN with the shape of a right-angled triangle (Fig. 13):

$$[R4, R3, R2, R1] \rightarrow [4.5, 3, 3, 3]$$

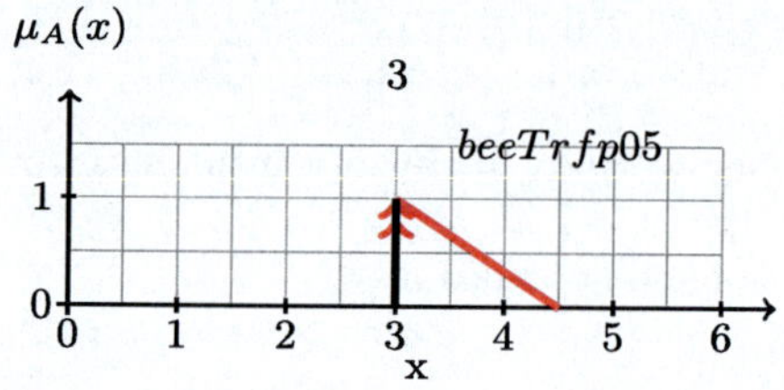

Fig. 12. Number 3 fuzzified with beeT fuzzifier – positive order, right-angled triangle shape

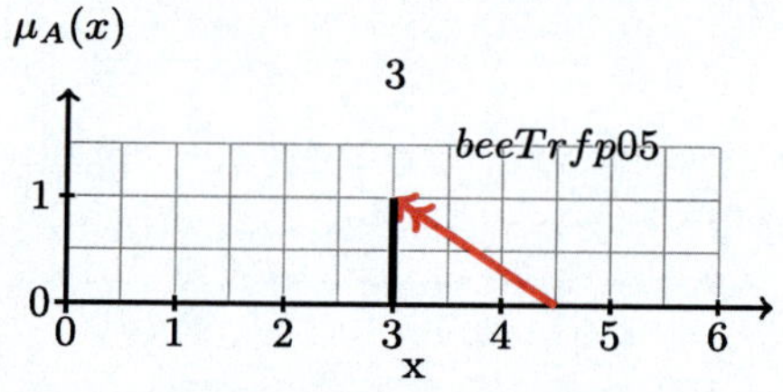

Fig. 13. Number 3 fuzzified with beeT fuzzifier – negative order, right-angled triangle shape

BeeT Fuzzification Functionals - Triangular Shape. The beeT fuzzy number of a triangular shape is created according to the following principle:

$$beeTrRf\alpha\beta\gamma = \begin{cases} R1 = number, \\ R2 = number + \mid number * \beta \mid, \\ R3 = number + \mid number * \beta \mid, \\ R4 = number + \mid number * \gamma \mid, \end{cases} \tag{24}$$

where:
$R1, R2, R3, R4$ – subsequent values of the ordered fuzzy number,
α – defines the order of the number and assumes the following values:

– p – number with positive order – $[R1, R2, R3, R4]$ (Fig. 14),
– n – number with negative order – $[R4, R3, R2, R1]$ (Fig. 15).

liczba – means a number we want to fuzzify into the form of the ordered fuzzy number acc. to beeT type,
β, γ – values specifying the level of fuzziness.

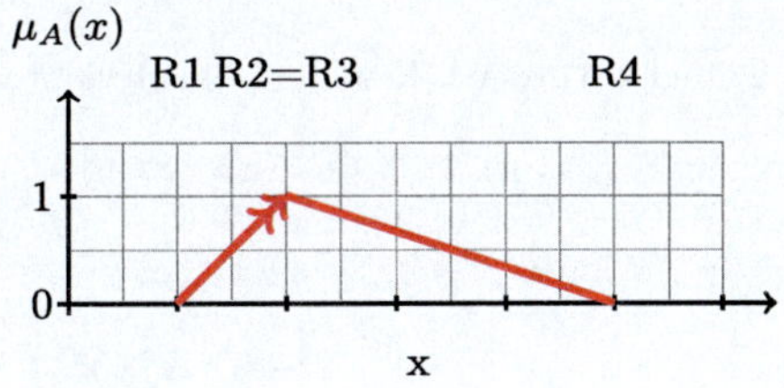

Fig. 14. Graphical interpretation of the beeT fuzzifier - positively ordered number of triangular shape

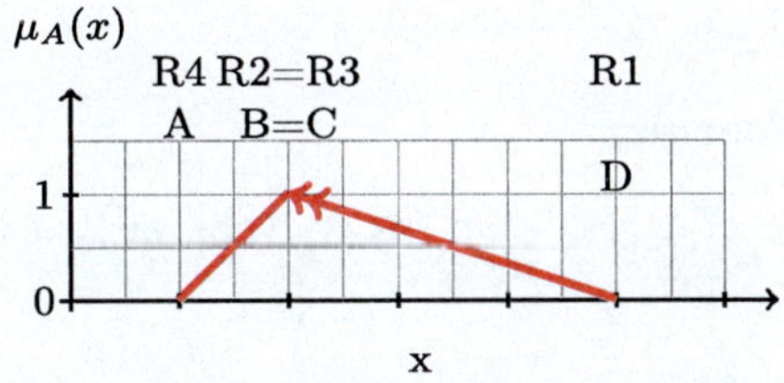

Fig. 15. Graphical interpretation of the beeT fuzzifier - negatively ordered number of triangular shape

Example:
Let's fuzzify number 3 using $beeTrRfp0511$ functional to the form of a positively ordered OFN with triangular shape (Fig. 16):

$$[R1, R2, R3, R4] \rightarrow [3, 4.5, 4.5, 4.1]$$

Let's fuzzify number 3 using the $beeTrRfn0511$ functional to the form of a negatively ordered OFN with triangular shape (Fig. 17):

$$[R4, R3, R2, R1] \rightarrow [4.1, 4.5, 4.5, 3]$$

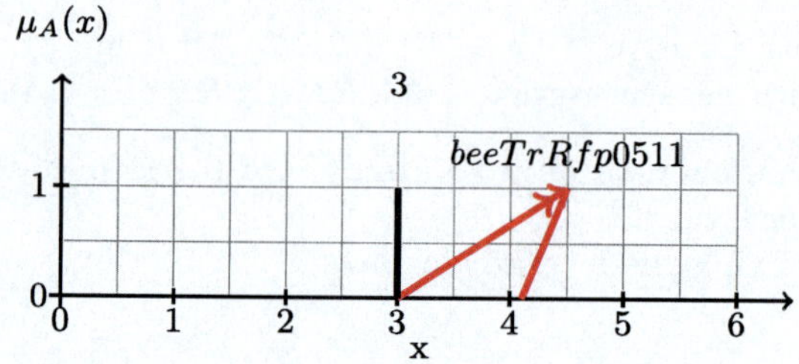

Fig. 16. Number 3 fuzzified with beeT fuzzifier - positive order, triangular shape

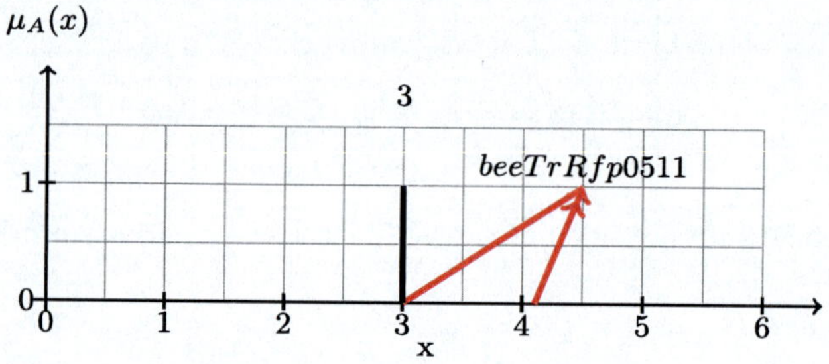

Fig. 17. Number 3 fuzzified with beeT fuzzifier - negative order, triangular shape

2.3 BeeM Type Fuzzifier

An important feature of OFN is an order, thus order-sensitive fuzzifiers are described later in the paper.

BeeM Fuzzification Functionals – Trapezoid Shape. The beeM fuzzy number of a trapezoidal shape and positive order is created according to the following principle:

$$beeMTp\beta\gamma = \begin{cases} R1 = number, \\ R2 = number, \\ R3 = number + \mid number * \beta \mid, \\ R4 = number + \mid number * \gamma \mid, \end{cases} \tag{25}$$

where:
$R1, R2, R3, R4$ – subsequent values of the ordered fuzzy number,
$[R1, R2, R3, R4]$ – positively ordered number (Fig. 18),
liczba – means a number we want to fuzzify into the form of the ordered fuzzy number acc. to beeM type,
β, γ – values specifying the level of fuzziness.

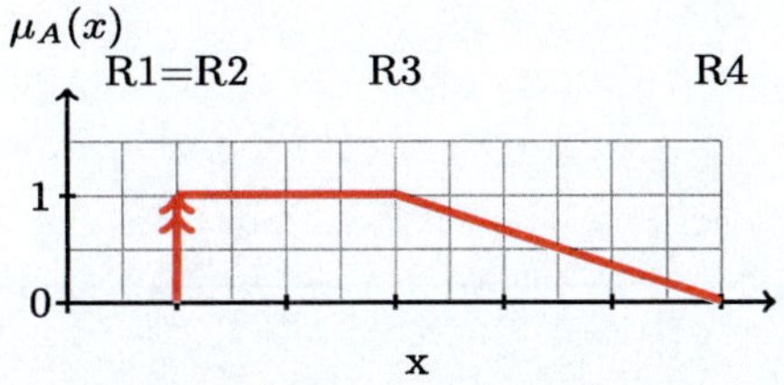

Fig. 18. Graphical interpretation of the beeM fuzzifier - positively ordered trapezoidal number

The beeM fuzzy number of trapezoid shape and negative order is created according to the following principle:

$$beeMTn\beta\gamma = \begin{cases} R1 = number, \\ R2 = number + \mid number * \beta \mid, \\ R3 = number + \mid number * \gamma \mid, \\ R4 = number + \mid number * \gamma \mid, \end{cases} \tag{26}$$

where:
$R1, R2, R3, R4$ – subsequent values of the ordered fuzzy number,
$[R1, R2, R3, R4]$ – negatively ordered number (Fig. 19),
liczba – means a number we want to fuzzify into the form of the ordered fuzzy number acc. to beeM type,
β, γ – values specifying the level of fuzziness.

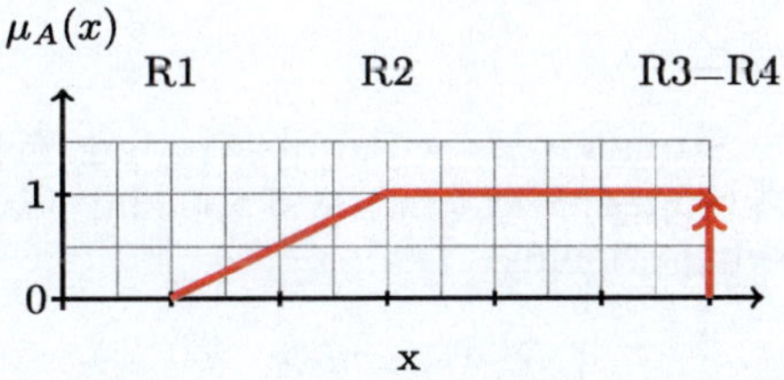

Fig. 19. Graphical interpretation of the beeM fuzzifier - negatively ordered trapezoidal number

Example:
Let's fuzzify number 3 using the *beeMTp*0207 functional to the form of a positively ordered OFN with trapezoid shape (Fig. 20):

$$[A, B, C, D] \rightarrow [3, 3, 3.6, 5.1]$$

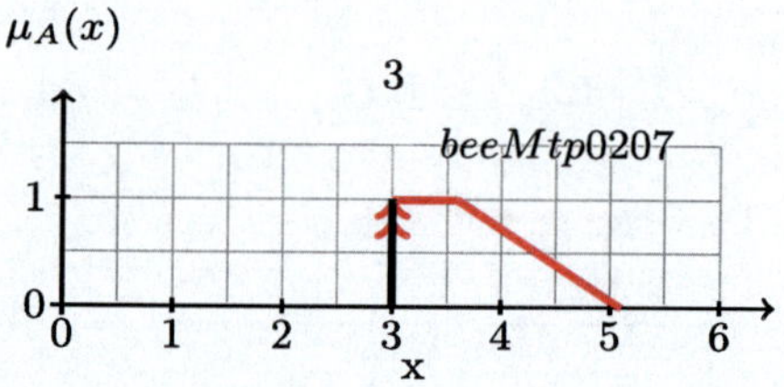

Fig. 20. Number 3 fuzzified with beeM fuzzifier - positive order, trapezoid shape

Let's fuzzify number 3 using the *beeMTp*0207 functional to the form of a negatively ordered OFN with trapezoid shape (Fig. 21):

$$[A, B, C, D] \rightarrow [3, 3.6, 5.1, 5.1]$$

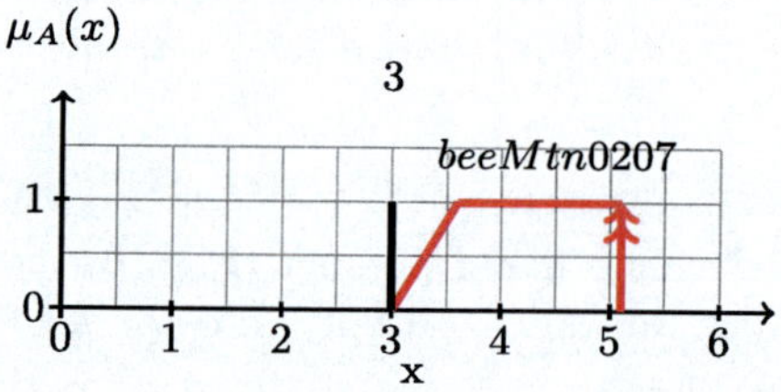

Fig. 21. Number 3 fuzzified with beeM fuzzifier - negative order, trapezoid shape

BeeM Fuzzification Functionals - Right-Angled Triangle Shape. The beeM fuzzy number of right–angled triangle shape and positive order is created according to the following principle:

$$beeMTrp\beta = \begin{cases} R1 = number, \\ R2 = number, \\ R3 = number, \\ R4 = number + \mid number * \beta \mid, \end{cases} \tag{27}$$

where:
$R1, R2, R3, R4$ – subsequent values of the ordered fuzzy number,
$[R1, R2, R3, R4]$ – positively ordered number (Fig. 22),
liczba – means a number we want to fuzzify into the form of the ordered fuzzy number acc. to beeM type,
β – value specifying the level of fuzziness.

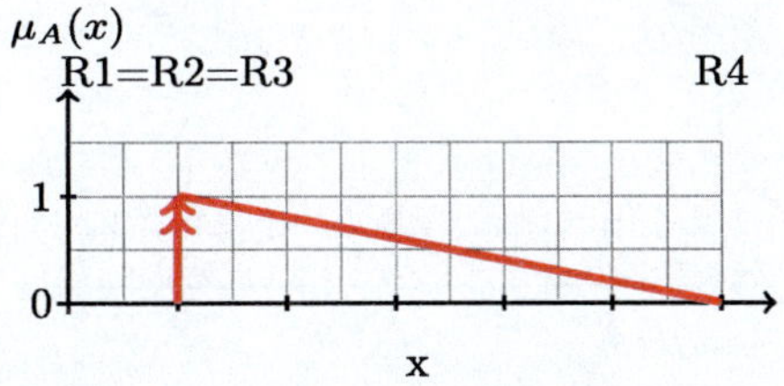

Fig. 22. Graphical interpretation of the beeM fuzzifier – positively ordered number of a right-angled triangle shape

The beeM fuzzy number of right-angled triangle shape and negative order is created according to the following principle:

$$beeMTrn\beta = \begin{cases} R1 = number, \\ R2 = number + \mid number * \beta \mid, \\ R3 = number + \mid number * \beta \mid, \\ R4 = number + \mid number * \beta \mid, \end{cases} \tag{28}$$

where:
$R1, R2, R3, R4$ – subsequent values of the ordered fuzzy number,
$[R1, R2, R3, R4]$ – negatively ordered number (Fig. 23),
liczba – means a number we want to fuzzify into the form of the ordered fuzzy number acc. to beeM type,
β – value specifying the level of fuzziness.

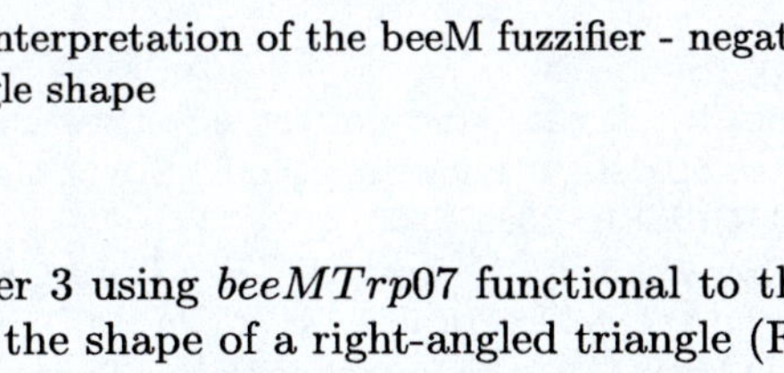

Fig. 23. Graphical interpretation of the beeM fuzzifier - negatively ordered number of a right-angled triangle shape

Example:
Let's fuzzify number 3 using *beeMTrp*07 functional to the form of a positively ordered OFN with the shape of a right-angled triangle (Fig. 24):

$$[A, B, C, D] \rightarrow [3, 3, 3, 5.1]$$

Let's fuzzify number 3 using the *beeMTrn*07 functional to the form of a negatively ordered OFN with the shape of a right-angled triangle (Fig. 25):

$$[A, B, C, D] \rightarrow [3, 5.1, 5.1, 5.1]$$

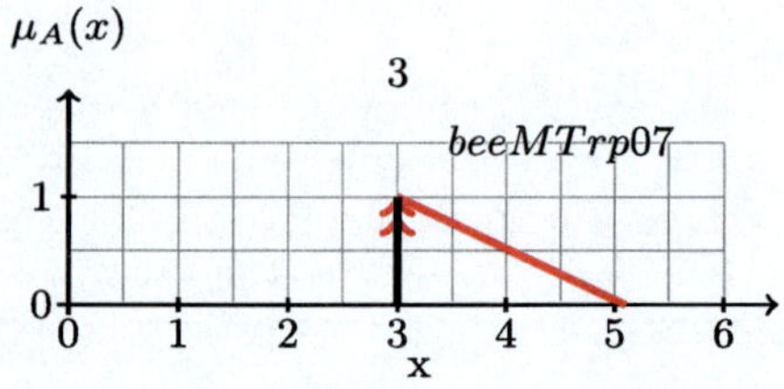

Fig. 24. Number 3 fuzzified with beeM fuzzifier - positive order, right-angled triangle shape

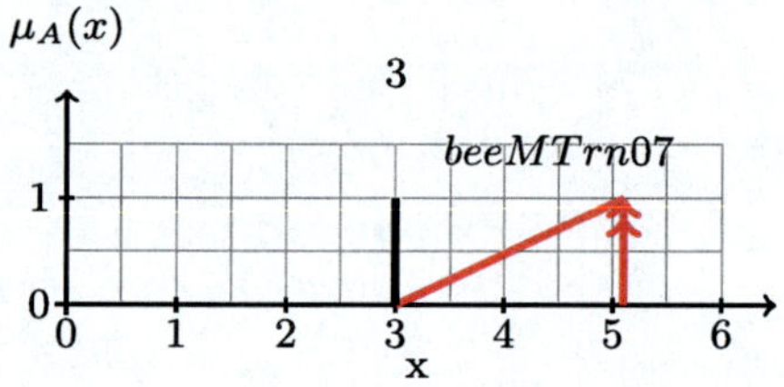

Fig. 25. Number 3 fuzzified with beeM fuzzifier - negative order, right-angled triangle shape

3 Summary

As there are few papers available in the literature on the fuzzification functionals dedicated to OFN, the proposals presented in the paper allow to fully exploit the potential of ordered fuzzy numbers. It is very important to systematize the knowledge on OFN because it would positively affect further development of the interest in this idea.

The selection of an appropriate fuzzification method can have a huge impact on the results obtained by algorithms or evaluation of processes in which OFNs are used. That is why it is so important to develop new fuzzification methods.

References

1. Baumgart, J., Sangho, B.: A case study of the effectiveness of new methods of swarm optimization compared to known methods. Stud. Mater. Appl. Comput. Sci. **13**(1), 47–50 (2021). ISSN: 1689–6300
2. Dobrosielski, W.T., Szczepański, J., Zarzycki, H.: A proposal for a method of defuzzification based on the golden ratio—GR. In: Atanassov, K.T., et al. (eds.) Novel Developments in Uncertainty Representation and Processing. AISC, vol. 401, pp. 75–84. Springer, Cham (2016). https://doi.org/10.1007/978-3-319-26211-6_7
3. Ewald, D., Czerniak, J.M., Paprzycki, M.: A new OFNBee method as an example of fuzzy observance applied for ABC optimization. In: Prokopowicz, P., Czerniak, J., Mikołajewski, D., Apiecionek, Ł, Ślęzak, D. (eds.) Theory and Applications of Ordered Fuzzy Numbers. SFSC, vol. 356, pp. 223–237. Springer, Cham (2017). https://doi.org/10.1007/978-3-319-59614-3_13

4. Ewald, D., Zarzycki, H., Czerniak, J.M.: Certain aspects of the OFNBee algorithm operation for different fuzzifiers. In: Atanassov, K.T., et al. (eds.) IWIFSGN BOS/SOR 2020 2020. LNCS, vol. 338, pp. 241–256. Springer International Publishing, Cham (2022). https://doi.org/10.1007/978-3-030-95929-6_19
5. Galas, K.: Drive unit as a replacementforthe platform. Stud. Mater. Appl. Comput. Sci. **12**(1), 10–14 (2020). ISSN: 1689–6300. https://doi.org/10.5281/zenodo.4362649
6. Jacko, P., et al.: Remote IoT education laboratory for microcontrollers based on the stm32 chips. Sensors **22**(4) (2022). https://doi.org/10.3390/s22041440, https://www.mdpi.com/1424-8220/22/4/1440
7. Kacprzak, D.: Analiza modelu Leontiewa z użyciem skierowanych liczb rozmytych. referat wygłoszony na II Konferencja Technologie Eksploracji i Reprezentacji Wiedzy (Białystok 2007)
8. Kosinski, W., Prokopowicz, P., Ślęzak, D.: On algebraic operations on fuzzy numbers. In: Kłopotek, M.A., Wierzchoń, S.T., Trojanowski, K. (eds.) Intelligent Information Processing and Web Mining. Advances in Soft Computing, vol. 22, pp. 353–362. Springer, Berlin (2003). https://doi.org/10.1007/978-3-540-36562-4_37
9. Kosinski, W., Prokopowicz, P., Ślęzak, D.: Ordered fuzzy numbers. Bull. Polish Acad. Sci.: Math. **51**(3), 327–338 (2003)
10. Kosiński, W., Koleśnik, R., Prokopowicz, P., Frischmuth, K.: On algebra of ordered fuzzy numbers. Soft Comput. Found. Theor. Aspects 291–302 (2004)
11. Kosiński, W., Prokopowicz, P.: Algebra liczb rozmytych. Matematyka Stosowana **5**, 37–63 (2004)
12. Livio, M.: The Golden Ratio: The Story of Phi. The World's Most Astonishing Number, p. 6 (2003)
13. Mikolajewska, E., Prokopowicz, P., Mikolajewski, D.: Computational gait analysis using fuzzy logic for everyday clinical purposes - preliminary findings. Bio-Algorithms Med-Syst. **13**(1), 37–42 (2017). https://doi.org/10.1515/bams-2016-0023
14. Namli, Ö.B., Tyburek, K.: Research of the efficiency of the reach of fire services to accidents in the city of Kocaeli on the basis of statistical data for the years 2013–2020. Stud. Mater. Appl. Comput. Sci. **14**(3), 1–5 (2022)
15. Piasecki, W.: Ordered fuzzy number defuzzyfication with the use of evolutionary algorithm. X International PhD Workshop OWD 2008 (2008)
16. Piszcz, A., Mikolajewski, D.: Application Forthe android platform with the system as a solution to the right problem horse nutrition. Stud. Mater. Appl. Comput. Sci. **12**(1), 5–9 (2020). ISSN: 1689–6300. https://doi.org/10.5281/zenodo.4362647
17. Prokopowicz, P.: Algorytmizacja działań na liczbach rozmytych i jej zastosowanie. Rozprawa doktorska IPPT PAN (2005)
18. Prokopowicz, P.: Flexible and simple methods of calculations on fuzzy numbers with the ordered fuzzy numbers model. In: Rutkowski, L., Korytkowski, M., Scherer, R., Tadeusiewicz, R., Zadeh, L.A., Zurada, J.M. (eds.) ICAISC 2013. LNCS (LNAI), vol. 7894, pp. 365–375. Springer, Heidelberg (2013). https://doi.org/10.1007/978-3-642-38658-9_33
19. Prokopowicz, P., Mikołajewski, D., Mikołajewska, E., Kotlarz, P.: Fuzzy system as an assessment tool for analysis of the health-related quality of life for the people after stroke. In: Rutkowski, L., Korytkowski, M., Scherer, R., Tadeusiewicz, R., Zadeh, L.A., Zurada, J.M. (eds.) ICAISC 2017. LNCS (LNAI), vol. 10245, pp. 710–721. Springer, Cham (2017). https://doi.org/10.1007/978-3-319-59063-9_64

20. Rojek, I., Macko, M., Mikolajewski, D., Saga, M., Burczynski, T.: Modern methods in the field of machine modelling and simulation as a research and practical issue related to industry 4.0. Bull. Polish Acad. Sci.-Tech. Sci. **69**(2) (2021). https://doi.org/10.24425/bpasts.2021.136717
21. Sangho, B.: Comparison of selected wolf pack algorithms used in solving optimization problems. Stud. Mater. Appl. Comput. Sci. **13**(1), 17–32 (2021). ISSN: 1689–6300. https://doi.org/10.5281/zenodo.4362647
22. Tyburek, K., Bora, Ö.: Comparison of the efficiency of time and frequency domain descriptors for the classification of selected wind instruments. Stud. Mater. Appl. Comput. Sci. **14**(3), 6–13 (2022)
23. Zarzycki, H.: Comparative study of the firefly algorithm and the whale algorithm. In: Kahraman, C., Tolga, A.C., Cevik Onar, S., Cebi, S., Oztaysi, B., Sari, I.U. (eds.) INFUS 2022. LNCS, vol. 504, pp. 999–1006. Springer International Publishing, Cham (2022). https://doi.org/10.1007/978-3-031-09173-5_114
24. Zarzycki, H., Dobrosielski, W.T., Czerniak, J.M., Ewald, D.: Use of OFN in the short-term prediction of exchange rates. In: Atanassov, K.T., et al. (eds.) IWIFSGN 2019 2019. AISC, vol. 1308, pp. 289–301. Springer, Cham (2021). https://doi.org/10.1007/978-3-030-77716-6_25
25. Zarzycki, H., Skubisz, O.: A new artificial bee colony algorithm approach for the vehicle routing problem. In: Kahraman, C., Cebi, S., Cevik Onar, S., Oztaysi, B., Tolga, A.C., Sari, I.U. (eds.) INFUS 2021. LNNS, vol. 307, pp. 562–569. Springer, Cham (2022). https://doi.org/10.1007/978-3-030-85626-7_66

Using Modified Canberra Distance as OFN Numbers Comparison Operator

Jacek M. Czerniak[1](✉), Jan Baumgart[1], Hubert Zarzycki[2], and Łukasz Apiecionek[1]

[1] Institute of Computer Science, Department of Intelligent Systems, Casimir the Great University in Bydgoszcz, ul. Kopernika 1, 85-064 Bydgoszcz, Poland
{jczerniak,jbaumgart,lapiecionek}@ukw.edu.pl

[2] General Tadeusz Kościuszko Military Academy of Land Forces in Wrocław, ul. Czajkowskiego 109, 51-147 Wrocław, Poland
hubert.zarzycki@awl.edu.pl

Abstract. This paper deals with the problem of fuzzy numbers comparison. There are a number of approaches to solving this problem. The authors tried to successfully avoid solutions whose parts are defuzzifiers. Their goal was to propose a comparison operator that would operate intuitively on fuzzy numbers similarly to operators in other number systems. As a result of trials and experiments, the final choice was the Canberra distance metric. To apply this metric to fuzzy numbers, it was necessary to modify it slightly, which was successfully done. The experimental part of the paper provides calculations performed on numbers in OFN notation. Preliminary experiments conducted by the authors show that the proposed approach capes well also with other fuzzy number notations. This is, however, to be fully confirmed in further research works.

Keywords: Fuzzy Logic · OFN · Canberra · JC · comparison

1 Introduction

Among various operators of every arithmetic that has been developed since Aristotle, there are always comparison operators. Each branch of science, depending on its peculiarities, tried to describe the nature of similarity in its own way. In fact, the problem of the similarity of objects is often the basis for description of relationships between those objects. Finding and describing relationships between objects is particularly valuable today when successful attempts are made to analyze mass data, i.e. so-called BigData [41,46]. Concepts such as similarity, dissimilarity or containment are classically needed to describe relationships between objects [5,25]. These terms have always been used in set arithmetic. Consequently, these terms can also be found in various fields of artificial intelligence [39,50]. As a natural consequence, these experiments and theories were transferred to fuzzy logic, which was created on the initiative of Lotfia Zadeh [54]. In recent decades, the fuzzy set theory has been applied in many fields of

K. T. Atanassov et al. (Eds.): BOS/SOR/IWIFSGN 2022, LNNS 793, pp. 67–82, 2023.
https://doi.org/10.1007/978-3-031-45069-3_7

science as well as everyday life. The need to compare fuzzy sets arose naturally already at the very beginning of the theory existence. The researchers originally developed a certain set of methods often referring to those used for classical sets. Intensive development of fuzzy logic and its applications often requires definition of new ways to compare objects. This issue is of particular importance in computer-aided decision making, classification or natural language processing. Although the issue of object comparison is crucial to many applications of the theory, it has still not been possible to explicitly formalize basic concepts such as similarity or containment [2,17,30]. While some fuzzy logic researchers strive to find precise definitions of terms, others question this approach, arguing that imposing a rigid framework limits the possibility of practical applications. Over the years of the fuzzy logic development, many researchers have created methods for comparison of fuzzy sets and numbers. Among them, it is absolutely necessary to mention research work that has been carried out since 1977 by Zadeh[54], Yager and Kaufman [53], Chang and Amado [1,7,27]. Already in the present century, Bortolan, Degani and Dadgostar reviewed some methods for comparison of fuzzy sets [44]. They selected well-known researchers and their methods. They included, among others, the first, second and third Yager index, Chang algorithm and Adamo methods. Baas and Kwakernaak methods, Baldwin and Guild method as well as Kerre method were also included in the list [4,49,52]. Another well-known approach worth mentioning is Jain method and, obviously, the four grades of dominance (NSD, ND, PSD, PD) proposed by the classics of that field of science: Dubois and Prade. In 2010, Dadgostar and Kerr published an article [2,17] where they proposed a consistent method called Partial Comparison Method (PCM). In subsequent years, Wang and Kerre were among those who undertook the task of formulating rules and axioms aimed at organizing the ways of fuzzy numbers comparison [7,20]. With the use of these criteria the authors were able to make a systematic comparison of a wide range of fuzzy ranking methods [26]. New approaches to the problem of fuzzy numbers comparison have emerged in the following years. These include the study by Dobrosielski, Czerniak et al. who proposed in 2018 the use of several original defuzzifiers in the process of comparing fuzzy numbers in OFN notation [19].

2 Similarity Measures

Most classification learning algorithms for both supervised and unsupervised cases require determining the measurement method of the similarity between feature vectors of objects. By determining the distance measure, or in other words, the similarity measure between objects or groups of objects, we can quantify the degree of similarity between them. This similarity can be understood directly, that is, the higher it is, the higher the similarity of the objects under consideration. Thus the similarity of objects is directly proportional to the value of the similarity measure. However, among the most commonly used measures are various distance measures, for which the larger the value, the less similar the objects are. Such measures should therefore be called dissimilarity measures.

Let us assume that each object of the learning set is represented in the form of a p-dimensional feature vector. Then suppose that we define the distance measure for the vectors xi, xj X RP. This measure is denoted by the symbol d(xi,xj). Formally, the distance is meant as a function that assigns a non-negative number to a pair of objects, satisfying the conditions of the metric. Thus for any xi, xj, xk X we have:

The above conditions are called reflexivity condition, symmetry condition and triangle condition, respectively. The most common distance measures include:

– Euclidean distance:

$$d^2(x_i, x_j) = (x_i - x_j)^T (x_i - x_j) \tag{1}$$

– Minkowski distance:

$$d_p(x_i, x_j) = \sqrt{\sum_{k=1}^{m} |x_{ik} - x_{jk}|^p} \tag{2}$$

– Mahalanobis distance:

$$d_M(x_i, x_j) = (x_i - x_j)^T \Sigma^{-1} (x_i - x_j) \tag{3}$$

– Sebestyen distance:

$$d_S(x_i, x_j) = (x_i - x_j)^T W (x_i - x_j) \tag{4}$$

Note that the Euclidean distance is a special case of the Minkowski distance for $p = 2$. When using it, it is recommended to normalize the data, because the scale of the attributes here significantly affects the value of this distance.

There are following methods intended for examining the distances between objects, which were developed later and gained popularity in such fields like econometrics, among others.

– Canberra distance:

$$C = \sum_{j=1}^{m} w_j \frac{v_{ik,j}}{(z_{ij} + z_{kj})} \tag{5}$$

– Bray-Curtis distance:

$$BC = \frac{\Sigma_{j=1}^{m} w_j v_{ik,j}}{\Sigma_{j=1}^{m} w_j (z_{ij} + z_{kj})} \tag{6}$$

– Clark distance

$$CL = \sqrt{\frac{1}{m} \sum_{j=1}^{m} w_j \left(\frac{(z_{ij} - z_{kj})}{(z_{ij} + z_{kj})} \right)^2} \tag{7}$$

– Jefferson-Matusita distance

$$JM = \sum_{j=1}^{m} w_j \left(\sqrt{z_{ij}} - \sqrt{z_{kj}}\right)^2 \tag{8}$$

Authors (B. Pawełek) point out the sensitivity of the Canberra distance metric to values close to zero. Modifications introduced by Bray-Curtis and by Clark were aimed, among others, at minimizing this effect. A modification of the Canberra distance metric that takes into account the specificity of ordered fuzzy numbers is proposed in the subsequent section.

3 Methodology

This section presents a new measure of similarity. It refers to the Carberry proposal mentioned in the previous section. However, the new measure removes the sensitivity of the metric to values close to zero, which was a significant inconvenience in the Canberra distance metric. The JC metric is dedicated to comparing ordered fuzzy numbers although, after ignoring the elements of the definition that relate to the specifics of OFN, it could be applied to other fuzzy number arithmetics.

3.1 The Ordered Fuzzy Numbers

The original approach to defining fuzzy numbers was taken in 1993 by Professor Witold Kosiński and his doctoral student P. Słysz [31,32,34,35]. Further publications by prof. W. Kośiński, in collaboration with P. Prokopowicz and D. Ślęzak, introduced the ordered fuzzy numbers (OFN) model [31,33,34].

The classical Kosinski's definition of ordered fuzzy numbers states that an OFN number is an ordered pair of a rising edge function f and a falling edge function g. The figure below shows two numbers with the start and end points of the mentioned functions marked, i.e. fA(0), fA(1), gA(1), gA(0) (Fig. 1).

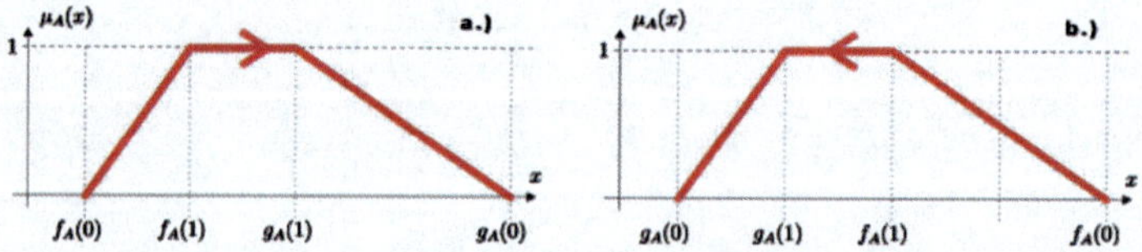

Fig. 1. Two OFN numbers with opposite orders

Definition 1. The ordered fuzzy number A is called the ordered pair of functions:

$$A = (x_{up}, x_{down})$$

where:
$x_{up}, x_{down} : [0,1] \longrightarrow R$ are continuous functions (Fig. 2).

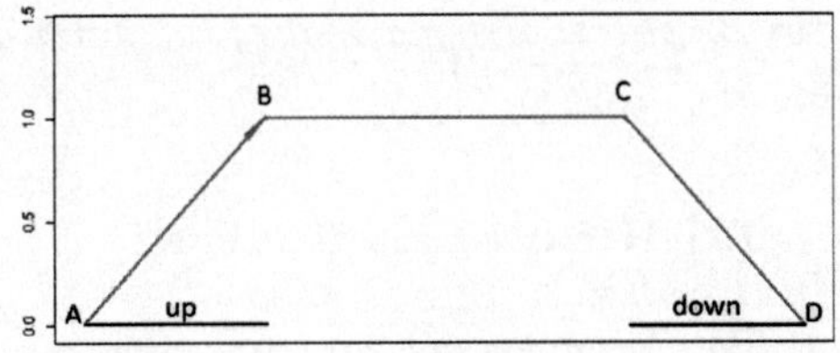

Fig. 2. Definition ordered fuzzy number

The characteristic feature of the directed fuzzy number is the orientation:

- If the number is oriented in the direction of the increase in the value of the axis OX - positive orientation,
- If the number is oriented in the opposite direction - negative orientation,

3.2 Lemma and Definitions

Lemma 1. *Let two OFN numbers A and B be given, each represented by a set of four boundary values of the rising and falling edge functions. This, for the number A, is the set* $\{f_{A(0)}, f_{A(1)}, g_{A(1)}, g_{A(0)}\}$, *and for the number* B $\{f_{B(0)}, f_{B(1)}, g_{B(1)}, g_{B(0)}\}$. *Thus, the comparison of the numbers A and B involves calculating the value of the JC metric expressed by the following formula.*

$$JC = \sum_{j=0}^{1} \frac{f_{A(i)} - f_{B(i)}}{|f_{A(i)}| + |f_{B(i)}|} + \sum_{j=0}^{1} \frac{g_{A(i)} - g_{B(i)}}{|g_{A(i)}| + |g_{B(i)}|} \tag{9}$$

where:
the condition that $|f_{A(i)}| + |f_{B(i)}| = 0$ and $|g_{A(i)}| + |g_{B(i)}| = 0$ should be taken into account if this condition is not met, an affine transformation (translation) of both OFN numbers in the rising direction of the OX axis should be applied. This is done by adding the value 1 to each coordinate of both numbers. This operation should be repeated until the zero value in each denominator disappears.

Definition 1. *If the value of the JC metric is 0 and both fuzzy numbers A and B are of the same order, it means that the numbers are equal.*

Definition 2. *If the value of the JC metric is 0 and fuzzy numbers A and B are reversely ordered, it means that the numbers are positively ordered.*

Definition 3. *If the value of the JC metric is <0 then the fuzzy number A is less than the number B. This is independent of the order of the fuzzy numbers A and B.*

Definition 4. *If the value of the JC metric is >0 then the fuzzy number A is greater than the number B. This is independent of the order of the fuzzy numbers A and B.*

4 Experiments and Results Evaluation

The literature on the subject provides several numbers commonly used for verification of comparison operators. These numbers, expressed in OFN notation, are shown in the figures below together with diagrams. The numbers were chosen intentionally, as this set includes both disjoint numbers and numbers containing one another. This may present some challenge as regards interpretation unlike in the domain of real numbers (Figs. 3, 4, 5, 6, 7, 8, 9 and 10).

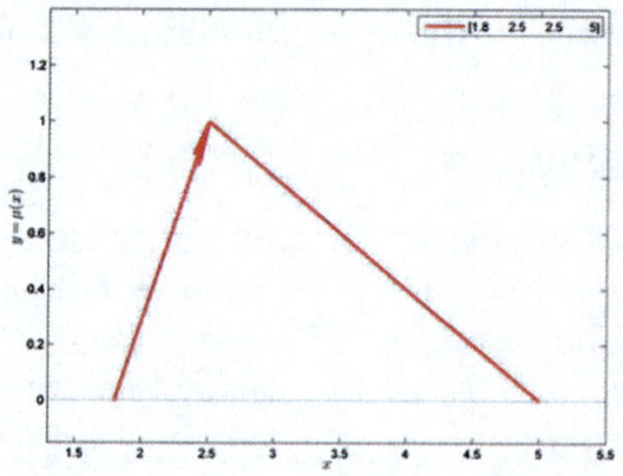

Fig. 3. Fuzzy number A[1.8, 2.5, 2.5, 5.0]

Thanks to the use of the JC metric described in the previous section it will be possible to compare the values of fuzzy numbers and to classify them. Each pair of numbers subjected to comparison will be assigned a number that is an outcome of the JC metric. After applying the similarity recognition rules formulated in the definitions specified in the previous section it will be possible to determine the relationships between individual pairs of fuzzy numbers.

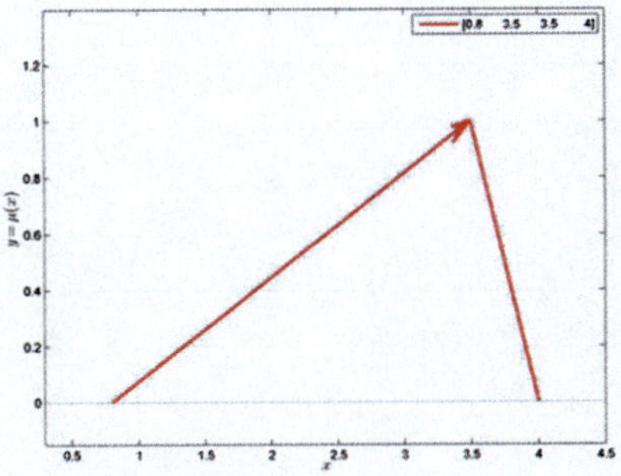

Fig. 4. Fuzzy number B[0.8, 3.5, 3.5, 4.0]

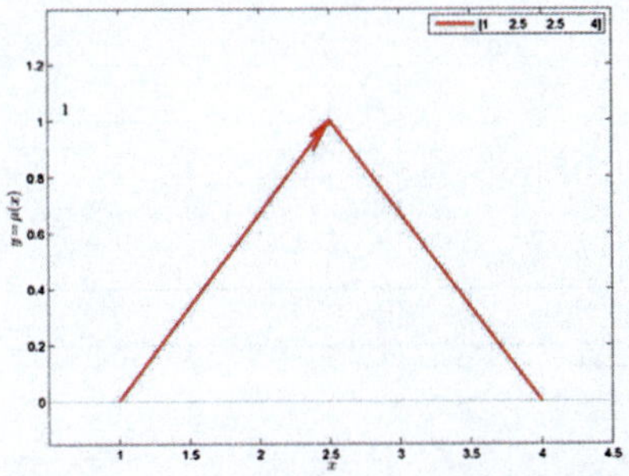

Fig. 5. Fuzzy number C[1.0, 2.5, 2.5, 4.0]

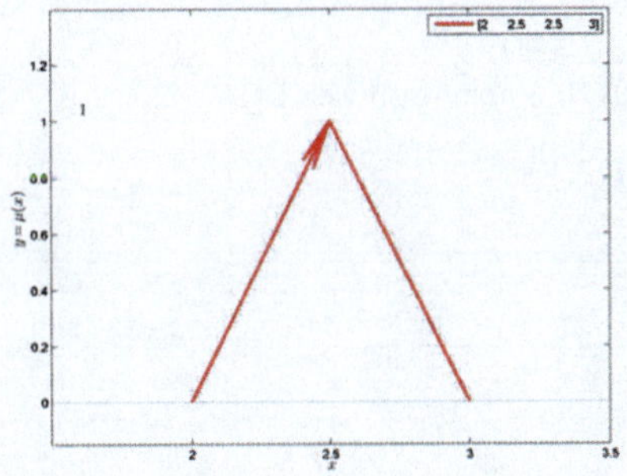

Fig. 6. Fuzzy number D[2.0, 2.5, 2.5, 3.0]

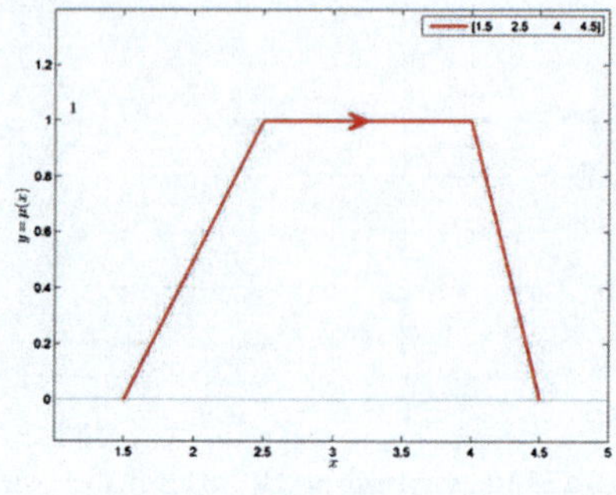

Fig. 7. Fuzzy number E[1.5, 2.5, 4.0, 4.5]

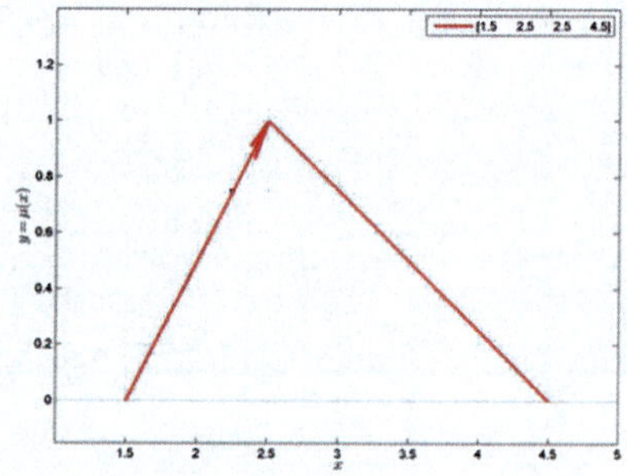

Fig. 8. Fuzzy number F[1.5, 2.5, 2.5, 4.5]

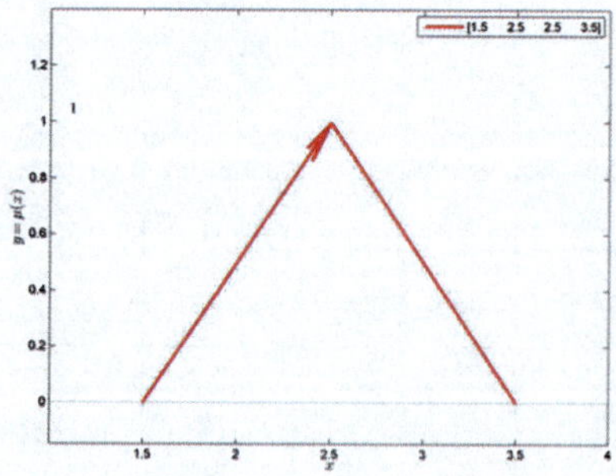

Fig. 9. Fuzzy number G[1.5, 2.5, 2.5, 3.5]

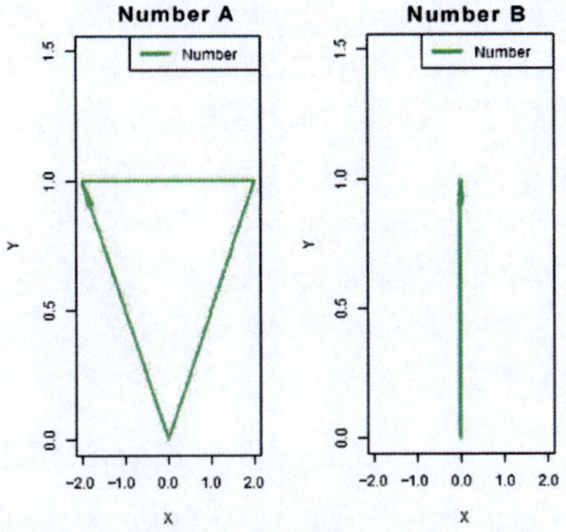

Fig. 10. Fuzzy numbers A[0,-2,2,0] and B[0,0,0,0]

5 Experiments and Results Evaluation

Below, values of the JC metric are calculated for the fuzzy numbers given in the previous section and for other numbers taken from the literature. Finally, a pair of numbers was also added, the values of which were chosen in a way to generate a divide-by-zero error. Such a scenario was anticipated in the definition and the relevant rule was applied there.

$$JC = \sum_{j=0}^{1} \frac{f_{A(i)} - f_{B(i)}}{|f_{A(i)}| + |f_{B(i)}|} + \sum_{j=0}^{1} \frac{g_{A(i)} - g_{B(i)}}{|g_{A(i)}| + |g_{B(i)}|} \tag{10}$$

$$A[3.0, 3.5, 3.5, 4.0] \tag{11}$$

$$B[4.0, 4.5, 4.5, 5.0] \tag{12}$$

$$JC_{AB} = -0,50 \rightarrow A < B \tag{13}$$

$$C[1.8, 2.5, 2.5, 5.0] \tag{14}$$

$$D[0.8, 3.5, 3.5, 4.0] \tag{15}$$

$$JC_{CD} = 0,16 \rightarrow C > D \tag{16}$$

$$E[2.0, 4.5, 4.5, 5.0] \tag{17}$$

$$F[2.0, 3.5, 3.5, 5.0] \tag{18}$$

$$G[2.0, 2.5, 2.5, 5.0] \tag{19}$$

$$JC_{EF} = 0,25 \rightarrow E > F \tag{20}$$

$$JC_{FG} = 0,33 \rightarrow F > G \tag{21}$$

$$JC_{GE} = -0,57 \rightarrow G < E \tag{22}$$

$$H[1.0, 2.5, 2.5, 4.0] \tag{23}$$

$$I[2.0, 2.5, 2.5, 3.0] \tag{24}$$

$$JC_{HI} = -0,19 \rightarrow H < I \tag{25}$$
$$J[1.5, 2.5, 4.0, 4.5] \tag{26}$$
$$K[1.5, 2.5, 2.5, 4.5] \tag{27}$$
$$L[1.5, 2.5, 2.5, 3.5] \tag{28}$$
$$JC_{JK} = 0,23 \rightarrow J > K \tag{29}$$
$$JC_{KL} = 0,13 \rightarrow K > L \tag{30}$$
$$JC_{LJ} = -0,36 \rightarrow L < J \tag{31}$$

$$P[0.0, -2.0, 2.0, 0.0] \tag{32}$$
$$Q[0.0, 0.0, 0.0, 0.0] \tag{33}$$
$$JC_{PQ} = divide - by - zero \tag{34}$$
$$P'[1.0, -1.0, 3.0, 1.0] \tag{35}$$
$$Q'[1.0, 1.0, 1.0, 1.0] \tag{36}$$
$$JC_{P'Q'} = -0,50 \rightarrow P < Q \tag{37}$$

As seen here, all the calculations made it possible to unambiguously determine the relationship between individual ordered fuzzy numbers. The last pair of compared numbers, which were P and Q, generated a divide-by-zero error. In accordance with the condition in Lemma 1, an affine translation with respect to the OX axis of both numbers was necessary. The new numbers P' and Q' no longer exhibit the error and the metric can be calculated.

6 Discussion, Limitations and Future Work

The results obtained in the previous section are unambiguous and there are no problems with their interpretation. Proper formulation of the Lemma 1 make it possible to resolve the indeterminacy caused by the divide-by-zero error that can occur during the calculation of the JC metric. Protection of this type was already included at the design stage of the metric. Although the JC metric works for equal values of fuzzy numbers located in different value ranges, it would be appropriate to compare the results obtained by its operation with the ideas of other researchers. The table below provides the comparison of the performance of the JC metric with other methods. In general, the interpretation of the results is as follows: a larger value of the factor indicates a number larger according to a given metric. This is a compilation of concepts whose operation is similar to comparing defuzzified values of fuzzy numbers, which was not the goal of the authors of this publication. As described above, the JC metric allows comparing two fuzzy numbers without their defuzzification or quasi-defuzzification. In other words, there is no change from the fuzzy numbers domain to real numbers domain in order to perform the comparison (Table 1).

Table 1. Summary of the results of different methods for comparing fuzzy numbers

Methods		A	B	C	D	E	F	G	H	I	J	K	L
Czerniak	JC	-0,5	A<B	0,16	C<D	E>F 0.25	F>G 0.33	G<E -0.57	-0,19	H<I	J>K 0.23	K>L 0.13	L<J -0.36
Yager	F1	0.70	0.90	0.61	0.53	0.76	0.70	0.63	0.50	0.50	0.62	0.56	0.50
	F2	0.72	0.90	0.66	0.69	0.90	0.76	0.66	0.61	0.54	0.81	0.64	0.58
	F3	0.70	0.90	0.58	0.56	0.80	0.70	0.60	0.50	0.50	0.62	0.54	0.50
Chang		0.14	0.18	0.40	0.34	0.46	0.41	0.38	0.29	0.10	0.56	0.33	0.20
Adamo	0.9M	0.71	0.91	0.55	0.66	0.91	0.73	0.55	0.53	0.51	0.81	0.54	0.52
	0.9m	0.71	0.91	0.55	0.66	0.91	0.73	0.55	0.53	0.51	0.81	0.54	0.52
	0.5	0.75	0.95	0.75	0.72	0.95	0.85	0.75	0.65	0.55	0.85	0.70	0.60
Baas-Kwakernaak		0	1	0.84	1	1	0.74	0.60	1	1	1	1	1
Baldwin-Guild	lap.	h	0.32	0.42	0.33	0.42	0.33	0.30	0.27	0.28	0.45	0.37	0.27
	g.	0	0.47	0.44	0.37	0.55	0.40	0.34	0.30	0.24	0.53	0.40	0.28
	r.a.	0	0.20	0.34	0.24	0.28	0.23	0.22	0.20	0.23	0.31	0.28	0.21
Kerre		0.80	1	0.96	0.89	1	0.86	0.76	0.91	0.91	1	0.85	0.75
Jain	k = 1	0.72	0.90	0.66	0.69	0.90	0.76	0.66	0.73	0.67	0.90	0.69	0.64
	k = 2	0.55	0.84	0.53	0.51	0.84	0.65	0.54	0.60	0.48	0.82	0.56	0.45
	k = ½	0.84	0.95	0.78	0.81	0.85	0.86	0.78	0.83	0.80	0.94	0.80	0.77
Dubois-Prade	PD	0	1	0.84	1	1	0.74	0.60	1	1	1	1	1
	PSD	0	1	0.54	0.46	0.74	0.23	0.16	0.73	0.24	0.80	0.20	0
	ND	0	1	0.54	0.46	0.63	0.38	0.18	0.27	0.76	0.50	0.50	0.50
	NSD	0	1	0	0.16	0.26	0	0	0	0	0	0	0
Lee-Li	U.m.	0.70	0.90	0.61	0.53	0.76	0.70	0.63	0.50	0.50	0.62	0.56	0.50
	U.g.	-	-	-	-	-	-	-	0.12	0.04	-	-	-
	P.m.	0.70	0.90	0.53	0.58	0.80	0.70	0.60	0.50	0.50	0.63	0.55	0.50
	P.g.	-	-	-	-	-	-	-	0.09	0.03	-	-	-
Dadgostar-Kerr	PCM	0	1	0.57	0.43	0.57	0.43	0.34	0.50	0.50	0.63	0.37	0.25
Dorohonceanu-Marin	B2	0	1	0.59	0.41	0.62	0.57	0.43	0.50	0.50	0.65	0.60	0.40
	B2 x	0	1	0.59	0.41	0.64	0.59	0.41	0.50	0.50	0.68	0.61	0.39

7 Conclusion

The research presented in this paper deals with a new method for solving the problem of fuzzy numbers comparison. As presented in the above sections, there are many approaches to solving this problem. Most of them explicitly or implicitly lead to leaving the fuzzy numbers domain and the comparison is performed in the domain of real numbers. The authors aimed to develop a comparison operator that does not change the domain in which relational operations are performed. The proposed comparison operator operates intuitively on fuzzy numbers similarly to operators in other number systems. The proposed JC metric was tested on the numbers used by other authors to verify their operators in the cited publications. The experimental part of the paper provides calculations performed on positively ordered numbers in OFN notation. The properties of OFN numbers demonstrate that these numbers behave identically to classic numbers. The results of the comparison of this set of fuzzy numbers confirm the good performance of the operator. No anomalies of the results were revealed. The numbers that most other authors considered to be in a certain relationship were also determined by the JC metric as being in the same relation. Therefore, it should be considered that the proposed JC comparison operator is an effective way of finding similarities and differences between fuzzy numbers without leaving the domain of fuzzy numbers. Moreover, the calculation is not complicated and even performing it repeatedly will not be computationally demanding. Preliminary experiments conducted by the authors show that the proposed approach capes well also with other fuzzy number notations. This is, however, to be fully confirmed in further research works. Further research and studies should concern tests of the JC metric over the entire spectrum of OFN numbers, although the example presented in the paper as well as other experiments carried out indicate that no problems are to be expected.

References

1. Abbasbandy, S., Asady, B.: Ranking of fuzzy numbers by sign distance. Inf. Sci. **176**, 2405–2416 (2006)
2. Adabitabar, F.M., Fath-Tabar, G., Eslampia, Z.: The similarity measure of generalized fuzzy numbers based on interval distance. Appl. Math. Lett. **25**(10), 1528–1534 (2012)
3. Apiecionek, L., Czerniak, J.M., Ewald, D.: NEW QoS CONCEPT for protecting network resources. In: Choraś, R.S. (ed.) IP&C 2016. AISC, vol. 525, pp. 234–239. Springer, Cham (2017). https://doi.org/10.1007/978-3-319-47274-4_28
4. Asady, B., Zendehnam, A.: Ranking fuzzy numbers by distance minimization. Appl. Math. Model. **31**, 2589–2598 (2007)
5. Baumgart, J., Sangho, B.: A case study of the effectiveness of new methods of swarm optimization compared to known methods. Stud. Mater. Appl. Comput. Sci. **13**(1), 47–50 (2021). ISSN 1689-6300
6. Bucko, R., Vince, T., Molnar, J., Dziak, J., Gladyr, A.: Safety system for intelligent building. In: International Conference on Modern Electrical and Energy Systems (MEES), pp. 252–255. IEEE (2017)
7. Cheng, C.: A new approach for ranking fuzzy numbers by distance method. Fuzzy Sets Syst. **95**, 307–317 (1998)
8. Czerniak, J., Ewald, D., Macko, M., Śmigielski, G., Tyszczuk, K.: Approach to the monitoring of energy consumption in eco-grinder based on ABC optimization. In: Kozielski, S., Mrozek, D., Kasprowski, P., Małysiak-Mrozek, B., Kostrzewa, D. (eds.) BDAS 2015. CCIS, vol. 521, pp. 516–529. Springer, Cham (2015). https://doi.org/10.1007/978-3-319-18422-7_46
9. Czerniak, J.M.: OFNAnt method based on TSP ant colony optimization. In: Prokopowicz, P., Czerniak, J., Mikołajewski, D., Apiecionek, Ł, Ślęzak, D. (eds.) Theory and Applications of Ordered Fuzzy Numbers. SFSC, vol. 356, pp. 207–222. Springer, Cham (2017). https://doi.org/10.1007/978-3-319-59614-3_12
10. Czerniak, J.M., Dobrosielski, W.T., Filipowicz, I.: Comparing fuzzy numbers using defuzzificators on OFN shapes. In: Prokopowicz, P., Czerniak, J., Mikołajewski, D., Apiecionek, Ł, Ślęzak, D. (eds.) Theory and Applications of Ordered Fuzzy Numbers. SFSC, vol. 356, pp. 99–132. Springer, Cham (2017). https://doi.org/10.1007/978-3-319-59614-3_6
11. Czerniak, J.M., Zarzycki, H.: Artificial acari optimization as a new strategy for global optimization of multimodal functions. J. Comput. Sci. **22**, 209–227 (2017)
12. Czerniak, J.M., Zarzycki, H., Apiecionek, Ł, Palczewski, W., Kardasz, P.: A cellular automata-based simulation tool for real fire accident prevention. Math. Probl. Eng. **2018**(3058241), 1–12 (2018)
13. Czerniak, J.M., Zarzycki, H., Ewald, D.: Aao as a new strategy in modeling and simulation of constructional problems optimization. Simul. Model. Pract. Theory **76**, 22–33 (2017). http://www.sciencedirect.com/science/article/pii/S1569190X17300709

14. Czerniak, J.M., Filipowicz, I., Ewald, D.: The novel shape normalization operator for fuzzy numbers in OFN notation. In: Kacprzyk, J., Szmidt, E., Zadrożny, S., Atanassov, K.T., Krawczak, M. (eds.) IWIFSGN/EUSFLAT -2017. AISC, vol. 641, pp. 548–562. Springer, Cham (2018). https://doi.org/10.1007/978-3-319-66830-7_49
15. Czerniak, J., Macko, M., Ewald, D.: The cutmag as a new hybrid method for multi-edge grinder design optimization. Adv. Intell. Syst. Comput. **401**, 327–337 (2016)
16. Czerniak, J., Smigielski, G., Ewald, D., Paprzycki, M.: New proposed implementation of ABC method to optimization of water capsule flight. In: Proceedings of the Federated Conference on Computer Science and Information Systems, IEEE Digital Library, ACSIS, vol. 5, pp. 489–493 (2015)
17. Debashree, G., Debjani, C.: A new approach to fuzzy distance measure and similarity measure between two generalized fuzzy numbers. Appl. Soft Comput. **10**(1), 90–99 (2010)
18. Dobrosielski, W.T., Czerniak, J.M., Zarzycki, H., Szczepański, J.: Fuzzy numbers applied to a heat furnace control. In: Prokopowicz, P., Czerniak, J., Mikołajewski, D., Apiecionek, Ł, Ślęzak, D. (eds.) Theory and Applications of Ordered Fuzzy Numbers. SFSC, vol. 356, pp. 269–288. Springer, Cham (2017). https://doi.org/10.1007/978-3-319-59614-3_16
19. Dobrosielski, W.T., Czerniak, J.M., Szczepański, J., Zarzycki, H.: Two new defuzzification methods useful for different fuzzy arithmetics. In: Atanassov, K.T., et al. (eds.) IWIFSGN 2016. AISC, vol. 559, pp. 83–101. Springer, Cham (2018). https://doi.org/10.1007/978-3-319-65545-1_9
20. Dorohonceanu, B., Marín, B.: A simple method for comparing fuzzy numbers (2002)
21. Dyczkowski, K.: A less cumulative algorithm of mining linguistic browsing patterns in the world wide web (2007)
22. Ewald, D., Czerniak, J.M., Zarzycki, H.: OFNBee method used for solving a set of benchmarks. In: Kacprzyk, J., Szmidt, E., Zadrożny, S., Atanassov, K.T., Krawczak, M. (eds.) IWIFSGN/EUSFLAT -2017. AISC, vol. 642, pp. 24–35. Springer, Cham (2018). https://doi.org/10.1007/978-3-319-66824-6_3
23. Ewald, D., Czerniak, J.M., Paprzycki, M.: A new OFNBee method as an example of fuzzy observance applied for ABC optimization. In: Prokopowicz, P., Czerniak, J., Mikołajewski, D., Apiecionek, Ł, Ślęzak, D. (eds.) Theory and Applications of Ordered Fuzzy Numbers. SFSC, vol. 356, pp. 223–237. Springer, Cham (2017). https://doi.org/10.1007/978-3-319-59614-3_13
24. Ewald, D., Zarzycki, H., Czerniak, J.M.: Certain aspects of the OFNBee algorithm operation for different fuzzifiers. In: Atanassov, K.T., et al. (eds.) IWIFSGN BOS/SOR 2020. LNNS, vol. 338, pp. 241–256. Springer, Cham (2022). https://doi.org/10.1007/978-3-030-95929-6_19
25. Galas, K.: Drive unit as a replacement for the platform. Stud. Mater. Appl. Comput. Sci. **12**(1), 10–14 (2020)
26. Guixiang, W., Jing, L.: Approximations of fuzzy numbers by step type fuzzy numbers. Fuzzy Sets Syst. **310**, 47–59 (2017)
27. Hajjari, T.: A new improved method for comparing fuzzy numbers by centroid point. J. Math. Ext. **10**(2), 1–23 (2016)
28. Jacko, P., Kovac, D., Bucko, R., Vince, T., Kravets, O.: The parallel data processing by nucleo board with STM32 microcontrollers. In: International Conference on Modern Electrical and Energy Systems (MEES), pp. 264–267 (2017)

29. Kacprzyk, J., Wilbik, A.: Using fuzzy linguistic summaries for the comparison of time series: an application to the analysis of investment fund quotations. In: IFSA/EUSFLAT Conference, pp. 1321–1326 (2009)
30. Khorshidi, H.A., Nikfalazar, S.: An improved similarity measure for generalized fuzzy numbers and its application to fuzzy risk analysis. Appl. Soft Comput. **52**, 478–486 (2017)
31. Kosinski, W.: On fuzzy number calculus. Int. J. Appl. Math. Comput. Sci. **16**(1), 51–57 (2006)
32. Kosinski, W.: On soft computing and modeling. Image Processing Communication, An International Journal with special section: Technologies of Data Transmission and Processing, held in 5th International Conference INTERPOR 2006, vol. 11, no. 1, pp. 71–82 (2006)
33. Kosiński, W., Frischmuth, K., Wilczyńska-Sztyma, D.: A new fuzzy approach to ordinary differential equations. In: Rutkowski, L., Scherer, R., Tadeusiewicz, R., Zadeh, L.A., Zurada, J.M. (eds.) ICAISC 2010. LNCS (LNAI), vol. 6113, pp. 120–127. Springer, Heidelberg (2010). https://doi.org/10.1007/978-3-642-13208-7_16
34. Kosinski, W., Prokopowicz, P., Slezak, D.: On algebraic operations on fuzzy numbers. In: Kłopotek, M.A., Wierzchoń, S.T., Trojanowski, K. (eds.) Intelligent Information Processing and Web Mining, pp. 353–362. Springer, Heidelberg (2003). https://doi.org/10.1007/978-3-540-36562-4_37
35. Kosinski, W., Słysz, P.: Fuzzy numbers and their quotient space with algebraic operations. Bull. Pol. Acad. Sci. Math. **41**(3), 285–295 (1993)
36. Macko, M., Szczepanski, Z., Mikolajewski, D., Mikolajewska, E., Listopadzki, S.: The method of artificial organs fabrication based on reverse engineering in medicine, pp. 353–365 (2017)
37. Marszalek, A., Burczynski, T.: Modeling and forecasting financial time series with ordered fuzzy candlesticks. Inf. Sci. **273**, 144–155 (2014)
38. Mikolajewska, E., Prokopowicz, P., Mikolajewski, D.: Computational gait analysis using fuzzy logic for everyday clinical purposes - preliminary findings. Bio-Algorithms Med-Syst. **13**(1), 37–42 (2017)
39. Namli, Ö.B., Tyburek, K.: Research of the efficiency of the reach of fire services to accidents in the City of Kocaeli on the basis of statistical data for the years 2013–2020. Stud. Mater. Appl. Comput. Sci. **14**(3), 1–5 (2022)
40. Piegat, A., Pluciński, M.: Computing with words with the use of inverse RDM models of membership functions. Int. J. Appl. Math. Comput. Sci. **25**(3), 675–688 (2015)
41. Piszcz, A., Mikolajewski, D.: Application for the android platform with the system as a solution to the right problem horse nutrition. Stud. Mater. Appl. Comput. Sci. **12**(1), 5–9 (2020). ISSN 1689-6300
42. Prokopowicz, P., Czerniak, J., Mikolajewski, D., Apiecionek, L., Slezak, D.: Theory and Applications of Ordered Fuzzy Numbers. Studies in Fuzziness and Soft Computing, vol. 356. Springer, Cham (2017). https://doi.org/10.1007/978-3-319-59614-3
43. Prokopowicz, P., Mikołajewski, D., Mikołajewska, E., Kotlarz, P.: Fuzzy system as an assessment tool for analysis of the health-related quality of life for the people after stroke. In: Rutkowski, L., Korytkowski, M., Scherer, R., Tadeusiewicz, R., Zadeh, L.A., Zurada, J.M. (eds.) ICAISC 2017. LNCS (LNAI), vol. 10245, pp. 710–721. Springer, Cham (2017). https://doi.org/10.1007/978-3-319-59063-9_64
44. Rezvani, S.: Ranking generalized fuzzy numbers with euclidian distance by the incentre of centroid. Mathematica Aeterna **3**, 103–114 (2013)

45. Rojek, I., Macko, M., Mikolajewski, D., Saga, M., Burczynski, T.: Modern methods in the field of machine modelling and simulation as a research and practical issue related to industry 4.0. Bull. Pol. Acad. Sci. Tech. Sci. **69**(2) (2021)
46. Sangho, B.: Comparison of selected wolf pack algorithms used in solving optimization problems. Stud. Mater. Appl. Comput. Sci. **13**(1), 17–32 (2021)
47. Stachowiak, A., Dyczkowski, K.: A similarity measure with uncertainty for incompletely known fuzzy sets. In: Proceedings of the 2013 Joint IFSA World Congress and NAFIPS Annual Meeting (IFSA/NAFIPS), pp. 390–394 (2013)
48. Szmidt, E., Kacprzyk, J.: Distances between intuitionistic fuzzy sets. Fuzzy Sets Syst. **114**, 505–518 (2000)
49. Trana, L., Ducksteinb, L.: Comparison of fuzzy numbers using a fuzzy distance measure. Fuzzy Sets Syst. **130**, 331–341 (2002)
50. Tyburek, K., Bora, Ö.: Comparison of the efficiency of time and frequency domain descriptors for the classification of selected wind instruments. Stud. Mater. Appl. Comput. Sci. **14**(3), 6–13 (2022)
51. Vince, T., Lukac, P., Schweiner, D., Tomcikova, I., Mamchur, D.: Android application supporting developed web applications testing. In: International Conference on Modern Electrical and Energy Systems (MEES), pp. 392–395. IEEE (2017)
52. Wang, Y., Luo, Y.: Area ranking of fuzzy numbers based on positive and negative ideal points. Comput. Math. Appl. **58**, 1763–1779 (2009)
53. Yager, R.: A procedure for ordering fuzzy subsets of the unit interval. Inf. Sci. **24**, 143–161 (1981)
54. Zadeh, L.: Outline of new approach to the analysis of complex systems and decision process. IEEE Syst. Man Cybern. SMC **3**, 28–44 (1973)
55. Zadrozny, S., Kacprzyk, J.: On the use of linguistic summaries for text categorization. In: Proceedings of IPMU, pp. 1373–1380 (2004)
56. Zarzycki, H.: Application of the finite difference CN method to value derivatives. Stud. Proc. Pol. Assoc. Knowl. Manag. **42**, 267–277 (2011)
57. Zarzycki, H., Dobrosielski, W., Apiecionek, Ł., Vince, T.: Center of circles intersection, a new defuzzification method for fuzzy numbers. Bull. Pol. Acad. Sci. Tech. Sci. **68**(2) 185–190 (2020). http://journals.pan.pl/Content/115177/PDF/02D_185-190_01346_Bpast.No.68-2_30.04.20_K4A_SS_TeX.pdf
58. Zarzycki, H.: Comparative study of the firefly algorithm and the whale algorithm. In: Kahraman, C., Tolga, A.C., Cevik Onar, S., Cebi, S., Oztaysi, B., Sari, I.U. (eds.) INFUS 2022. LNNS, vol. 504, pp. 999–1006. Springer, Cham (2022). https://doi.org/10.1007/978-3-031-09173-5_114
59. Zarzycki, H., Apiecionek, Ł, Czerniak, J.M., Ewald, D.: The proposal of fuzzy observation and detection of massive data DDOS attack threat. In: Atanassov, K.T., et al. (eds.) IWIFSGN 2018. AISC, vol. 1081, pp. 363–378. Springer, Cham (2021). https://doi.org/10.1007/978-3-030-47024-1_34
60. Zarzycki, H., Czerniak, J.M., Dobrosielski, W.T.: Detecting nasdaq composite index trends with OFNs. In: Prokopowicz, P., Czerniak, J., Mikołajewski, D., Apiecionek, Ł, Ślęzak, D. (eds.) Theory and Applications of Ordered Fuzzy Numbers. SFSC, vol. 356, pp. 195–205. Springer, Cham (2017). https://doi.org/10.1007/978-3-319-59614-3_11
61. Zarzycki, H., Dobrosielski, W.T., Czerniak, J.M., Ewald, D.: Use of OFN in the short-term prediction of exchange rates. In: Atanassov, K.T., et al. (eds.) IWIFSGN 2019 2019. AISC, vol. 1308, pp. 289–301. Springer, Cham (2021). https://doi.org/10.1007/978-3-030-77716-6_25

62. Zarzycki, H., Ewald, D., Skubisz, O., Kardasz, P.: A comparative study of two nature-inspired algorithms for routing optimization. In: Atanassov, K.T., et al. (eds.) IWIFSGN BOS/SOR 2020. LNNS, vol. 338, pp. 215–228. Springer, Cham (2022). https://doi.org/10.1007/978-3-030-95929-6_17
63. Zarzycki, H., Skubisz, O.: A new artificial bee colony algorithm approach for the vehicle routing problem. In: Kahraman, C., Cebi, S., Cevik Onar, S., Oztaysi, B., Tolga, A.C., Sari, I.U. (eds.) INFUS 2021. LNNS, vol. 307, pp. 562–569. Springer, Cham (2022). https://doi.org/10.1007/978-3-030-85626-7_66

Advanced Data Analysis and Machine Learning

Parameterized Interval-Valued Aggregation Functions in Classification of Data with Large Number of Missing Values

Urszula Bentkowska(✉) and Marcin Mrukowicz

Institute of Computer Science, University of Rzeszów, Rzeszów, Poland
{ubentkowska,mmrukowicz}@ur.edu.pl

Abstract. In this contribution parameterized interval-valued aggregation functions are applied in an algorithm based on k nearest neighbours algorithm which uses interval modelling to improve the quality of binary classification in the case of large number of missing values in datasets. Missing values are common in real-life problems and as a result performance of classifiers may be lowered significantly. One of the approach to cope with this problem is to use diverse methods of filling missing values. In this contribution we proposed to apply a kind of multiple imputation method involved in interval modelling approach. The considered interval-valued aggregation functions belong to diverse families of aggregations considered with respect to monotonicity conditions based on classical partial order for intervals and also other comparability relations for intervals. Moreover, these interval-valued aggregation functions are defined with diverse parameters which makes possibility to fit a given aggregation to the data and considered algorithm.

Keywords: Aggregation functions · Interval-valued fuzzy sets · Classifiers · kNN · Missing values · Multiple imputation

1 Introduction

Missing values are common in practice in datasets. There are various reasons for this situation, "such as manual data entry procedures, equipment errors and incorrect measurements" [8]. The missing data could be described as intended to collect, but not actually collected for various reasons. For example, respondent refuses to answer a question in questionnaire or the sensor breaks down and do not record data. It is also possible that some data was anonymized so some data need to be deleted to protect the privacy of some person. The missing values could be represented as well-known NULL in databases, or other dedicated symbol. The missing data causes a few problems. First of all it is a loss of efficiency, because dealing with missing data requires some additional time compared to non-missing data. The second problem is that not all AI algorithms are able to

K. T. Atanassov et al. (Eds.): BOS/SOR/IWIFSGN 2022, LNNS 793, pp. 85–94, 2023.
https://doi.org/10.1007/978-3-031-45069-3_8

process datasets with missing data, which leads to "complications with handling and analyzing the data" [8]. Finally, missing values add some bias to dataset; even if the missing data will be for example imputed, they are different and not equivalent to the complete data [8]. However, since the missing data are common, there is usually no other choice than working with them, despite the mentioned problems.

The literature described diverse solutions to handle missing data. The simplest way is elimination of such data. It can be both reducing the whole attributes or whole objects. However this approach could be used effectively only if the number of missings in dataset is small and the dataset is large. This method is waste-full, since we reject some information present in dataset. The second possible solution is to treat missing values as special values and prepare AI algorithms in the way, they can recognize them and process them in a special way. The last common used approach is to replace the missing values with some concrete values, which is believed to be closed to real value of an object if it will be collected or at least imputing this value will not influence negatively future analysis of the dataset. The basic methods of this family is replacing missing values with average or most frequent values for given attributes. Other most sophisticated solutions assume that we can find objects similar (or somehow related) to those with missing values and then if they do not contain missings, we can impute the missing value based on those values. The literature knows many methods of this kind of imputation, for example using k-nearest neighbours, neural nets, association rules and regression [7,8].

The other possible method of dealing with missing values is to fill missing values with values "generated at random from the variable distribution observed" ([10], p. 31). This method has one advantage over the mean imputation: it preserves the shape of the distribution. When there are many missing values, imputing for example a mean value, will artificially change the original shape of the distribution of attribute for known, non-missing data [7,10]. However, there is hard to state how much imputed value is close to the concrete value in hypothetically complete dataset. There will always be uncertainty about the correctness of imputation, independently of the used method. The situation, when the missing value is filled only once is called single imputation [7]. To overcome the limits of single imputation, multiple imputation was introduced. As the name suggest, this strategy is based on doing imputation multiple times and hence produce more than one dataset with imputed missings. As a result each of this newly created dataset need to be processed independently and at the end the results is merged into one [7]. This method is believed to better handle uncertainty of the corectness of imputation than a single imputation. In [13] there is comprehensive analysis between single and multiple imputation. Both methods produce unbiased estimates. "The single imputation procedure commonly results in an underestimation of the standard errors or too small P-values, i.e., overestimation of the precision of the study associations" [13]. As opposite, the multiple imputation results in correctly estimated standard errors and confidence intervals.

Recently, the interval modelling was applied to the problem of handling missing values [2,4,14]. The interval modelling seems to be able to process uncertain knowledge in efficient way. In [2] the algorithm, based on the multiple random filling from observed distribution, was proposed. The main idea of this algorithm is to aggregate main class confidence, represented as intervals, derived from applying multiple kNN algorithm on multiple datasets, created at random filling missing values. The generalization of this algorithm to the multi-class case was proposed in [4]. The further details of this algorithm is descried in Sect. 3. In this contribution we will apply and compare parameterized interval-valued aggregation functions to aggregate main class confidence.

The paper is constructed as follows. In Sect. 2, there are provided some preliminary notions related to interval-valued aggregation functions. In Sect. 3, details of the algorithms and performed experiments are given and in Sect. 4, results of the experiments are listed.

2 Basic Notions and Properties Related to IV Aggregation Functions

We recall now some notions applied in our further work. The family of intervals in the unit interval is denoted by L^I, i.e. $L^I = \{[\underline{x}, \overline{x}] : \underline{x}, \overline{x} \in [0,1], \underline{x} \leqslant \overline{x}\}$. Classical partial order in L^I is of the form

$$[\underline{x}, \overline{x}] \preceq [\underline{y}, \overline{y}] \Leftrightarrow \underline{x} \leqslant \underline{y},\ \overline{x} \leqslant \overline{y}.$$

Basic operations in L^I are the following

$$[\underline{x}, \overline{x}] \vee [\underline{y}, \overline{y}] = [\max(\underline{x}, \underline{y}), \max(\overline{x}, \overline{y})], \quad [\underline{x}, \overline{x}] \wedge [\underline{y}, \overline{y}] = [\min(\underline{x}, \underline{y}), \min(\overline{x}, \overline{y})].$$

$(L^I, \vee, \wedge)$ is a complete lattice (with respect to $\preceq$) with bounds $\mathbf{0} = [0,0]$, $\mathbf{1} = [1,1]$. Other examples of comparability relations between intervals ([5,11,12]) are the possible $\preceq_\pi$ and the necessary relation $\preceq_\nu$, respectively, where

$$[\underline{x}, \overline{x}] \preceq_\pi [\underline{y}, \overline{y}] \Leftrightarrow \underline{x} \leqslant \overline{y}$$

$$[\underline{x}, \overline{x}] \preceq_\nu [\underline{y}, \overline{y}] \Leftrightarrow \overline{x} \leqslant \underline{y}.$$

Below we recall some basic properties of these relations.

Proposition 1. *Relation $\preceq_\pi$ is an interval-order in L^I. Relation $\preceq_\nu$ is antisymmetric and transitive in L^I. Moreover, $\preceq_\nu \Rightarrow \preceq \Rightarrow \preceq_\pi$.*

Thanks to the diversity of comaparability relations (some of them are orders) we may consider interval-valued aggregation functions with respect to these relations (shortly called IV-aggregation functions).

Definition 1 (cf. [1,9]). *Let be given $\mathcal{A} : (L^I)^n \to L^I$ and boundary conditions $\mathcal{A}(\underbrace{\mathbf{0}, ..., \mathbf{0}}_{n\times}) = \mathbf{0}$, $\mathcal{A}(\underbrace{\mathbf{1}, ..., \mathbf{1}}_{n\times}) = \mathbf{1}$ be fulfilled. $\mathcal{A}$ is called:*

– *interval-valued aggregation function if it is increasing, i.e.*

$$\forall_{\boldsymbol{x}_i,\boldsymbol{y}_i\in L^I} \boldsymbol{x}_i \preceq \boldsymbol{y}_i \Rightarrow \mathcal{A}(\boldsymbol{x}_1,\ldots,\boldsymbol{x}_n) \preceq \mathcal{A}(\boldsymbol{y}_1,\ldots,\boldsymbol{y}_n),$$

– *possible aggregation function (pos–aggregation function), if*

$$\forall_{\boldsymbol{x}_i,\boldsymbol{y}_i\in L^I} \boldsymbol{x}_i \preceq_\pi \boldsymbol{y}_i \Rightarrow \mathcal{A}(\boldsymbol{x}_1,\ldots,\boldsymbol{x}_n) \preceq_\pi \mathcal{A}(\boldsymbol{y}_1,\ldots,\boldsymbol{y}_n),$$

– *necessary aggregation function (nec–aggregation function), if*

$$\forall_{\boldsymbol{x}_i,\boldsymbol{y}_i\in L^I} \boldsymbol{x}_i \preceq_\nu \boldsymbol{y}_i \Rightarrow \mathcal{A}(\boldsymbol{x}_1,\ldots,\boldsymbol{x}_n) \preceq_\nu \mathcal{A}(\boldsymbol{y}_1,\ldots,\boldsymbol{y}_n).$$

Among these aggregation functions we may distinguish some subfamilies.

Definition 2 **([6]).** *$\mathcal{A} : (L^I)^n \to L^I$ is called decomposable (representable) if there exist functions $A_1, A_2 : [0,1]^2 \to [0,1]$ such that*

$$\mathcal{A}(\boldsymbol{x}_1,\ldots,\boldsymbol{x}_n) = [A_1(\underline{x}_1,\ldots,\underline{x}_n), A_2(\overline{x}_1,\ldots,\overline{x}_n)], \quad \boldsymbol{x} = [\underline{x}_1,\overline{x}_1],\ldots,\boldsymbol{x}_n = [\underline{x}_n,\overline{x}_n] \in L^I.$$

Under the assumption of decomposability we have some special properties of the mentioned IV-aggregation functions, e.g. $\mathcal{A}_\nu \subset \mathcal{A} \subset \mathcal{A}_\pi$ (cf. [1]).

Proposition 2 **([1]).** *Let $\mathcal{A} : (L^I)^n \to L^I$ be decomposable, $\mathcal{A} = [A_1, A_2]$. $\mathcal{A}$ is a nec-aggregation function if and only if $A_1 = A_2$, A_1 is an aggregation function on $[0,1]$.*

Example 1 (cf. [3]). Let $q \leqslant s$, $q, s \in \mathbb{R}_0$

$$\mathcal{A}_I(\mathbf{x}_1,\mathbf{x}_2,\ldots,\mathbf{x}_n) = \left[\left(\frac{\underline{x}_1^q + \underline{x}_2^q + \ldots + \underline{x}_n^q}{n}\right)^{\frac{1}{q}}, \left(\frac{\overline{x}_1^s + \overline{x}_2^s + \ldots + \overline{x}_n^s}{n}\right)^{\frac{1}{s}}\right]$$

$\mathcal{A}_I$ is decomposable, it is a necessary aggregation for $q = s$ (and then it is an interval-valued aggregation and possible aggregation).

Let $q \in \mathbb{R}_0$

$$\mathcal{A}_{II}(\mathbf{x}_1,\mathbf{x}_2,\ldots,\mathbf{x}_n) = \left[\left(\frac{\underline{x}_1^q + \underline{x}_2^q + \cdots + \underline{x}_n^q}{n}\right)^{\frac{1}{q}}, \right.$$

$$\left. \max\left(\left(\frac{\underline{x}_1^q + \overline{x}_2^q + \cdots + \overline{x}_n^q}{n}\right)^{\frac{1}{q}},\ldots,\left(\frac{\overline{x}_1^q + \cdots + \overline{x}_{n-1}^q + \underline{x}_n^q}{n}\right)^{\frac{1}{q}}\right]$$

$\mathcal{A}_{II}$ is not decomposable. It is a necessary aggregation and interval-valued aggregation function.

Let $q, s \in \mathbb{R}_0$, $p \in [0,1]$

$$\mathcal{A}_{III}(\mathbf{x}_1,\mathbf{x}_2,\ldots,\mathbf{x}_n) = \left[p\min(\underline{x}_1,\underline{x}_2,\ldots,\underline{x}_n) + (1-p)\left(\frac{\underline{x}_1^q + \underline{x}_2^q + \ldots + \underline{x}_n^q}{n}\right)^{\frac{1}{q}}, \right.$$

$$\left. p\left(\frac{\overline{x}_1^s + \overline{x}_2^s + \ldots + \overline{x}_n^s}{n}\right)^{\frac{1}{s}} + (1-p)\max(\overline{x}_1,\overline{x}_2,\ldots,\overline{x}_n)\right]$$

A_{III} is a decomposable interval-valued aggregation function and it is a pos-aggregation.

Let $q \in \mathbb{R}$, $q \geqslant 1$, $p \in [0,1]$

$$\mathcal{A}_{IV}(\mathbf{x}_1, \mathbf{x}_2, \ldots, \mathbf{x}_n) = \left[p \sqrt[n]{\prod_{k=1}^{n} \underline{x}_k} + (1-p)\frac{1}{n}\sum_{k=1}^{n} \underline{x}_k, \right.$$
$$\left. p\frac{\overline{x}_1^2 + \overline{x}_2^2 + \cdots + \overline{x}_n^2}{\overline{x}_1 + \overline{x}_2 + \cdots + \overline{x}_n} + (1-p)\frac{\overline{x}_1^q + \overline{x}_2^q + \cdots + \overline{x}_n^q}{\overline{x}_1^{q-1} + \overline{x}_2^{q-1} + \cdots + \overline{x}_n^{q-1}} \right]$$

A_{IV} is a decomposable pos-aggregation function.

Diverse construction methods allow to obtain many examples of the mentioned families of IV-aggregation functions (cf. Definition 1, [3]). We consider here four parameterized aggregation functions presented in Example 1.

3 Algorithm and Details of Experiments

Algorithm 1 which uses interval modelling and IV-aggregation functions to cope with missing values was introduced in [2]. The main idea of this algorithm is to find objects with at least one missing value, then make a fixed number of copies of this object. Then for each copy and for each attribute, which value is missing, the missing value is filled with a value randomly selected, based on the range of all possible values of this attribute, delivered from the training dataset. Then for each object a collection of k-NN classifiers is applied and vectors of main class confidences is determined for each object, respectively. Based on these vectors, the intervals are generated, where the lower bound is the minimum value of corresponding vector and analogously the upper bound is the maximum value of corresponding vector. Then the intervals are aggregated and finally the received interval is reduced to a scalar value, used as a threshold, based on which the main or subordinate class is assigned to a test object. In this contribution the impact of parameterized IV-aggregations on received classification quality (AUC) is examined. In Table 2 we present the parameters of IV-aggregation functions which were tested in our experiments. As we may see there were tested diverse sets of parameters for each aggregation, i.e. positive, negative, integers, and rational numbers.

We tested our algorithm using a few datasets form UCI Irvine Repository (Table 1). The collection of k-NN classifiers was used with the Euclidean distance and parameters $k \in \{3, 5, 7, 9, 11, 13, 15, 30\}$. We decided to make some assumptions about the methodology of experiments. The binary classification datasets with no missing values were taken into account, and then the various levels of missing data was imputed to the original datasets. However, the imputation of missing values was combined with the 10-fold cross-validation, with the assumption that, there was no missing values in a training dataset and some amount of missing values was randomly imputed into test data with the same proportion. For example, if the input parameter is 0.2, then 20% of the missing value is randomly entered into every attribute (column of the dataset). Seeds were used in every applicable place, i.e., cross-validation, imputation of missing values and finally the random sampling fill to maintain repeatability of experiments and provide exactly the same final dataset to all parameterized IV-aggregations. Average value of AUC and standard deviation on the basis of different folds were determined for every combination of parameters given in Table 2. Implementation of the algorithm and Python scripts to reproduce performed experiments are available on Github [15].

Algorithm 1. Binary classifier

Input:

1. data set represented by $\mathbf{T} = (U, A, d)$,
2. collection $C_1, ..., C_m$ of k-NN classifiers, e.g., $k \in \{5, 10, 20, 30\}$,
3. fixed parameter r, e.g., $r = 10$,
4. aggregation function $\mathcal{A}$,
5. test object u

Output:

1. The certainty coefficient representing the probability of belonging the object u to the "main class"

if exists at least one missing value in the object u **then**
  for $i := 1$ **to** m **do**
    Choose randomly with the Monte Carlo method r objects $u_1, ..., u_r$ on the basis of object u, where any object u_j is constructed as follows, $j \in \{1, ..., r\}$: copy values of attributes from u to u_j
    for all attribute whose value in u_j is missing **do**
      replace it with a randomly selected value from the range of possible values for this attribute (from the training data)
    Compute certainty coefficient for objects $u_1, ..., u_r$ using the classifier C_i and assign these values to $p_1, ..., p_r$
    Compute $min\{p_1, ..., p_r\}$ and assign it to min_i
    Compute $max\{p_1, ..., p_r\}$ and assign it to max_i
    Determine the uncertainty interval $[down(u), up(u)]$ for the object u by aggregating (with the use of $\mathcal{A}$) the intervals $[min_1, max_1]$, ..., $[min_m, max_m]$
    Determine the final certainty coefficient $\mathbf{p} = \frac{down(u)+up(u)}{2}$ for the object u
else
  for $i := 1$ **to** m **do**
    Compute certainty coefficient ("main class" membership probability) for the given test object u using the classifier C_i and assign it to p_i
    Determine the uncertainty interval $[down(u), up(u)]$ for the object u by aggregating (with the use of $\mathcal{A}$) the intervals $[p_1, p_1]$, ..., $[p_m, p_m]$
    Determine the final certainty coefficient $\mathbf{p} = \frac{down(u)+up(u)}{2}$ for the object u
if $\mathbf{p} > 0.5$ **then**
  return u belongs to the "main class"
else
  return u belongs to the "subordinate class"

Table 1. Datasets details

Dataset	Objects	Attributes	Classes
Biodeg	1,055	41	2
German	1,000	24	2
Ozone	1,847	72	2
Parkinson	1,040	27	2
Spam	4,601	57	2

Table 2. Parameteres of IV-aggregation functions

Agg	q	s	Agg	q	Agg	q	s	p	Agg	q	p
A_I	−2.0	−0.5	A_{II}	−1.5	A_{III}	−1.5	−0.5	0.5	A_{IV}	1.0	1.0
A_I	−0.5	2.0	A_{II}	−0.5	A_{III}	−0.5	0.5	0.5	A_{IV}	2.0	0.0
A_I	0.5	1.0	A_{II}	0.5	A_{III}	0.5	2.0	0.5	A_{IV}	3.0	0.0
A_I	1.0	1.0	A_{II}	1.0	A_{III}	1.0	2.0	1.0	A_{IV}	3.0	0.2
A_I	1.5	3.0	A_{II}	1.5	A_{III}	2.0	1.0	0.0	A_{IV}	3.0	0.5
A_I	2.0	2.0	A_{II}	2.0	A_{III}	3.0	3.0	0.2	A_{IV}	3.0	0.8
A_I	2.0	3.0			A_{III}	3.0	3.0	0.8	A_{IV}	3.0	1.0
A_I	3.0	4.0							A_{IV}	5.0	0.5
A_I	4.0	5.0									

Table 3. Biodeg and German datasets

Level of missing values	Biodeg dataset				German dataset			
	max AUC				max AUC			
	A_I	A_{II}	A_{III}	A_{IV}	A_I	A_{II}	A_{III}	A_{IV}
0.00	**0.915**	0.914	0.914	0.914	**0.746**	**0.746**	**0.746**	**0.746**
0.01	**0.909**	0.908	0.908	0.907	**0.740**	**0.740**	0.738	**0.740**
0.05	**0.892**	0.887	0.890	0.886	**0.753**	0.750	**0.753**	0.748
0.10	**0.861**	0.851	0.860	0.854	**0.740**	0.739	0.734	0.736
0.20	0.815	0.813	**0.817**	0.813	**0.719**	0.718	0.715	0.713
0.30	0.774	0.774	**0.778**	0.771	**0.714**	**0.714**	0.706	0.708
0.40	0.714	0.713	**0.716**	0.707	0.692	**0.693**	0.684	0.688
0.50	**0.696**	**0.696**	0.693	0.686	0.662	**0.663**	0.652	0.656

4 Results and Statistical Tests

If we look at results in Tables 3, 4 and 5, we can see that parameterized IV-aggregation functions provide comparable results and there is no a clear leader in all examined datasets. The quality of classification decreases with the growth

Table 4. Ozone and Parkinson datasets

Level of missing values	Ozone dataset				Parkinson dataset			
	max AUC				max AUC			
	A_I	A_{II}	A_{III}	A_{IV}	A_I	A_{II}	A_{III}	A_{IV}
0.00	**0.878**	**0.878**	0.877	**0.878**	**0.976**	**0.976**	**0.976**	**0.976**
0.01	**0.879**	0.877	0.877	0.878	0.970	0.969	**0.971**	0.970
0.05	0.875	**0.876**	0.872	0.873	**0.968**	0.963	**0.968**	0.966
0.10	0.874	0.873	**0.875**	0.869	**0.957**	0.956	**0.957**	0.953
0.20	0.854	**0.855**	0.854	0.847	**0.925**	0.923	**0.925**	0.916
0.30	**0.843**	**0.843**	0.841	0.824	**0.892**	**0.892**	0.889	0.874
0.40	**0.812**	0.811	0.804	0.779	0.835	**0.836**	0.834	0.818
0.50	**0.750**	**0.750**	0.739	0.713	0.804	**0.807**	0.802	0.785

Table 5. Spam dataset

Level of missing values	**Spam dataset**			
	max AUC			
	A_I	A_{II}	A_{III}	A_{IV}
0.00	**0.963**	0.962	**0.963**	**0.963**
0.01	**0.957**	**0.957**	0.953	0.953
0.05	**0.930**	**0.930**	0.923	0.921
0.10	0.887	**0.888**	0.881	0.875
0.20	0.811	**0.812**	0.807	0.798
0.30	**0.760**	**0.760**	0.756	0.746
0.40	**0.713**	**0.713**	0.711	0.703
0.50	**0.670**	**0.670**	**0.670**	0.660

of the level of missing values in a dataset, as expected. There is also no pattern, that some aggregation is better or worse depending on the level of the missing values in a dataset. In Table 7, we can see that family A_{II} seems to have both the lowest mean and median globally, computed for all dataset results combined. On the other hand, the family A_{III} has both the greatest mean and median globally. For statistical analysis, first of all, we checked if computed results are normally distributed with Shapiro-Wilk test for each examined dataset and all results combined. Due to this test, none of the received results was normally distributed. As a result, we performed a non-parametric Mann-Whitney U test between pairs of parameterized IV-aggregation families and we did not obtain any statistically significant difference between pairs. However, it also means, that we cannot state, that any of the examined parameterized family of IV-aggregation functions is not appropriate to the problem of classifying datasets with large percentage of missing values. Additionally, the comparison with aggregations $A_1 - A_{10}$ from [2] was made. For small levels of missing values, there

Table 6. Max AUC obtained for aggregations A_1-A_{10} as parameters of Algorithm 1

Level of missing values	Aggregation, maximum AUC				
	Biodeg	German	Ozone	Parkinson	Spam
0.1	A_9, 0.858	A_7, 0.739	A_1, A_7, 0.874	A_8, 0.953	A_2, **0.888**
0.2	A_6, 0.815	A_1, A_2, A_7, 0.718	A_2, A_{10}, 0.854	A_8, 0.916	A_2, **0.812**
0.3	A_5, 0.774	A_1, A_7, 0.713	A_1, A_2, **0.843**	A_8, 0.874	A_1, A_2, **0.760**
0.4	A_6, 0.714	A_2, A_7, **0.693**	A_7, **0.812**	A_8, 0.818	A_{10}, **0.713**
0.5	A_1, A_2, A_7, **0.696**	A_1, 0.661	A_1, A_2, A_7, **0.750**	A_8, 0.785	A_6, **0.670**

Table 7. Summary of the results for all considered datasets and missing values

		AUC					
	count	max	mean	median	std	sum	var
A_{III}	240	0.976	0.820	0.832	0.097	196.705	0.009
A_I	360	0.976	0.816	0.813	0.098	293.792	0.010
A_{IV}	320	0.976	0.815	0.818	0.098	260.947	0.010
A_{II}	200	0.976	0.802	0.801	0.106	160.374	0.011

is no significant differences between aggregations, however for levels provided in Table 6 we may observe, that parameterized IV-aggregation functions outperformed previously proposed examples. For Biodeg and German datasets only for one level of missings, aggregations $A_1 - A_{10}$ has maximum AUC equal to maximum AUC of families $A_I - A_{IV}$. In Ozone dataset such equality is for three levels. In Parkinson dataset the parameterized families were better for all levels. However, for Spam dataset none of the parameterized families outperformed previous examples from [2]. The conclusion is that the families $A_I - A_{IV}$ has better or equal AUC than aggregations $A_1 - A_{10}$ for the given datasets. Especially in the case of one dataset (Parkinson) the difference seems to be significant. This leads to another conclusion, that concrete aggregation could be better suited to concrete dataset than other.

5 Conclusions and Future Plans

Diverse types of IV-aggregation functions proved to be the best in the case of concrete datasets and levels of missing values so it is worth to consider them to improve the performance of classifiers. In future study more examples of parameterized families of IV-aggregations will be studied considering also the problem of fitting IV-aggregations to datasets (learning parameters of IV-aggregations). Moreover, classification algorithm for missing values will be adjusted to the case of different percentage of missing values of the conditional attributes and real (not simulated) datasets with missing values of the conditional attributes. Furthermore, since received results are not normally distributed, it will be valuable to examine the distributions of the results and their possible impact on comparison between received quality measures of parameterized IV-aggregation families.

Acknowledgements. This work was partially supported by the Centre for Innovation and Transfer of Natural Sciences and Engineering Knowledge of University of Rzeszów, Poland, the project RPPK.01.03.00-18-001/10.

References

1. Bentkowska, U.: New types of aggregation functions for interval-valued fuzzy setting and preservation of pos-B and nec-B-transitivity in decision making problems. Inf. Sci. **424**, 385–399 (2018)
2. Bentkowska, U., Bazan, J.G., Rząsa, W., Zaręba, L.: Application of interval-valued aggregation to optimization problem of k-NN classifiers for missing values case. Inf. Sci. **486**, 434–449 (2019)
3. Bentkowska, U.: Interval-Valued Methods in Classifications and Decisions. SFSC, vol. 378. Springer, Cham (2020). https://doi.org/10.1007/978-3-030-12927-9
4. Bentkowska, U., Bazan, J.G., Mrukowicz, M., Zaręba, L., Molenda, P.: Multi-class classification problems for the k-NN algorithm in the case of missing values. In: 2020 IEEE International Conference on Fuzzy Systems (FUZZ-IEEE), pp. 1–8 (2020)
5. De Baets, B.: Aggregation of structured objects. In: Lecture during International Symposium on Aggregation on Bounded Lattices (ABLAT) 2014, Karadeniz Technical University, Trabzon, Turkey, 16–20 June (2014)
6. Drygaś, P., Pękala, B.: Properties of decomposable operations on some extension of the fuzzy set theory. In: Atanassov, K.T., Hryniewicz, O., Kacprzyk, J., et al. (eds.) Advances in Fuzzy Sets, Intuitionistic Fuzzy Sets, Generalized Nets and Related Topics, pp. 105–118. EXIT, Warsaw (2008)
7. Jadhav, A., Pramod, D., Ramanathan, K.: Comparison of performance of data imputation methods for numeric dataset. Appl. Artif. Intell. **33**(10), 913–933 (2019)
8. Kaiser, J.: Dealing with missing values in data. J. Syst. Integr. **5**(1), 1804–2724 (2014)
9. Komorníková, M., Mesiar, R.: Aggregation functions on bounded partially ordered sets and their classification. Fuzzy Sets Syst. **175**, 48–56 (2011)
10. Larose, D.T.: Discovering Knowledge In Data: An Introduction to Data Mining. John Wiley & Sons Inc, Hoboken, New Jersey (2005)
11. Pękala, B., Bentkowska, U., De Baets, B.: On comparability relations in the class of interval-valued fuzzy relations. Tatra Mountains Math. Publ. **66**, 91–101 (2016)
12. Pękala, B., De Baets, B.: Structures of the class of interval-valued fuzzy relations created by different comparability relations. Submitted
13. Rogier, A., Donders, T., Van der Heijden, G.J.M.G., Stijnen, T., Moons, K.G.M.: Review: a gentle introduction to imputation of missing values. J. Clin. Epidemiol. **59**(10), 1087–1091 (2006)
14. Wójtowicz, A., Zywica, P., Stachowiak, A., Dyczkowski, K.: Solving the problem of incomplete data in medical diagnosis via interval modeling. Appl. Soft Comput. **47**, 424–437 (2016)
15. https://github.com/furoDMGroup/IWIFSGN2022 . Accessed 31 Oct 2022

Rough Set Flow Graphs for Information Systems over Ontological Graphs

Krzysztof Pancerz[1(✉)] and Jaromir Sarzyński[2]

[1] Institute of Technology and Computer Science, Academy of Zamość, Zamość, Poland
krzysztof.pancerz@akademiazamojska.edu.pl

[2] College of Natural Sciences, Institute of Computer Science, University of Rzeszów, Rzeszów, Poland
jsarzynski@ur.edu.pl

Abstract. In the paper, we present an approach in which information flow over time is represented by rough set flow graphs. We assume that the underlying information consists of concepts embedded in ontologies representing knowledge of the domains under consideration. Moreover, the Rough Set Flow Graph Visualizer module, implemented as a part of the Classification and Prediction Software System (CLAPSS), is depicted. An example of using the module to model sample data containing sequential information about the educational paths and job careers of people is also discussed.

Keywords: rough sets · flow graphs · information systems · ontological graphs

1 Introduction

One of the intensively developed trends of data processing and knowledge discovery from data, inspired by Zadeh's *computing with words* (cf. [12]), is processing words, concepts, and natural language statements. The main idea is that words and concepts are used in place of numbers for computing and reasoning. Over time, a lot of methodologies were proposed to process data of such character. The main topic of our research concerns information or decision systems (in Pawlak's sense [9]), whose attribute values are words or concepts that describe objects or phenomena. To effectively carry out data mining processes, we require to take into consideration the domain knowledge related to data semantics [11]. The domain knowledge delivers important information about different aspects of mined data. Some of the significant aspects are relations between values of attributes describing objects, especially, semantic relations. Data mining with the domain knowledge was an extensively studied research area in the past. Different forms of the domain knowledge have been used. In our case, we use domain ontologies. The domain knowledge is expressed as a set of concepts together with

K. T. Atanassov et al. (Eds.): BOS/SOR/IWIFSGN 2022, LNNS 793, pp. 95–105, 2023.
https://doi.org/10.1007/978-3-031-45069-3_9

the relationships, which have been defined between them, comprising the vocabulary from a given area (cf. [3]). In [5], we proposed to incorporate directly the domain knowledge in a form of ontology into information systems. In this case, values of a given attribute are concepts from the domain described by this attribute. The domain is modelled using an ontology presented, in a simplified way, by means of the graph structure, called the ontological graph. In such a graph, each node represents one concept from the ontology, whereas each edge represents a relation between two concepts.

In [6], we presented a possibility of application of Pawlak's flow graphs [10] (called here rough set flow graphs) to modelling information flow described by information or decision systems over ontological graphs. In each vertical layer of the graph, there are nodes corresponding to all concepts appearing in the ontological graph associated with a given attribute. Edges of the graph are labelled with three coefficients, certainty, strength, and covering. The graph allows us to characterize the possibility of the emergence of a given flow in the described system. By taking into consideration semantic relations, it is possible to characterize flows containing concepts at different levels of abstraction defined in ontological graphs.

The presented approach has been implemented in Classification and Prediction Software System (CLAPSS) [4]. CLAPSS is a tool in which some specialized approaches based on fuzzy sets and rough sets are implemented (cf. [7,8]). The tool is equipped with a user-friendly graphical interface. One of the new functionalities, recently added to the CLAPSS, is visualizing rough set flow graphs for information systems over ontological graphs. On the basis of rough set flow graphs, it is also possible in CLAPSS to perform the temporal inference supported by fuzzy set operators.

As an example, we give in the paper modelling paths of young people's education and professional careers. The model includes the sequential information about the education career path, and next, about the professional career.

2 Basic Notions

Before proceeding to the rest of the paper, it is recommended to get familiar with the selected notions like: information and decision systems IS/DS [9], rough sets [9], Rough Set Flow Graphs $RSFG$ [10], ontological graphs OG [5], simple information systems over ontological graphs SIS^{OG} [5]. To explain these notions, a simple example is demonstrated.

Example 1. Let $SIS^{OG} = (U, A, \{OG_a\}_{a \in A}, f_{inf})$ be an information system over ontological graphs, where: $U = \{u_1, u_2, ..., u_{16}\}$ is the set of objects, $A = \{a_1, a_2\}$ is the set of attributes, $\{OG_{a_1}, OG_{a_2}\}$ is the family of ontological graphs assigned to the attributes, and f_{inf} is the information function represented by the following tables:

U/A	a_1	a_2
u_1	A	C
u_2	A	C
u_3	A	C1
u_4	A	C2
u_5	A1	C1
u_6	A2	C1

U/A	a_1	a_2
u_7	A	D
u_8	A	D1
u_9	A1	D
u_{10}	B	C
u_{11}	B	C
u_{12}	B1	C1

U/A	a_1	a_2
u_{13}	B1	C1
u_{14}	B2	C2
u_{15}	B2	C1
u_{16}	B	C1
u_{17}	B2	C1
u_{18}	B2	C2

The ontological graph $OG_{a_1} = (\mathcal{C}_1, E_1, \mathcal{R}_1, \rho_1)$ assigned to attribute a_1 has the form:

- $\mathcal{C}_1$ ={"A", "A1", "A2", "B", "B1", "B2"},
- E_1 = {("A", "A1"), ("A", "A2"), ("B", "B1"), ("B", "B2")},
- $\mathcal{R}_1$ = {"has subclass"},
- ρ_1(("A", "A1")) = "has subclass", ρ_1(("A", "A2")) = "has subclass", ρ_1(("B", "B1")) = "has subclass", ρ_1(("B", "B2")) = "has subclass".

In any ontological graph OG: $\mathcal{C}$ is the nonempty, finite set of nodes representing concepts in the ontology, E is the finite set of edges representing semantic relations between concepts, $\mathcal{R}$ is the family of semantic descriptions (in a natural language) of types of relations (represented by edges) between concepts, and ρ is the function assigning a semantic description of the relation to each edge (Fig. 1).

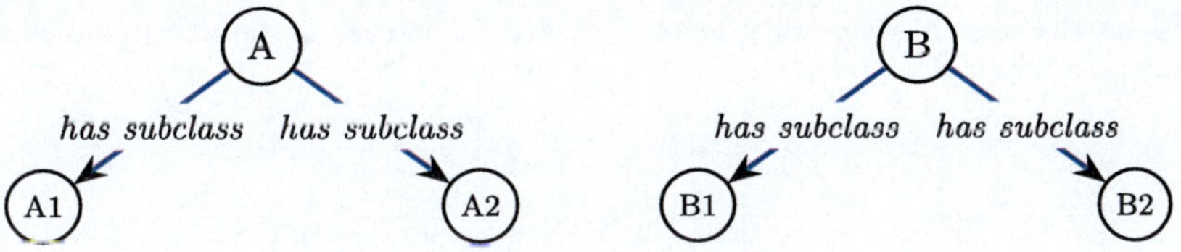

Fig. 1. Ontological graph OG_{a_1}.

The ontological graph $OG_{a_2} = (\mathcal{C}_2, E_2, \mathcal{R}_2, \rho_2)$ assigned to attribute a_2 has the form (Fig. 2):

- $\mathcal{C}_2$ ={"C", "C1", "C2", "D", "D1", "D2"},
- E_1 = {("C", "C1"), ("C", "C2"), ("D", "D1"), ("D", "D2")},
- $\mathcal{R}_2$ = {"has subclass"},
- ρ_2(("C", "C1")) = "has subclass", ρ_2(("C", "C2")) = "has subclass", ρ_2(("D", "D1")) = "has subclass", ρ_2(("D", "D2")) = "has subclass".

2.1 Rough Set Flow Graphs for Information Systems Over Ontological Graphs

Instead of a formal definition of rough set flow graphs for information systems over ontological graphs, we can explain them in a graphical way at different levels of details.

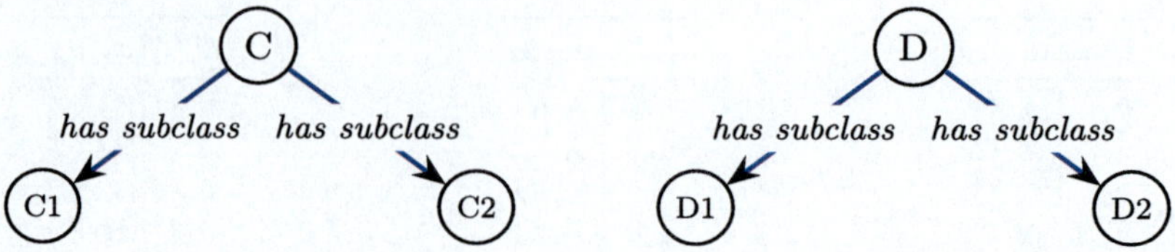

Fig. 2. Ontological graph OG_{a_2}.

Example 2. Using SIS^{OG} from Example 1 and the methods of calculation of the coefficients *cer*, *str*, *cov* [10], the rough set flow graph $\mathcal{RSFG}(SIS^{OG})$ for the information system SIS^{OG} over ontological graphs is given as follows:

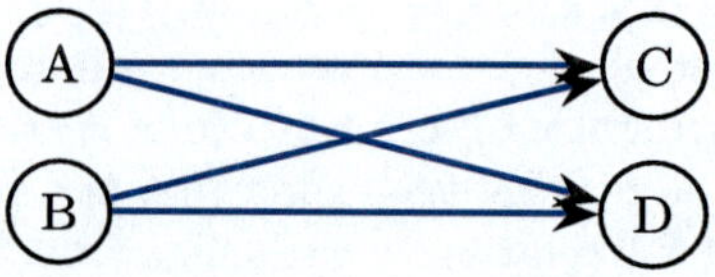

Fig. 3. The rough set flow graph $\mathcal{RSFG}$ on the highest generality level according to ontological graphs.

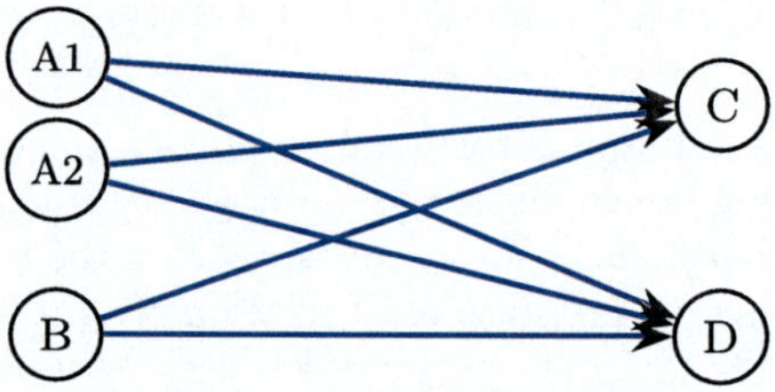

Fig. 4. The rough set flow graph $\mathcal{RSFG}$ in case of moving to subclasses of A.

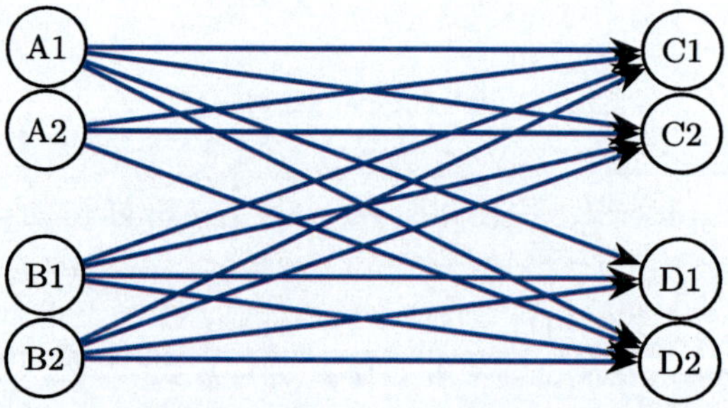

Fig. 5. The rough set flow graph $\mathcal{RSFG}$ on the lowest generality level according to ontological graphs.

As we can see in Figs. 3, 4, and 5, we can consider rough set flow graphs for information systems over ontological graphs at different levels of generality both locally and globally. For each view of the rough set flow graph, we can calculate coefficients *cer*, *str*, and *cov*. In each case, attribute values representing concepts from ontological graphs at given levels of generality are taken into consideration.

Calculation of coefficients *cer*, *str*, and *cov*, for the $A \to C$ edge has the form (it is worth noting that, according to the subclass relation, $A1$ is A, $A2$ is A, $C1$ is C, and $C2$ is C):

- $cer(A, C) = \dfrac{\text{card}(\{u_1, u_2, u_3, u_4, u_5, u_6\})}{\text{card}(\{u_1, u_2, u_3, u_4, u_5, u_6, u_7, u_8, u_9\})} = \dfrac{6}{9} = 0.6666...$
- $str(A, C) = \dfrac{\text{card}(\{u_1, u_2, u_3, u_4, u_5, u_6\})}{\text{card}(\{u_1, u_2, ..., u_{18}\})} = \dfrac{6}{18} = 0.3333...$
- $cov(A, C) = \dfrac{\text{card}(\{u_1, u_2, u_3, u_4, u_5, u_6\})}{\text{card}(\{u_1, u_2 ..., u_6, u_{10}, u_{11}, ..., u_{18}\})} = \dfrac{6}{15} = 0.4$

Calculation of coefficients *cer*, *str*, and *cov*, for the $A2 \to D$ edge has the form:

- $cer(A2, D) = \dfrac{\text{card}(\varnothing)}{\text{card}(\{u_6\})} = \dfrac{0}{1} = 0$
- $str(A2, D) = \dfrac{\text{card}(\varnothing)}{\text{card}(\{u_1, u_2, ..., u_{18}\})} = \dfrac{0}{18} = 0$
- $cov(A2, D) = \dfrac{\text{card}(\varnothing)}{\text{card}(\{u_7, u_8, u_9\})} = \dfrac{0}{3} = 0$

Calculation of coefficients *cer*, *str*, and *cov*, for the $B \to D$ edge has the form:

- $cer(B, D) = \dfrac{\text{card}(\varnothing)}{\text{card}(\{u_{10}, u_{11}, ..., u_{18}\})} = \dfrac{0}{9} = 0$
- $str(B, D) = \dfrac{\text{card}(\varnothing)}{\text{card}(\{u_1, u_2, ..., u_{18}\})} = \dfrac{0}{18} = 0$
- $cov(B, D) = \dfrac{\text{card}(\varnothing)}{\text{card}(\{u_7, u_8, u_9\})} = \dfrac{0}{3} = 0$

Calculation of coefficients *cer*, *str*, and *cov*, for the $B2 \to C1$ edge has the form:

- $cer(B2, C1) = \dfrac{\text{card}(\{u_{15}, u_{17}\})}{\text{card}(\{u_{14}, u_{15}, u_{17}, u_{18}\})} = \dfrac{2}{4} = 0.5$
- $str(B2, C1) = \dfrac{\text{card}(\{u_{15}, u_{17}\})}{\text{card}(\{u_1, u_2, ..., u_{18}\})} = \dfrac{2}{18} = 0.1111...$
- $cov(B2, C1) = \dfrac{\text{card}(\{u_{15}, u_{17}\})}{\text{card}(\{u_3, u_5, u_6, u_{12}, u_{13}, u_{15}, u_{16}, u_{17}\})} = \dfrac{2}{8} = 0.25$

3 RSFG Visualizer Module

In Example 2, it could be seen that working with rough set flow graphs for information systems over ontological graphs is not easy when they have to be considered at different levels of generality. Even an information system with a small number of objects, a small number of attributes, and uncomplicated ontological graphs, makes it necessary to draw many graphs and perform many calculations, at which it is necessary to pay attention to whether given objects are generalized/specialized by others. For this reason, the Rough Set Flow Graph Visualizer module (see Fig. 11 in the next section) was implemented as a part of the Classification and Prediction Software System (CLAPSS) [4]. The possibility of interactive visualization is very useful, due to the complicated structure of these graphs for human analysis. The module is suitable for use with information systems with symbolic attribute values that are concepts embedded in ontological graphs.

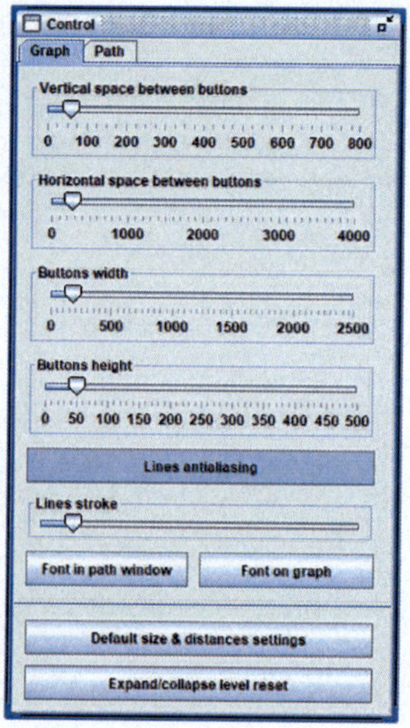

Fig. 6. *Graph* control tab.

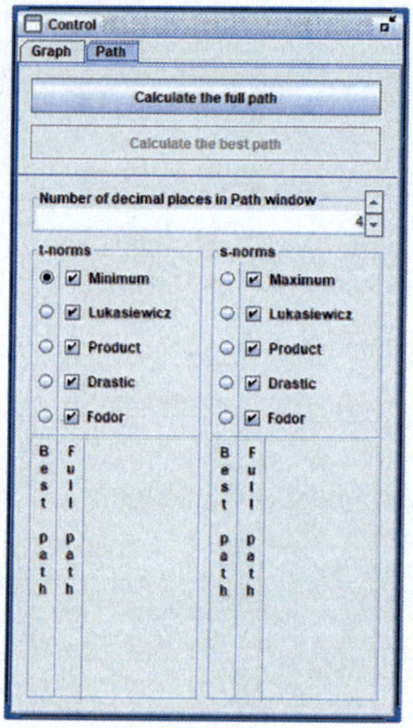

Fig. 7. *Path* control tab.

The control panel of the module contains two tabs *Graph* and *Path* (Figs. 6 and 7). *Graph* is used to control the graphic settings of the panel, e.g. changing the vertical/horizontal distance between nodes in the graph panel, adjusting the line thickness, restoring the default values of graphic settings. *Path* is used to manage and open the results. It contains, for example, opening the window for the *Full Path* mode and *Best Path* mode, a group of check boxes to select t-norms and s-norms to make calculations in the *Full Path* mode, a group of radio buttons to select one t-norm or s-norm for calculations in the *Best Path* mode.

4 Modeling with RSFG Visualizer

In this section, we discuss the basic elements (ontologies, data, graphs) that are used to model sequential information about the educational paths and job careers of individuals.

4.1 Ontological Graphs for Educational Paths and Job Careers

The first layer considered in the model is education in secondary schools: high schools and technical schools. On the basis of available information, the structure presented in Fig. 8 has been developed as well as a new ontology has been created.

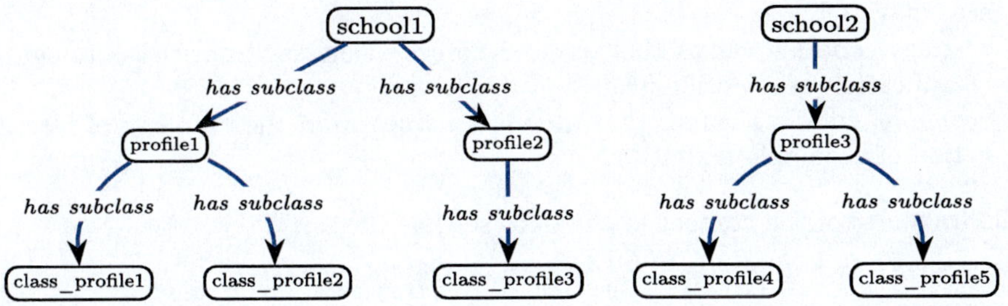

Fig. 8. An ontological graph which presents the structure of the ontology of secondary schools.

The second stage is the current classification of scientific fields and disciplines in Poland. It is a two-level classification. According to the structure presented in Fig. 9 as well as available information a new ontology has been created.

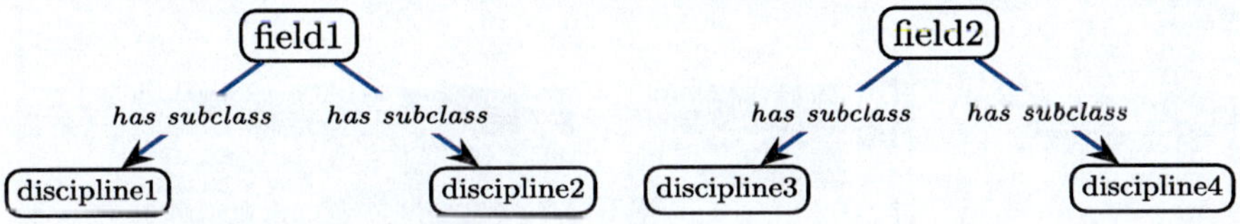

Fig. 9. An ontological graph which presents the structure of the ontology of classification of scientific fields and disciplines.

The third stage is the classification of economic activities. The most known classification is Polish Classification of Activities (*PKD*) [1]. The analogical European classification is also available [2]. According to the structure shown in Fig. 10 and selected information from the classification scheme, a new ontology has been created.

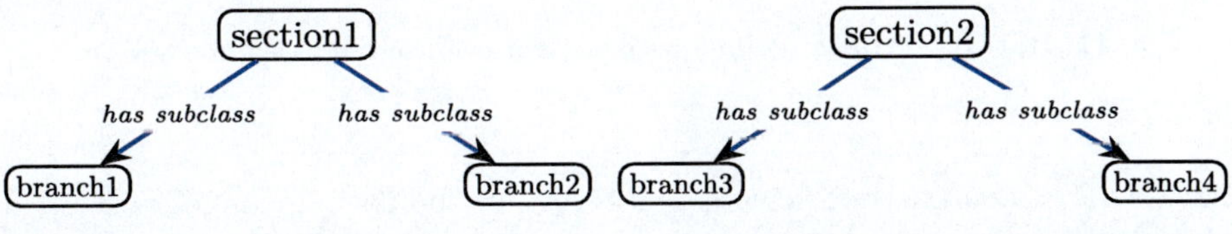

Fig. 10. An ontological graph which presents the structure of the ontology of classification of economic activities.

4.2 RSFG Graph with Assigned Ontological Graphs

To build a model, a sample data set has been prepared. The objects represent the educational and job career paths of individuals. The information system contains three attributes:

- *school*, contains values that are the names of classes from the ontology of secondary schools,
- *science*, contains values that are the names of classes from the ontology of classification of scientific fields and disciplines,
- *economy*, contains values that are class names from the ontology of classification of economic activities.

The fragment of the content of the data set has the form:

```
science_profile mathematics FINANCIAL_AND_INSURANCE_ACTIVITIES
mathematics_english mathematics FINANCIAL_AND_INSURANCE_ACTIVITIES
mathematics_physics_informatics astronomy Scientific_research_and_development
geography_mathematics_english mathematics Scientific_research_and_development
```

To start modeling with the RSFG Visualizer module, the first step is to load the considered data set into the CLAPSS system, and then to assign the ontological graphs to particular attributes.

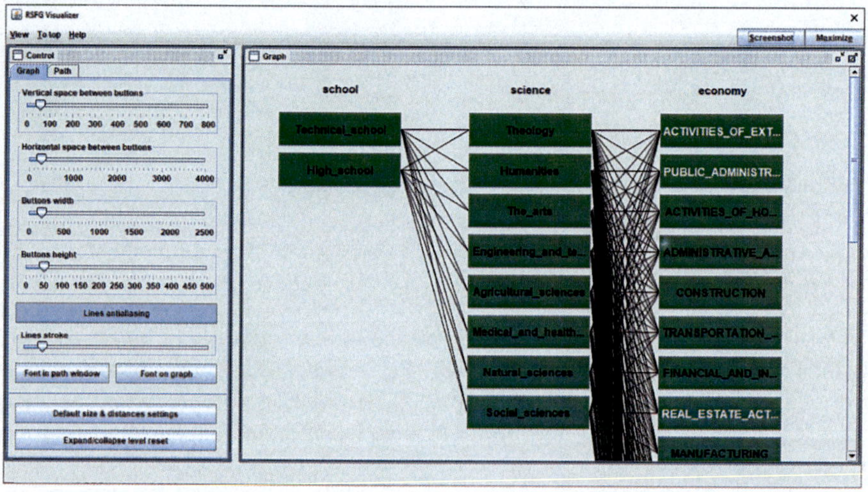

Fig. 11. RSFG Visualizer module with the view of the discussed flow graph.

Example 3. Calculated coefficients and norms for the path are as follows:

$High\_school \rightarrow Natural\_sciences \rightarrow$
$PROFESSIONAL\_SCIENTIFIC\_AND\_TECHNICAL\_ACTIVITIES$

High_school → *Natural_sciences*

- *cer* – out of 105 people who attended high school, 40 studied in the field of natural sciences,
- *str* – 40 out of 150 people took this path,
- *cov* – 40 out of 50 people who studied in the field of natural sciences were in high school.

Natural_sciences →
PROFESSIONAL_SCIENTIFIC_AND_TECHNICAL_ACTIVITIES

- *cer* – out of 50 people who studied in the field of natural sciences, 20 work in professional, scientific and technical activities,
- *str* – 20 out of 150 people took this path,
- *cov* – 20 out of 27 people who work in professional, scientific and technical activities have previously studied in the field of natural sciences.

Figure 12 shows that CLAPSS enables us to calculate, for a selected full path, the total certainty of this path. Calculation is made using a set of different t-norms and s-norms.

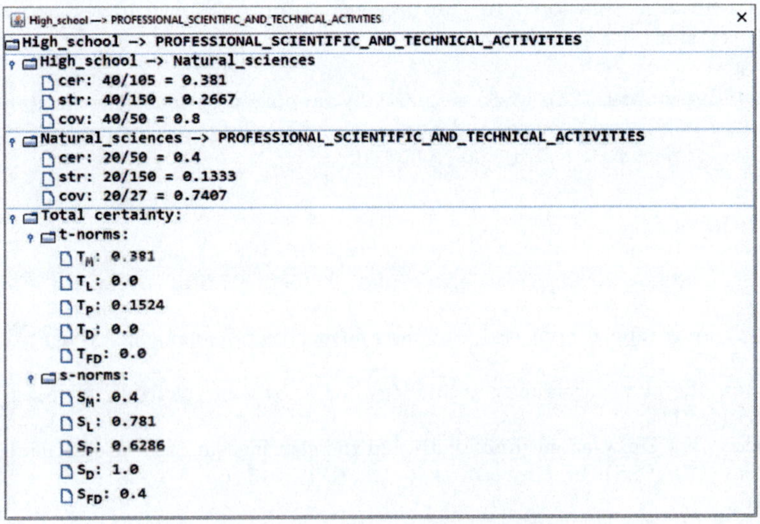

Fig. 12. *Full path* mode for the discussed path.

Example 4. The best path is found for nodes (Fig. 13):

High_school →
PROFESSIONAL_SCIENTIFIC_AND_TECHNICAL_ACTIVITIES

turned out to be the same path as the one discussed earlier:

High_school → *Natural_sciences* →
PROFESSIONAL_SCIENTIFIC_AND_TECHNICAL_ACTIVITIES

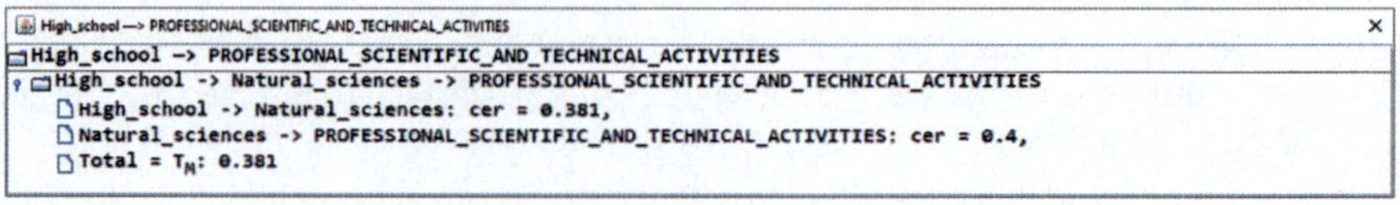

Fig. 13. *Best path* mode for the discussed path.

5 Conclusions

In the paper, we have presented the topic of rough set flow graphs for information systems over ontological graphs. Basic information about the RSFG Visualizer module has also been presented. An example of using a module with generated data containing sequential information about the educational path and job career of people has been discussed. In the future, it will be possible to create model for real-life data, e.g. data collected in a survey, which will allow to characterize the path of educational and job career development of the surveyed people. Moreover, it will be possible to create views of rough set flow graphs according to other semantic relations. In the presented approach, we have used the *has subclass* relation.

Acknowledgements. This work was partially supported by the Centre for Innovation and Transfer of Natural Sciences and Engineering Knowledge of University of Rzeszów, Poland, the project RPPK.01.03.00-18-001/10.

References

1. https://stat.gov.pl/Klasyfikacje/doc/pkd_07/pkd_07.htm . Accessed 15 Sept 2022
2. https://ec.europa.eu/eurostat/ramon/nomenclatures/index.cfm . Accessed 15 Sept 2022
3. Neches, R., et al.: Enabling technology for knowledge sharing. AI Mag. **12**(3), 36–56 (1991)
4. Pancerz, K.: On selected functionality of the classification and prediction software system (CLAPSS). In: Proceedings of the IDT 2015, pp. 278–285. Zilina, Slovakia (2015)
5. Pancerz, K.: Toward information systems over ontological graphs. In: Yao, J.T., et al. (eds.) RSCTC 2012. LNCS (LNAI), vol. 7413, pp. 243–248. Springer, Heidelberg (2012). https://doi.org/10.1007/978-3-642-32115-3_29
6. Pancerz, K.: Paradigmatic and syntagmatic relations in information systems over ontological graphs. Fund. Inform. **148**(1–2), 229–242 (2016)
7. Pancerz, K., Lewicki, A., Sarzyński, J.: Discovering flow graphs from data tables using the classification and prediction software system (CLAPSS). In: Mihálydeák, T., et al. (eds.) IJCRS 2019. LNCS (LNAI), vol. 11499, pp. 356–368. Springer, Cham (2019). https://doi.org/10.1007/978-3-030-22815-6_28
8. Pancerz, K., Sarzynski, J.: A fuzzy set tool in the classification and prediction software system (CLAPSS) (short paper). In: Ropiak, K., Polkowski, L., Artiemjew, P. (eds.) Proceedings of the CS&P 2019. CEUR Workshop Proceedings, vol. 2571 (2019)

9. Pawlak, Z.: Rough Sets. Theoretical Aspects of Reasoning about Data. Kluwer Academic Publishers, Dordrecht (1991)
10. Pawlak, Z.: Flow graphs and data mining. In: Peters, J.F., Skowron, A. (eds.) Transactions on Rough Sets III. LNCS, vol. 3400, pp. 1–36. Springer, Heidelberg (2005). https://doi.org/10.1007/11427834_1
11. Witten, I.H., Frank, E.: Data Mining: Practical Machine Learning Tools and Techniques. Morgan Kaufmann (2005)
12. Zadeh, L.: Fuzzy logic = computing with words. IEEE Trans. Fuzzy Syst. **4**(2), 103–111 (1996)

Aggregation Functions in Researching Connections Between Bio-Markers and DNA Micro-arrays

Jan G. Bazan[1], Stanislawa Bazan-Socha[2], Urszula Bentkowska[1(✉)], Wojciech Gałka[1], Marcin Mrukowicz[1], and Lech Zaręba[1]

[1] Institute of Computer Science, University of Rzeszów, Rzeszów, Poland
{jbazan,ubentkowska,wgalka,mmrukowicz,lzareba}@ur.edu.pl

[2] Department of Internal Medicine, Jagiellonian University Medical College, Kraków, Poland
stanislawa.bazan-socha@uj.edu.pl

Abstract. In this contribution a combination of diverse types of regression models in the case of microarray datasets is considered. There is a proposition of ensemble algorithm where datasets for training constituent regression models are created in a deterministic way. Moreover, aggregation functions are used here to combine the output values of the constituent regression models. Proposed model were tested with Regression Tree, K-Nearest Neighbours and Support Vector Regression as individual models. Some known families of aggregation functions defined on arbitrary $[a, b]$ interval are applied and compared. The applied aggregations are arithmetic mean, exponential mean, olimpic aggregation, arithmetic-min average, arithmetic-max average and median. Moreover, the proposed approach is compared with the bagging regression model with the optimized parameters based on Grid Search. Typical measure such as RMSE is applied to evaluate the proposed ensemble model. The proposed ensemble regression model with some set of parameters outperforms significantly the corresponding single models which was proved using a statistical test.

Keywords: Regression model · Aggregation function · Microarary dataset · Ensembling of regression models

1 Introduction

Regression problems for the microarray datasets derive from the so-called *large p, small n* problem where the number p of available predictors (genes) vastly exceeds the number n of samples. An additional consideration is that, by virtue of pathway and gene network relationships, there will likely be strong and complex correlations between expression levels of various genes across samples [15].

K. T. Atanassov et al. (Eds.): BOS/SOR/IWIFSGN 2022, LNNS 793, pp. 106–115, 2023.
https://doi.org/10.1007/978-3-031-45069-3_10

In our contribution we consider an ensemble regression model. According to the literature most of works on ensemble learning focus on classification problems. However, techniques that are successful for classification are often not directly applicable for regression [12]. Therefore, although both classification and regression ensemble approaches are related, they have been developed independently. In our ensemble regression model we apply aggregation functions (cf. [4]). Aggregation functions [5,17] are often applied with a great success in diverse areas also in machine learning models and ensemble learning methods (cf. [7,13]).

Ensemble learning is a method applied to provide better predictive performance by combining the predictions from a collection of input constituent models (cf. [8,14,18]). Main types of ensemble learning methods are bagging, stacking, and boosting. The algorithm proposed in this paper may be treated as a bagging (Bootstrap AGGregatING) ensemble learning. It is based on several subsets of ensemble input members and the varying training data. A single regression model is applied (diverse versions) and each model is trained on a different sample of the same training dataset (sampling method). Usually microarrays datasets consist of large number of features and small number of instances. Because of that in our model instead of drawing samples, we apply method similar to drawing features (random sub-spaces method) [9]. The predictions delivered by the group of ensemble regression models are then combined using aggregation functions. In such model we need to apply aggregation functions defined on arbitrary interval $[a, b]$. In the literature such aggregation functions are called sometimes $[a, b]$-aggregation functions and appeared recently in the following papers [1,2,11,16] where both theoretical results and applications are provided.

The aim of our work is to apply ensemble approach involving aggregation functions in order to improve predictions in our proposition of ensemble regression model. We provide a discussion of aggregation functions impact on regression quality. As base regression models we consider a few ones, i.e. Decision Trees, K-Nearest Neighbours, Support Vector Regression (base regression models are tested with diverse parameters). We propose to use a few classes of known aggregation functions such as quasi-arithmetic means, ordered weighted averages, some well-known convex combinations of means. We compare the performance of the applied aggregation functions in the proposed model. As a result an aggregation function is a parameter of a model. Moreover, a statistical test is applied to show that our ensemble model performs significantly better than single models.

The proposed regression model is dedicated to datasets with large number of conditional attributes and especially we are concentrated on microarray datasets since microarrays are used for researching expression of thousands of genes. The presented algorithm is focused on a non public-available dataset called Asthma which was gathered as a result of the scientific project dedicated to the disease of asthma and one distinguished biomarker, namely the amount of Red Blood Cells (RBC) which may be treated as biomarker of oxygen supply in asthma.

The paper is organized as follows. In Sect. 2, we recall the notion of an aggregation function. In Sect. 3, we provide information about regression models and adequate measure of quality. Details of the discussed dataset and experiments are given in Sect. 4. Finally, in Sect. 5 discussion on the results is presented providing also the results of statistical test.

2 Aggregation Functions

In the context of regression problems we may need to aggregate values also outside the unit interval $[0,1]$, including negative values. This is why we consider the so-called $[a,b]$-aggregation functions on arbitrary interval $[a,b] \subset \mathbb{R}$.

Definition 1 (cf. [1]). *Let $[a,b] \subset \mathbb{R}$, $a < b$. Function $A : [a,b]^n \to [a,b]$ is called an $[a,b]$-aggregation function if it is increasing with respect to any variable and fulfils boundary conditions $A(a,...,a) = a$ and $A(b,...b) = b$.*

Not all aggregation functions which are well-known on the unit interval may be applied when working on negative numbers. This is due to the restricted domain of the component functions or not satisfied monotonicity or boundary conditions.

Example 1 (cf. [4]). The following are examples of $[a,b]$-aggergation functions applied in our experiments:

- arithmetic mean

$$A_{ar}(x_1,...,x_n) = \frac{1}{n}\sum_{k=1}^{n} x_k$$

- exponential mean

$$A_{ex}^{(r)}(x_1,...,x_n) = \frac{1}{r}\ln\left(\frac{1}{n}\sum_{k=1}^{n} e^{rx_k}\right), \quad r \in \mathbb{R}, \quad r \neq 0$$

- arithmetic-min average

$$A_{amn}^{(p)}(x_1,...,x_n) = \frac{p}{n}\sum_{k=1}^{n} x_k + (1-p)\min_{1\leqslant k\leqslant n} x_k, \quad p \in [0,1]$$

- arithmetic-max average

$$A_{amx}^{(p)}(x_1,...,x_n) = \frac{p}{n}\sum_{k=1}^{n} x_k + (1-p)\max_{1\leqslant k\leqslant n} x_k, \quad p \in [0,1]$$

- median

$$A_{md}(x_1,\ldots,x_n) = \begin{cases} y_{(n+1)/2}, & \text{if } n \text{ is odd} \\ \frac{y_{n/2}+y_{(n/2)+1}}{2}, & \text{if } n \text{ is even} \end{cases},$$

– *olympic* aggregation

$$A_{ow1}(x_1, \ldots, x_n) = \frac{1}{n-2}\sum_{k=2}^{n-1} y_k, \quad n > 2$$

– *modified olympic* aggregation

$$A_{ow2}(x_1, \ldots, x_n) = \frac{1}{n-4}\sum_{k=3}^{n-2} y_k, \quad n > 4.$$

Aggregation functions A_{ar} and A_{ex} are examples of quasi-arithmetic means. Arithmetic-min average A_{amn} is a convex combination of the arithmetic mean and minimum and arithmetic-max average A_{amx} is a convex combination of the arithmetic mean and maximum. Median A_{md} is an OWA (Ordered Weighted Average) operator, similarly the so-called olympic aggregation A_{ow1}, where there are essentially disregarded the highest and the lowest scores and A_{ow2} are OWA operators.

3 Regression

We consider a typical regression problem with continuous decision. Data consist of n examples of the form $\{(x_1, f(x_1)), (x_2, f(x_2)), ..., (x_n, f(x_n))\}$. The aim is to induce the function $\hat{f}$ from the data, $\hat{f} : X \to \mathbb{R}$, where $\hat{f}(x) = f(x)$ for all $x \in X$, where f represents the unknown true function, $x = (x_1, ..., x_n)$. The algorithm used to obtain the $\hat{f}$ function is called *induction algorithm or learner*. The $\hat{f}$ function is called *model* or *predictor* [12]. X is a set of objects, so x is a vector of attributes and $f(x)$ is a mapping from attributes to continuous decision.

Obtaining exact true function is very hard for example due to software limitations (computer systems usually store real numbers with some precision) or measurement accuracy. Usually $\hat{f}$ function is only approximation of true f function so that there is a need to calculate how accurate is $\hat{f}$ function compared to true f function. The goal for regression is to minimize a prediction error. It may be for example Mean Absolute Error, Mean Squared Error or Root Mean Squared Error given by the following formula (cf. [12])

$$RMSE = \sqrt{\frac{1}{n}\sum_{i=1}^{n}(\hat{f}(x_i) - f(x_i))^2}.$$

Machine learning regression generally involves plotting a curve of best fit through the data points. The distance between each point and the curve is minimised to achieve the best fit line. Mean Absolute Error is a sum of these distances. MSE and RMSE measures unequally penalize errors larger than 1 than smaller than 1. If errors are evenly distributed then RMSE is equal to MAE. If variance of errors are small then RMSE is a bit grater than MAE. It is worth to pay attention to the scale of the problem which we try to solve. When we try to predict age error tolerance could be greater than when we try to predict birth date. RMSE is considered to be a good metric to evaluate numerical values [6] (cf. [20–22]). MSE penalizes larger errors more while the advantage of RMSE is working on the natural scale of the data.

It is worth noting, that error definitions are used both for training a model and evaluating a model quality. For regression tasks it is impossible to obtain a well-known confusion matrix. Instead the prediction error, computed on test dataset is a quality metric. Because of that, the quality metric is dependent on concrete dataset and its decision attribute scale, which means that it is hard to compare a quality of regression between different datasets. For this reason we are focused on one dataset in this paper.

4 Experiments

Details of the performed experiments are the following:

- Leave One Out Cross-validation was applied
- Different base regression models were considered (Decision Trees, K-Nearest Neighbours, Support Vector Regression)
- Different number of models in ensemble algorithm were tested (3, 6, 8, 10, 12, 14, 16, 19, 25, 30, 40, 50, 75, 100, 200, 400, 800, 1600, 3200,750, 850, 1500, 1700)

The results were compared with Bagging Regression model where Grid Search method to find optimal parameters was applied. Parameters like number of models, base models and cross validation method were the same as mentioned. We also force Bagging algorithm to draw features to obtain model similar to our approach.

Since the distance between our values are small we decided to use RMSE as it converts MSE to original scale. Moreover, when performing Leave one out method, Root Mean Squared Error is equal to Mean Absolute Error.

The study was performed on the dataset Asthma. The data were collected during the scientific project on airway remodelling in asthma [3,10]. The study included 46 non-smoking moderate to severe asthma patients aged 20–70 years with at least 10-year history of asthma diagnosed by a physician following Global Initiative for Asthma (GINA) guidelines [19]. Original dataset has 48,010 conditional attributes, additionally some biomarkers data were obtained from patients. In this research we consider RBC (Red Blood Cells) biomarker of oxygen supply in asthma, as a continuous decision.

We cannot use all of the 46 instances due to missing decision values so in experiments we consider 43 instances. Moreover, we tried to reduce number of features by applying Recursive Feature Elimination with Cross Validation method searching linear (SVR with linear kernel) correlations. The number of features was reduced to 38,438. Figure 1 shows distribution of our decision attribute.

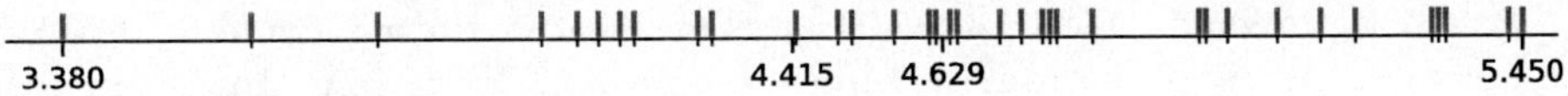

Fig. 1. The distribution of RBC values in the dataset Asthma

We can observe that distance between the smallest value and the greatest value is equal to 2.07. Maximum distance between two values is equal to 0.27 which is 13% of the whole range. It is worth to mention that this distance is between two smallest values.

Algorithm 1 shows details of training procedure of our ensemble model. Input to the algorithm is a training matrix X which contains measurements for n instances. Each instance have also known measured decision value y. In our model we also specify number of constituent models s. As our model consist of s the same models we can call it homogeneous [12]. First step in training procedure depends on calculating features importance's using Support Vector Regression and sort features using this ranking. We also store this ranking to use it during predictions to sort features in the same way. Next step is dividing dataset into s datasets and create s regression models (find $\hat{f}$ functions). Datasets are indexed from 0 to s and also features (columns) in X matrix are indexed starting from 0. Features are taken one by one and assigned to the tables until nothing left. We can describe it by function that calculates to which table feature with a given index will be assigned by the following formula: $f(i) = i \mod s$ where i is a feature index, s is a number of datasets (also number of models). As the output of this function we get a number of the dataset to which we assign the given feature. Last step in training procedure is to prepare regression models using for training each model one of the created dataset. As a model we can use any regression model. Moreover, unlike classic ensemble method our proposition is deterministic. Datasets are not created using draws.

Algorithm 1. Prediction model training procedure

Input:
X training dataset, consist of rows and columns
y continuous decision
s number of constituent models
M constituent regression model
Output:
G collection of $\hat{f}$ functions discovered by constituent models
i $\overline{\overline{X}}$-dimensional vector of features importance

$i \leftarrow$ determine features importance of X by SVR algorithm.
Sort X according to i
$G =$ allocate empty collection
for $datasetIndex = 0, 1, 2 \ldots, s - 1$ **do**
  $X' =$ allocate empty dataset
  for $indexColumn = 0, 1, 2, \ldots, card(X.columns) - 1$ **do**
    if $indexColumn \mod s == datasetIndex$ **then**
      X'.append(X.getColumn(indexColumn))
  determine $\hat{f}_{datasetIndex}$ using M on X' and y
  G.append($\hat{f}_{datasetIndex}$)
return G, i

Predict procedure (test procedure) uses, created during training, feature importance's to sort features of input test object. Then test object features are divided into s sets like in the training procedure. Next, prediction models trained before are used to obtain predictions. Last step is to aggregate predictions of models using aggregation functions and the result is the final prediction as shown in Algorithm 2.

Algorithm 2. Prediction model test procedure

Input:
- X test dataset, consist of rows and columns
- G collection of $\hat{f}$ functions discovered by constituent models
- A aggregation function
- i $\overline{\overline{X}}$-dimensional vector of features importance

Output:
- v prediction value

```
Sort X according to i.
D = allocate empty collection
for indexModel = 0, 1, 2, ..., card(G) − 1 do
    X' = allocate empty dataset
    for indexColumn = 0, 1, 2, ..., card(X.columns) − 1 do
        if indexColumn mod s == indexModel then
            X'.append(X.getColumn(i))
    Read f̂_indexModel from G[indexModel]
    y' = Apply f̂_indexModel function on X'
    D.append(y')
v = A(D)
return v
```

5 Results

In the proposed ensemble model (with various parameters) we observe that ensembling decreases model error by 0.1 on average which is 4.83% of RBC scale. Exception is KNNRegressor for which RMSE is better for a single model but difference is very small. The best obtained result is equal to 0.32 which makes 15.5% of the value range. The examined dataset has both very small number of instances and problematic distribution of decision attribute. Due to distribution on RBC we tried to remove some samples or carry out duplication. Removing instances does not have positive influence on prediction quality. Duplicating records, as expected, improved quality of predictions to 9% but it is not a good method as small number of instances cause that artificiality of dataset increases very fast.

Bagging Regression was applied to compare our approach. The best results for Bagging Regression were obtained with the given parameters BaggingRegressor(base_estimator = SVR(kernel = 'poly'), bootstrap = False, max_features = 45, n_estimators = 850) and are comparable to our proposed model, i.e.:

- Bagging Regression - RMSE mean = 0.315774531, RMSE std = 0.259003142
- Proposed Model - RMSE mean = 0.320414676, RMSE std = 0.2537746

Dependence of RMSE on s - the number of subtables, for different models that we propose is presented in Table 1.

Values of RMSE depending on aggregation functions are provided in Table 2.

Table 1. The best RMSE results obtained by tested single regression model with comparison to our proposed model

Proposed model			Single model	
s	Regressor	RMSE mean	RMSE mean	Difference
850	DecisionTree(max_depth=3)	0.367112963	0.667268274	0.300155
750	DecisionTree()	0.36340031	0.660697674	0.297297
30	KNN()	0.357182946	0.356651163	−0.00053
400	SVR(kernel='linear')	0.330228744	0.356236102	0.026007
850	SVR(kernel='poly')	0.320735242	0.356334489	0.035599

Table 2. The best results with RMSE smaller than 0.6

s	Regression model	Aggregation	Params	Proposed model RMSE	Single model RMSE
850	SVR(kernel='poly')	A_{ow1}	{}	0.320415	0.356334
850	SVR(kernel='poly')	A_{ow2}	{'r': 2}	0.320475	0.356334
850	SVR(kernel='poly')	A_{ar}	{}	0.320735	0.356334
800	SVR(kernel='poly')	A_{md}	{}	0.322243	0.356334
800	SVR(kernel='poly')	A_{ow2}	{'r': 2}	0.322322	0.356334
...	...	...	...	...	...
1600	KNN()	A_{amx}	{'p': 0.05}	0.598088	0.356651
1600	SVR(kernel='linear')	A_{amn}	{'p': 0.8}	0.598705	0.356236
3200	SVR(kernel='linear')	A_{amn}	{'p': 0.8}	0.599006	0.356236
10	DecisionTree()	A_{amn}	{'p': 0.5}	0.599349	0.660698
3	DecisionTree()	A_{amx}	{'p': 0.1}	0.599651	0.660698

Table 3. The best results for Support Vector Regression and given kernel with RMSE mean smaller than 0.4 for proposed model vs single model

Kernel	Agg	Prop. mean	95% CI lower	95% CI upper	Sing. mean	p- value
'poly'	A_{ar}	0.344352	0.339641	0.349063	0.356334489	< 0.0001
'poly'	A_{md}	0.345571	0.341026	0.350117	0.356334489	< 0.0001
'poly'	A_{ow1}	0.344311	0.339568	0.349053	0.356334489	< 0.0001
'poly'	A_{ow2}	0.344232	0.339490	0.348975	0.356334489	< 0.0001
'linear'	A_{ar}	0.347686	0.343950	0.351422	0.356236102	< 0.0001
'linear'	A_{md}	0.348147	0.344313	0.351982	0.356236102	0.000300
'linear'	A_{ow1}	0.347669	0.343906	0.351431	0.356236102	0.000100
'linear'	A_{ow2}	0.347605	0.343832	0.351378	0.356236102	0.000100

In Table 3 there are given the best results of RMSE mean for Support Vector Regression and the statistical test which proved that the proposed model is statistically significantly better than the single model. Tested hypothesis is that RMSE mean of Proposed model is equal to the Single model mean (one-sample test). We can conclude that both models tend to yield better results when constituent model is Support Vector Regression with polynomial kernel. The Decision Trees and K-Nearest Neighbours were significantly the worst as a base model, respectively.

6 Conclusions

In our research we obtained significantly better results of ensemble regression approach comparing to respective single regression models. We applied an $[a, b]$-aggregation function as a parameter to obtain possibly the best ensemble model result. It turned out that in our model the type of feature distribution is very problematic for models like K-NN regression model or Decision Tree regression model. It seems that Decision Trees and K-NN can only predict values used in training procedure. If we do LeaveOne Out Cross-validation, records which are on both ends (the last one and the one before last) and very far from others will cause high errors. Our future plans are to optimize the obtained ensemble regression model by applying ensemble pruning to select from a set of models a proper subset in order to optimize a given objective function and as a consequence to reduce computational complexity and, if possible, to improve the performance of the model [12]. Furthermore, applying oversampling/undersampling methods should improve the quality of our proposed ensemble regression model.

Acknowledgements. This work was partially supported by the Centre for Innovation and Transfer of Natural Sciences and Engineering Knowledge of University of Rzeszów, Poland, the project RPPK.01.03.00-18-001/10.

Medical data acquisition was funded by the Polish National Science Centre based on decision No: DEC-2013/09/B/NZ5/00758 (to Stanislawa Bazan-Socha)

References

1. Asmus, T., et al.: A constructive framework to define fusion functions with floating domains in arbitrary closed real intervals. Inf. Sci. **610**, 800–829 (2022)
2. Asmus, T., et al.: Negations and dual aggregation functions on arbitrary closed real intervals. In: 2022 IEEE International Conference on Fuzzy Systems, pp. 1–8, FUZZ-IEEE, Padova (2022). https://doi.org/10.1109/FUZZ-IEEE55066.2022.9882708
3. Bazan-Socha, S., et al.: Reticular basement membrane thickness is associated with growth- and fibrosis-promoting airway transcriptome profile-study in asthma patients. Int. J. Mol. Sci. **22**(3), 998 (2021)
4. Bazan, J.G., Bazan-Socha, S., Bentkowska, U., Gałka, W., Mrukowicz, M., Zaręba, L.: Comparison of aggregation classes in ensemble classifiers for high dimensional datasets. In: 2022 IEEE International Conference on Fuzzy Systems, pp. 1–10. Padova (2022). https://doi.org/10.1109/FUZZ-IEEE55066.2022.9882768

5. Beliakov, G., Bustince Sola, H., Calvo Sánchez, T.: A Practical Guide to Averaging Functions. SFSC, vol. 329. Springer, Cham (2016). https://doi.org/10.1007/978-3-319-24753-3
6. Chai, T., Draxler, R.R.: Root mean square error (RMSE) or mean absolute error (MAE)? - Arguments against avoiding RMSE in the literature. Geosci. Model Dev. **7**, 1247–1250 (2014)
7. Costa, V.S., Farias, A.D.S., Bedregal, B., Santiago, R.H.N., Canuto, A.M.D.P: Combining multiple algorithms in classifier ensembles using generalized mixture functions. Neurocomputing **313**, 402–414 (2018)
8. González, S., García, S., Del Ser, J., Rokach, L., Herrera, F.: A practical tutorial on bagging and boosting based ensembles for machine learning: algorithms, software tools, performance study, practical perspectives and opportunities. Inf. Fusion **64**, 205–237 (2020)
9. Ho, T.K.: The random subspace method for constructing decision forests. IEEE Trans. Pattern Anal. Mach. Intell. **20**(8), 832–844 (1998)
10. Kozlik et al., P.: The relationship of airway structural changes to blood and bronchoalveolar lavage biomarkers, and lung function abnormalities in asthma. Clin. Exp. Allergy **50**(1), 15–28 (2020)
11. Matusiewicz, Z, Homenda, W.: Balanced relations equations. In: 2022 IEEE International Conference on Fuzzy Systems, pp. 1–8. FUZZ-IEEE, Padova (2022). https://doi.org/10.1109/FUZZ-IEEE55066.2022.9882798
12. Mendes-Moreira, J., Soares, C., Jorge, A.M., Freire De Sousa, J.: Ensemble approaches for regression: a survey. ACM Comput. Surv. **45**(1), 1–40 (2012). Article No.: 10
13. Mohammed, A.M., Onieva, E., Woźniak, M., Martínez-Muñoz, G.: An analysis of heuristic metrics for classifier ensemble pruning based on ordered aggregation. Pattern Recogn. **124**, 108493 (2022)
14. Schapire, R., Freund, Y.: Boosting: Foundations and Algorithms. MIT Press, Cambridge (2012)
15. Segal, M.R., Dahlquist, K.D., Conklin, B.R.: Regression approaches for microarray data analysis. J. Comput. Biol. **10**(6), 961–980 (2003)
16. Špirková, J., Bustince, H., Fernandez, J., Sesma-Sara, M.: New classes of the moderate deviation functions. In: Joint Proceedings of the 19th World Congress of the International Fuzzy Systems Association (IFSA), the 12th Conference of the European Society for Fuzzy Logic and Technology (EUSFLAT), and the 11th International Summer School on Aggregation Operators (AGOP), pp. 661–666, Atlantis Studies in Uncertainty Modelling, vol. 3 (2022)
17. Torra, V., Narukawa, Y.: Modelling Decisions. Information Fusion and Aggregation Operators. Springer, Heidelberg (2007). https://doi.org/10.1007/978-3-540-68791-7
18. Zhihua, Z.: Ensemble Methods: Foundations and Algorithms. Chapman and Hall/CRC, Boca Raton (2012)
19. Global Initiative for Asthma - GINA. https://ginasthma.org/. Accessed 25 Oct 2022
20. https://scikit-learn.org/stable/modules/generated/sklearn.metrics.mean_absolute_error.html . Accessed 25 Oct 2022
21. https://scikit-learn.org/stable/modules/generated/sklearn.metrics.mean_squared_error.html . Accessed 25 Oct 2022
22. https://www.oreilly.com/library/view/machine-learning-with/9781789343700/f4faf2f6-06e5-4c06-ad17-d350b3d13e83.xhtml . Accessed 25 Oct 2022

A Comparison of Client Weighting Schemes in Federated Learning

Anna Wilbik[1], Barbara Pękala[2,3](✉), Krzysztof Dyczkowski[4], and Jarosław Szkoła[2]

[1] Maastricht University, Maastricht, The Netherlands
a.wilbik@maastrichtuniversity.nl
[2] University of Rzeszów, Rzeszów, Poland
bpekala@ur.edu.pl
[3] University of Information Technology and Management, Rzeszów, Poland
[4] Adam Mickiewicz University, Poznań, Poland
chris@amu.edu.pl

Abstract. Data is the new oil of the digital economy. Many business organizations are gathering and using their data to optimize their business performance. However, in some cases, an individual organization may not have a sufficient amount of data or data quality to build a well-performing model, especially in a dynamic environment. In some cases, the companies, which collect similar data, may be willing to exchange their knowledge, yet without sharing their data, e.g. due to privacy or legal issues. In such a case, federated learning is a solution. In horizontal federated learning, each client (organization) iteratively improves its model, so that it can be regularly aggregated and shared with all clients participating in the federation for further improvements. In the aggregation mechanism based on the weighted average, weights depend on the model's quality for each client. In this paper, we extend our previous approach to federated learning with missing information, and we investigate different weighting schemes, that depend on the effectiveness of the local models. The initial results are promising.

Keywords: First keyword · Second keyword · Another keyword

1 Introduction

In this paper, we extend the federated learning model proposed in [1] with interval-valued fuzzy sets to use imprecise and/or incomplete data for training the models, moreover, we propose the use of the effectiveness of local models to another. We test our method on medical data on breast cancer diagnosis. A precisely, our studies concentrated on the situations of non-well-balanced data (distortion of data) and the impact assessment of federated learning on effectiveness of diagnostic (medical) processes for them. Interval-valued fuzzy sets (IVFSs) ([2,3]) are one of the many extensions of fuzzy sets (FSs) and occurred very

K. T. Atanassov et al. (Eds.): BOS/SOR/IWIFSGN 2022, LNNS 793, pp. 116–128, 2023.
https://doi.org/10.1007/978-3-031-45069-3_11

useful in their flexibility [4]. Interval arithmetic is successfully used in scientific fields such as uncertainty theory, or fuzzy systems, to determine the uncertain data and modeling of uncertain systems. So, diverse applications of IVFSs for solving real-life problems involving, for example, pattern recognition, medical diagnosis, or image thresholding were successfully proposed.

The discussed problem is not limited to healthcare but can be seen also in other industry domains, such as smart industry (e.g., predictive maintenance when a single party does not have sufficient data about failures), finance (e.g., fraud detection, money laundering), but also societal welfare (e.g., fighting the poverty). Federated learning is an approach that enables parties to share their knowledge, without sharing the actual data. Even though federated learning works successfully in some applications [5–7], still further development is needed.

Thus, the paper is focused on two important aspects of contemporary decision support systems, i.e. federated learning (FL) and modeling various forms of uncertainty. Specifically, the purpose of our research is to:

1. Propose a federated learning method that can handle uncertain and/or incomplete data (in the form of interval fuzzy sets);
2. Improve the performance of local models using the fusion of third-party models by construction weights based on the effectiveness each model;

2 Background and Related Work

2.1 Federated Learning

Federated learning enables collaboration between multiple parties to jointly train a machine learning model without exchanging the local data [8]. The federated learning model was originally proposed by google researchers [9–11]. Their main idea was to build machine learning models based on datasets that are distributed across multiple devices. (cf. [12,13] or [14]).

Federated learning is a learning paradigm seeking to address the problem of data governance and privacy by training algorithms collaboratively without exchanging the data itself [12,15]. The core challenges associated with solving the optimization problem during federated learning, make the federated setting distinct from other classical problems, such as distributed learning in data center settings or traditional private data analyses. These challenges are: communication, heterogeneity, and privacy. Generally, FL can be divided into different scenarios based on how the data is partitioned or distributed among the data owners, i.e., horizontally or vertically. Horizontal federated learning is used when different parties collect the same features but from different subjects. A common example of horizontal federated learning is a group of hospitals collaborating to build a model that can predict a health risk for their patients, based on agreed data. Vertical federated learning is used when multiple parties share not the features, but the subjects, like e.g., a telecom company collaborating with a home entertainment company (cable tv provider), or an airline collaborating with a car rental agency.

In this paper, we consider a horizontal federated learning scenario. Figure 1 shows the general architecture of the federated model. The assumption is that all clients have the same local data structure and use a common machine learning model. They exchange with the server only coefficients describing the learned local models and parameters describing the classification quality, which is only used to stop the moment of the iteration process. The server performs model aggregation, i.e. appropriate aggregation of coefficients. The server then returns the new coefficients to the clients.

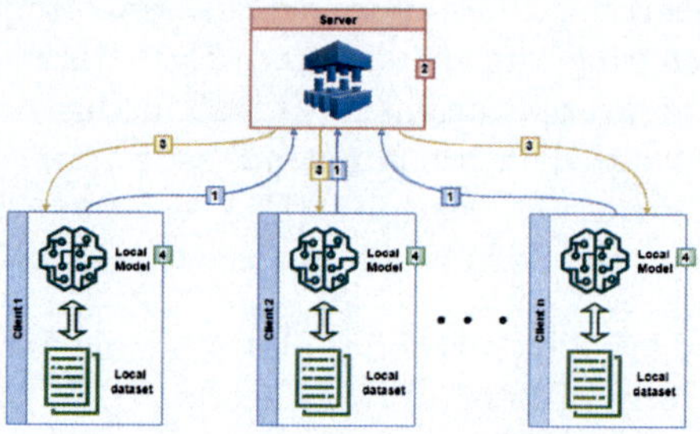

Fig. 1. Proposed federated model

The federated learning scheme was extended to include a cross-validation step for local models [1]. This allows the use of foreign data sets to make decisions in subsequent iterations of the model without having direct access to such data and avoiding model overfitting (see Fig. 2):

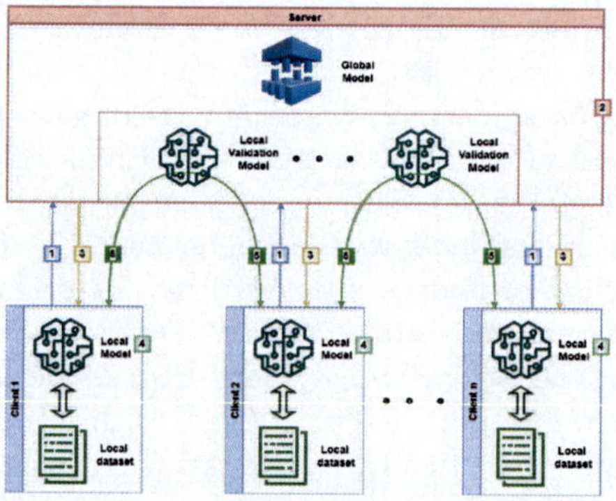

Fig. 2. Proposed federated model

Originally, in Federated averaging [11] the aggregation uses weighted averaging, where weights depend on the amount of data available to each client.

However, this is not the only solution. For instance, FedMA [16] combines the weights of neurons with similar characteristics based on the permutation invariant of the neural network and makes efficient use of the communication rounds to improve the convergence speed. Another approach is inspired by active learning and assumes that the central server could evaluate every client data and indicate the potential utility of training before aggregation and select only the most useful clients to participate in the training [17]. Zhao [18] proposed that the central server collects a small amount of data from each client to macro-distribute them, resulting in more homogeneous data. There were also some experiments with the introduction of different weighting schemes, e.g. [19], where the authors assigned weights of the clients according to a fairness measure.

2.2 Interval-Valued Fuzzy Set Theory

Many approaches and theories for investigating and modeling imprecision have been proposed since fuzzy sets were originally introduced by Zadeh [20]. For example, interval-valued fuzzy sets [2,3] are an effective tool for uncertainty modeling in many practical issues. In $L^I = \{[\underline{p}, \overline{p}] : \underline{p}, \overline{p} \in [0,1],\ \underline{p} \leq \overline{p}\}$, i.e., in a family of intervals belonging to the unit interval for $X \neq \emptyset$ and according to the following papers [2,3,21,22] we define an **interval-valued fuzzy set** (IVFS) S in X as a mapping $S : X \to L^I$ such that for each $x \in X$

$$S(x) = [\underline{S}(x), \overline{S}(x)]$$

means the degree of membership of an element x into S. The family of all IVFSs in X we denoted by IVFS(X). We assume, reflect an aspect of applications on a finite set $X = \{x_1, \ldots, x_n\}$. In many areas, data aggregation is needed, which is the process of gathering data and presenting it in a summarized format. The data may be gathered from multiple data sources with the intent of combining these data sources into a summary for data analysis. This is a crucial step since the accuracy of insights from data analysis depends heavily on the amount and quality of data used. It is important to gather high-quality accurate data and a large enough amount to create relevant results. An aggregate function takes as input a set, a multiset (bag), or a list from some input domain and outputs an element of an output domain.

Especially, the notion of an aggregation function on L^I being a significant concept in numerous applications (e.g. [23,24] or [25]). For the input data in the form of interval-valued fuzzy sets, we can define aggregations as follows

Definition 1 ([25–27]). *Let $n \in \mathbf{N}$, $n \geq 2$. An operation $\mathcal{A} : (L^I)^n \to L^I$ is called an interval-valued (I-V) aggregation function if it is increasing with regard to the order $\leq$ (partial or linear), i.e.*

$$\forall x_i, y_i \in L^I \quad x_i \leq y_i \Rightarrow \mathcal{A}(x_1, ..., x_n) \leq \mathcal{A}(y_1, ..., y_n) \tag{1}$$

$$\text{and } \mathcal{A}(\underbrace{[0,0], ..., [0,0]}_{n\times}) = [0,0], \quad \mathcal{A}(\underbrace{[1,1], ..., [1,1]}_{n\times}) = [1,1].$$

The fundamental examples of I-V aggregation functions are the following:

- the representable arithmetic mean
  $\mathcal{A}_{mean}([\underline{x},\overline{x}],[\underline{y},\overline{y}]) = [\frac{\underline{x}+\underline{y}}{2}, \frac{\overline{x}+\overline{y}}{2}]$,
- the α mean
  $\mathcal{A}_{\alpha}(x,y) = [\alpha\underline{x} + (1-\alpha)\underline{y}, \alpha\overline{x} + (1-\alpha)\overline{y}]$
  is an I-V aggregation function on L^I with regard to the lexicographical and Xu-Yager order [26] for $x, y \in L^I$.

Often discussed in the literature and applied in practice, are OWA operators introduced by Yager in 1988. The concept of OWA has been extended to the interval-valued setting (or more generally, to the type-2 fuzzy sets setting or in the real set), which was called the uncertain OWA operator and which is the next generalization of $\mathcal{A}_{mean}$ aggregation and is crucial in our novelty method in this paper, where we create weights for each local models used in the aggregation process.

OWA operators are a particular case of more general aggregation functions called Choquet integrals. In [28] was introduced discrete interval-valued Choquet integrals of interval-valued fuzzy sets based on admissible orders.

In [28], the class of linear orders on L^I is used to extend the definition of OWA operators for interval-valued fuzzy setting in following way.

Definition 2 **([28]).** *Let $\leq$ be an admissible order on L^I, and $w = (w_1,\ldots,w_n) \in [0,1]^n$, with $w_1 + \cdots + w_n = 1$. The interval-valued ordered weighted averaging (OWA) operator (IVOWA) associated with $\leq$ and w is a mapping $IVOWA_{\leq,w} : (L^I)^n \to L^I$, given by*

$$IVOWA_{\leq,w}([\underline{x}_1,\overline{x}_1],\ldots,[\underline{x}_n,\overline{x}_n]) = \sum_{i=1}^{n} w_i \cdot [\underline{x}_{(i)},\overline{x}_{(i)}],$$

where $[\underline{x}_{(i)},\overline{x}_{(i)}]$, $i = 1,\ldots,n$, denotes the i-th greatest of the inputs with respect to the order $\leq$ and $w \cdot [\underline{x},\overline{x}] = [w\underline{x}, w\overline{x}]$, $[\underline{x}_1,\overline{x}_1] + [\underline{x}_2,\overline{x}_2] = [\underline{x}_1 + \underline{x}_2, \overline{x}_1 + \overline{x}_2]$.

Moreover, interval arithmetic was deemed as necessary with the development of the theory of uncertainty. It was realized that the use of uncertain parameters and uncertain data is very important for the description of reality in the form of a mathematical model. The most common and most frequently used interval arithmetic is Moore arithmetic [29,30]. In Moore arithmetic basic operations on intervals $X = [\underline{x},\overline{x}]$ and $Y = [\underline{y},\overline{y}]$ are realized by formulas for sum, difference and product:

$$[\underline{x},\overline{x}] + [\underline{y},\overline{y}] = [\underline{x}+\underline{y}, \overline{x}+\overline{y}]$$

$$[\underline{x},\overline{x}] - [\underline{y},\overline{y}] = [\underline{x}-\overline{y}, \overline{x}-\underline{y}]$$

$$a * [\underline{x},\overline{x}] = [a\underline{x}, a\overline{x}], \quad a \in R^+$$

$$a * [\underline{x},\overline{x}] = [a\overline{x}, a\underline{x}], \quad a \in R^-$$

$$[\underline{x},\overline{x}] * [\underline{y},\overline{y}] =$$

$$[\min(\underline{x} * \underline{y}, \overline{x} * \overline{y}, \underline{x} * \overline{y}, \overline{x} * \underline{y}), \max(\underline{x} * \underline{y}, \overline{x} * \overline{y}, \underline{x} * \overline{y}, \overline{x} * \underline{y})]$$

for $\underline{x}, \overline{x}, \underline{y}, \overline{y} \in R$ and $\underline{x} \leq \overline{x}, \underline{y} \leq \overline{y}$. Some limitations and drawbacks have been found in the Moore interval arithmetic such as the excess width effect problem so the alternative for Moore arithmetic we may used multidimensional interval arithmetic. The idea of multidimensional arithmetic was developed by A. Piegat [31], where given value x from interval $X = [\underline{x}, \overline{x}]$ is described using variable γ_x, $\gamma_x \in [0, 1]$, as shown:

$$Rep_\gamma(x) = \underline{x} + \gamma_x(\overline{x} - \underline{x}). \tag{2}$$

In this notation the interval $X = [\underline{x}, \overline{x}]$ is described in the form:.

$$X = \{Rep_\gamma(x) : Rep_\gamma(x) = \underline{x} + \gamma_x(\overline{x} - \underline{x}), \gamma_x \in [0, 1]\}.$$

The variable γ_x gives the possibility to obtain any value between the left border $\underline{x}$ and right border $\overline{x}$ of interval X.

3 Proposed Method

We consider a horizontal federated learning scenario, where each client has its own independent data set $\{Y_i, x_{i1}, ...x_{ip}\}$ and $x_{ip} \in L^I$, $Y_i \in \{0, 1\}$ for $i = 1, ..., n$, n is the number of instances, and p number of attributes.

Each client trains a set model on its data (n_k observations) in a specified number of internal iterations and provides the training result in the form of a result vector according to the selected Machine Learning model.

In our earlier paper [1], we proposed a federated learning approach, that could deal with missing data (Fig. 2). There we used the arithmetic average in the aggregation process.

However, in this paper, we investigate different weighted aggregation schemes, that depend on the quality of the local models. The new learning model which after the server initializes the empty model sent to local models, consists of iterative executions (after initialization) of the following steps (see Fig. 3, with the new 6-th element (6-th line on the picture)):

1. Each client performs a few training steps of its own model on its local data and passes it to the server (parameters of models (line 1) and their efficiency: Accuracy, Sensitivity, Specificity, Precision, AUC (line 5));
2. The server aggregates the models (line 2);
3. The server returns the new model to the clients (line 6);
4. Local models are updated (line 4);
5. In the aggregation mechanism based on the weighted average, weights depend on the model's effectiveness for each client (line 6).

Thus, we realize our process in general two aspects:

(1) In the first, for initial validation, new models are distributed to other clients on the server where they are tested and model quality metrics are returned to the client in question. The client decides whether to continue updating its model for better quality.

(2) In the second, approach, based on the effectiveness of local models, we build the weights used to aggregate the local models in the server;

The process continues until the obtained quality of local models is high enough and it is impossible to improve them (subsequent iterations do not reduce the error function). As can be seen, the federated learning scheme has been extended to include a cross-validation step for local models. This allows the use of foreign data sets to make decisions in subsequent iterations of the model without having direct access to such data and avoiding model overfitting.

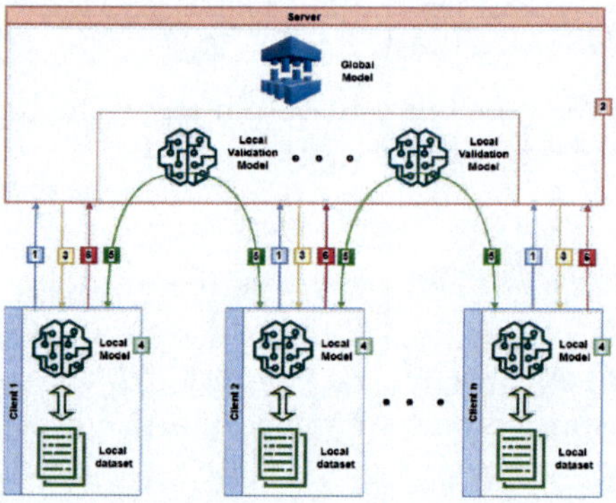

Fig. 3. Proposed federated mode

In our model, we allow the data to be in interval form and for there to be gaps in the data. We assume that the data is normalized and the missing data are presented in the form of intervals [0, 1].

As mentioned above the federated learning scheme thus constructed is independent of the choice of a particular machine learning model. For simplicity, we chose logistic regression with stochastic gradient descent. For the experiment, we modify it to operate on interval data, as in [1].

One iteration of the local learning process follows the scheme:

1. calculation of the model response for each training sample according to the sigmoid function:

$$f(y_i) = \frac{1}{1 + e^{-Rep_\gamma(\beta_0 + \beta_1 \cdot x_{i1} + \ldots + \beta_p \cdot x_{ip} + \epsilon_i)}}$$

for $\gamma \in [0, 1]$ and $f : L^I \to R$.
From this step, in every single iteration, we switch from the interval calculus to the real model using the defined *Rep* function in (2). This allows us to operate on data in the form of interval-valued fuzzy sets while receiving the model in the form of a vector of real numbers.

2. For the computation of an error (loss function) between the computed value and the actual value we assumed:

$$\mathcal{L}(y_i) = -\log(f(y_i)) \cdot Y_i - \log(1 - f(y_i)) \cdot (1 - Y_i),$$

where Y_i - actual output value,
3. Finally, we update of learning coefficients in steps:

$$\beta_j = \beta_j + \alpha \cdot \bigtriangledown_{\beta_j} \mathcal{L}(y_i) \cdot x_{ij},$$

$$\beta_0 = \beta_0 + \alpha \cdot \bigtriangledown_{\beta_0} \mathcal{L}(y_i),$$

α is the learning coefficient and $\bigtriangledown$ is the gradient for $i = 1, .., n_k,\ j = 1, .., p.$

4 Experiment and Results

In this section we describe our initial evaluation of the proposed method.

4.1 Structure of Dataset

The dataset is a Wisconsin (diagnostic) breast cancer dataset. This is one of the popular datasets from UCI Machine Learning Repository [32]. Data contains information on 569 medical cases. Features are calculated from a digitized image of a fine needle aspirate (FNA) of a breast mass. They describe the characteristics of the cell nuclei present in the image.
Ten real-valued features are computed for each cell nucleus:

- radius (mean of distances from center to points on the perimeter),
- texture (standard deviation of gray-scale values),
- perimeter,
- area,
- smoothness (local variation in radius lengths),
- compactness,
- concavity (severity of concave portions of the contour),
- concave points (number of concave portions of the contour),
- symmetry,
- fractal dimension (*coastline approximation* – 1).

For each value, the standard deviation and mean value of the trait measurements for the patient are given. On the basis of both of these values, the value of the interval is constructed:

$$[mean\ -\ standard\ deviation,\ mean\ +\ standard\ deviation]$$

after earlier fuzzyfing both values: "mean-standard deviation" and "mean+ standard deviation", by normalization.

The decision attribute stores information about the diagnosis: malignant (0) or benign (1). The dataset consists of 212 malignant objects and 357 benign objects. Since the dependent variable, the explained variable, takes two dichotomous values of 0 and 1, the optimal model choice for decision prediction turned out to be the logistic regression model, which determines the probability of a given event occurring for the values of the predictors entered into the model.

To simulate the data sets of a group of clients (three in this case), the data were randomly divided into three groups with the decision-balance and unbalanced behavior. The data in each client is then randomly split into a training set and a test set in a ratio of 90% to 10%.

4.2 Comparison and Analysis of Different Models

We checked our model in various real-life scenarios, dealing with uncertain and many local models, where we concentrated on the situations for non-well-balanced data (bad data qualitatively) and compared it with:

I. **Benchmark model.** As a benchmark model, we chose a situation in which the data (complete) without uncertainty. The model was trained on a 90% training set (sum of customer sets) and tested on a 10% test set. To ensure the correctness of the learning process, we conducted a 10-fold cross validation. That is, a standard logistic regression model without a federated learning model was used;

Reference performance of Benchmark model are presented on Table 1

Table 1. Performance of benchmark model

	ACC	SENS	SPEC	PREC
Complete data	0,965	0,972	0,935	0,965

II. The Baseline models (no federal learning, learning only locally);

III. (Case I) The standard model, i.e., when during aggregation of parameters of local models we used arithmetic mean.

III. (Case II) and (Case III) Aggregation by weights, shows, influence of effectiveness local models to other, is as follows: For each i-th local model, we determine the average efficiency parameter based on its parameters: ACC, SENS, SPEC, PREC and AUC - w(i). So for this model in Case II:

$$w(i) = ACC.$$

Let us assume that $i = 1, ..z$ (z local models). Let

$$w(i) := w(i)/(\sum_{i=1}^{z} w(i).$$

Then for every $k = 0, ..., p$

$$\beta_{k(i)} := \beta_{k(i)} * w(i).$$

Ultimately:

$$\beta_k := A_{mean}{}_{i=1}^{z}\beta_{k(i)}.$$

In Case III, we also apply proceed from Case II but we create adequate $w(i)$ based on AUC, i.e.,

$$w(i) = AUC.$$

During the research we assumed $\epsilon_i = 0$, $\alpha = 0.01$ and $\gamma = 0.5$ (optimal results) of the logistic regression algorithm described above with 50 learning epochs and 100 aggregation cycles.

The above cases of federal learning (case I - III) compared to the Baseline models (no federal learning, learning only locally) for the following two scenarios (different methods of building a local dataset - in all scenarios, we assume the relatively equal local sets):

- Scenario I.
  All three local models have decision-balanced data (Tables 2, 3 and 4).

Table 2. Performance of baseline model in Scenario I

Dataset	ACC	SENS	SPEC	PREC	AUC
Client 1	0,94	0,96	0,89	0,94	0,938
Client 2	0,92	0,95	0,86	0,92	0,831
Client 3	0,92	0,92	0,94	0,96	0,867

Table 3. Performance of model in Case I of Scenario I

Dataset	ACC	SENS	SPEC	PREC	AUC
Client 1	0,94	0,96	0,89	0,94	0,938
Client 2	0,921	0,95	0,86	0,92	0,832
Client 3	0,92	0,922	0,94	0,96	0,868

Table 4. Performance of model in Case II of Scenario I

Dataset	ACC	SENS	SPEC	PREC	AUC
Client 1	0,94	0,96	0,89	0,94	0,938
Client 2	0,92	0,95	0,86	0,92	0,932
Client 3	0,92	0,92	0,94	0,96	0,957

- Scenario II.
  The first local set has balanced data, but the second has a decision distribution of 20% to 80% for 0 and 1 respectively, in the third duality (Tables 5, 6, 7, 8 and 9).

Table 5. Performance of model in Case III of Scenario I

Dataset	ACC	SENS	SPEC	PREC	AUC
Client 1	0,94	0,96	0,89	0,94	0,919
Client 2	0,93	0,95	0,86	0,92	0,933
Client 3	0,92	0,922	0,94	0,96	0,949

Table 6. Performance of baseline model in Scenario II

Dataset	ACC	SENS	SPEC	PREC	AUC
Client 1	0,94	0,96	0,89	0,94	0,938
Client 2	0,81	0,86	0,89	0,89	0,802
Client 3	0,70	0,89	0,78	0,88	0,826

Table 7. Performance of model in Case I of Scenario II

Dataset	ACC	SENS	SPEC	PREC	AUC
Client 1	0,94	0,96	0,94	0,96	0,954
Client 2	0,916	0,89	0,98	0,98	0,943
Client 3	0,890	0,89	0,78	0,88	0,912

Table 8. Performance of model in Case II of Scenario II

Dataset	ACC	SENS	SPEC	PREC	AUC
Client 1	0,94	0,96	0,94	0,96	0,954
Client 2	0,922	0,89	0,99	0,97	0,988
Client 3	0,801	0,97	0,48	0,76	0,943

Table 9. Performance of model in Case III of Scenario II

Dataset	ACC	SENS	SPEC	PREC	AUC
Client 1	0,94	0,96	0,94	0,96	0,954
Client 2	0,93	0,89	0,99	0,99	0,991
Client 3	0,92	0,99	0,72	0,88	0,953

Let us note that each of the local sets consists of the training and test sets, which for each scenario are balanced in terms of decisions.

Note that the weighted average of case III, based on building weights taking into account the AUC parameter for aggregating the β parameters of the global model, turned out to be the most effective and resistant to distortion of data in local sets.

5 Concluding Remarks

In this paper, we have extended our earlier approach to federated learning that could handle the missing data by using the arithmetic mean. We investigated namely different weighting schemes, other than the number of available data for each client, as in the case of federated averaging. The initial results on one data set show that including the quality of the model on locally available data can improve the overall quality of the model, especially in the context of not independent and identically distributed (non-IID) data. In future work, we plan to extensively test this approach.

References

1. Dyczkowski, K., Pękala, B., Szkoła, J., Wilbik, A.: Federated learning with uncertainty on the example of a medical data. In: 2022 IEEE International Conference on Fuzzy Systems (FUZZ-IEEE), pp. 1–8. IEEE (2022)
2. Zadeh, L.A.: The concept of a linguistic variable and its application to approximate reasoning-I. Inf. Sci. **8**(3), 199–249 (1975)
3. Sambuc, R.: Fonctions ϕ-floues: application á l'aide au diagnostic en pathologie thyroidienne, Ph.D. dissertation, Faculté de Médecine de Marseille (1975) (in French)
4. Bustince, H., et al.: A historical account of types of fuzzy sets and their relationships. IEEE Trans. Fuzzy Syst. **24**(1), 179–194 (2016)
5. Hard, A., et al.: Federated learning for mobile keyboard prediction (2018). https://arxiv.org/abs/1811.03604
6. Deist, T.M., et al.: Infrastructure and distributed learning methodology for privacy-preserving multi-centric rapid learning health care: euroCAT. Clin. Transl. Radiat. Oncol. **4**, 24–31 (2017)
7. Zheng, W., Yan, L., Gou, C., Wang, F.Y.: Federated meta-learning for fraudulent credit card detection. In: Proceedings of the Twenty-Ninth International Joint Conference on Artificial Intelligence (IJCAI-20) (2020)
8. Kairouz, P., et al.: Advances and open problems in federated learning. Found. Trends® Mach. Learn. **14**, 1–210 (2021)
9. Konecný, J., McMahan, H.B., Ramage, D., Richtárik, P.: Federated optimization: distributed machine learning for on-device intelligence. ArXiv, vol. 1610.02527 (2016)
10. Konečný, J., McMahan, H.B., Yu, F.X., Richtárik, P., Suresh, A.T., Bacon, D.: Federated learning: strategies for improving communication efficiency. ArXiv, vol. 1610.05492 (2017)
11. McMahan, B., Moore, E., Ramage, D., Hampson, S., y Arcas, B.A.: Communication-efficient learning of deep networks from decentralized data. In: AISTATS 2017 (2017)

12. Yang, Q., Liu, Y., Chen, T., Tong, Y.: Federated machine learning: concept and applications. ACM Trans. Intell. Syst. Technol. **10**(2), 1–19 (2019)
13. Wilbik, A., Grefen, P.: Towards a federated fuzzy learning system. IEEE International Conference on Fuzzy Systems (FUZZ-IEEE), pp. 1–6 (2021)
14. Yan, H., Hu, L., Xiang, X., Liu, Z., Yuan, X.: Privacy-preserving collaborative learning for mitigating indirect information leakage. Inf. Sci. **548**, 423–437 (2021)
15. Li, T., Sahu, A.K., Talwalkar, A., Smith, V.: Federated learning: challenges, methods, and future directions. IEEE Signal Process. Mag. **37**(3), 50–60 (2020)
16. Wang, H., Yurochkin, M., Sun, Y., Papailiopoulos, D., Khazaeni, Y.: Federated learning with matched averaging. arXiv preprint arXiv:2002.06440 (2020)
17. Goetz, J., Malik, K., Bui, D., Moon, S., Liu, H., Kumar, A.: Active federated learning. arXiv preprint arXiv:1909.12641 (2019)
18. Zhao, Y., Li, M., Lai, L., Suda, N., Civin, D., Chandra, V.: Federated learning with non-IID data. arXiv preprint arXiv:1806.00582 (2018)
19. Li, T., Sanjabi, M., Beirami, A., Smith, V.: Fair resource allocation in federated learning. arXiv preprint arXiv:1905.10497 (2019)
20. Zadeh, L.A.: Fuzzy sets. Inf. Control **8**(3), 338–353 (1965)
21. Turksen, I.B.: Interval valued fuzzy sets based on normal forms. Fuzzy Sets Syst. **20**(2), 191–210 (1986)
22. Gorzałczany, M.B.: A method of inference in approximate reasoning based on interval-valued fuzzy sets. Fuzzy Sets Syst. **21**(1), 1–17 (1987)
23. Dyczkowski, K., Wójtowicz, A., Żywica, P., Stachowiak, A., Moszyński, R., Szubert, S.: An intelligent system for computer-aided ovarian tumor diagnosis. In: Filev, D., et al. (eds.) Intelligent Systems'2014. AISC, vol. 323, pp. 335–343. Springer, Cham (2015). https://doi.org/10.1007/978-3-319-11310-4_29
24. Pękala, B.: Uncertainty Data in Interval-Valued Fuzzy Set Theory. SFSC, vol. 367. Springer, Cham (2019). https://doi.org/10.1007/978-3-319-93910-0
25. Beliakov, G., Bustince Sola, H., Calvo Sánchez, T.: A Practical Guide to Averaging Functions. SFSC, vol. 329. Springer, Cham (2016). https://doi.org/10.1007/978-3-319-24753-3
26. Zapata, H., et al.: Interval-valued implications and interval-valued strong equality index with admissible orders. Int. J. Approximate Reason. **88**, 91–109 (2017)
27. Komorníková, M., Mesiar, R.: Aggregation functions on bounded partially ordered sets and their classification. Fuzzy Sets Syst. **175**(1), 48–56 (2011)
28. Bustince, H., Galar, M., Bedregal, B., Kolesárová, A., Mesiar, R.: A new approach to interval-valued Choquet integrals and the problem of ordering in interval-valued fuzzy sets applications. IEEE Trans. Fuzzy Syst. **21**(6), 1150–1162 (2013)
29. Moore, R.E.: Interval Analysis. Prentice Hall, Upper Saddle River (1966)
30. Moore, R.E.: Methods and Applications of Interval Analysis. SIAM, Philadelphia (1979)
31. Piegat, A., Landowski, M.: Multidimensional approach to interval uncertainty calculations. In: Atanassov, K., et al. (eds.) New Trends in Fuzzy Sets, Intuitionistic: Fuzzy Sets, Generalized Nets and Related Topics, Volume II: Applications, pp. 137–151. IBS PAN - SRI PAS, Warsaw (2013)
32. Dua, D., Graff, C.: UCI machine learning repository (2017). http://archive.ics.uci.edu/ml

InterCriteria Analysis of the Supervisory Statistic Data for Selected 8 EU Countries During the Period 2020–2021

Slaviiana Danailova-Veleva[1], Lyubka Doukovska[1](✉), and Atanas Dukovski[2]

[1] Intelligent Systems Department, Institute of Information and Communication Technologies, Bulgarian Academy of Sciences, Acad. Georgi Bonchev Str., 1113 Sofia, Bulgaria
lyubka.doukovska@iict.bas.bg

[2] Modelling and Optimization Department, Institute of Information and Communication Technologies, Bulgarian Academy of Sciences, Acad. Georgi Bonchev Str., 1113 Sofia, Bulgaria

Abstract. In this paper, we will analyze the correlation between some indicators for financial condition of the significant banks in eight European countries. Banking system in European Union is under supervision of European Central Bank (ECB) and national supervisory authorities. ECB supervised directly significant banks in EU. For analysis purposes, we will use multicriteria decision-making method - InterCriteria Analysis. The presented multicriteria decision making method is based on two fundamental concepts: Intuitionistic Fuzzy Sets (IFSs) and Index Matrices (IMs).

Keywords: Intelligent systems · InterCriteria analysis · Intuitionistic fuzzy sets · Banking system

1 Introduction

In Europe, the countries that are part of the European Union, the activity of their banks is controlled by the supervisory institution of the given country and by the European Central Bank (ECB), [5]. This constitutes the so-called Single Supervisory Mechanism, [10]. The European Central Bank is an independent institution of the European Council.

Every year the European Central Bank reviews the criteria one institution to be classified as a significant. They also reviews the list of the banks already classified as a significant to be sure that the institution is still fulfilling the criteria.

The European Central Bank publish the criteria for significant on their internet page. To be classified as a significant the institutions should fulfill at least these criteria: Size, Economic importance, Cross-border activities and Direct public financial assistance.

A supervised bank can also be considered as a significant if it is one of the three most significant banks established in a particular country, [6, 9].

In this regard, the European Central Bank publishes a review report on its website. In it, it reports for the next reporting year how many significant credit institutions it will monitor and which are the new ones, as well as by what criteria they are determined

K. T. Atanassov et al. (Eds.): BOS/SOR/IWIFSGN 2022, LNNS 793, pp. 129–137, 2023.
https://doi.org/10.1007/978-3-031-45069-3_12

to be significant. It also announces whether there is a change for any institution in the materiality criteria. The European Central Bank regulates the activity of credit institutions by means of issued Implementing Regulations. The main Regulation is Regulation 575/2013, [13].

The European Central Bank, the Banking supervision, published every quarter banking statistic. In this report, there is financial information for significant banks in EU. This document consist of 62 tables, separate in six main sections. For our research, we use the information for the banks in eight countries. These countries are – Germany, Spain, France, Italy, Luxemburg, Netherlands, Austria and Portugal. We made this decision because for these countries there is complete information on all indicators. The information that we use is as follow - profit and loss figures, key performance indicators and non-performing loans and advances by country.

We already published the paper where we analyzed the same information but the period was for the fourth quarter of 2020. We believe that there are dependencies between these indicators and when using the InterCriteria Analysis, which is based on two concepts: Intuitionistic Fuzzy Sets (IFSs), [1] and Index Matrices (IMs), [2], to analyze them, they will manifest. The analysis will confirm the conclusions already drawn in the previous publications, [3, 4, 7].

2 Data Analysis

In this paper, the analysis was conducted for the following indicators/criteria:

Profit and loss indicators by country:

1. *Net interest income;*
2. *Net fee and commission income;*
3. *Net gains or losses on financial assets and liabilities held for trading;*
4. *Net gains or losses on financial assets and liabilities at fair value through profit and loss;*
5. *Net gains or losses from hedge accounting;*
6. *Exchange differences, net;*
7. *Net other operating income;*
8. *Operating income;*
9. *Administrative expenses and depreciation;*
10. *Net income before impairment, provision and taxes;*
11. *Impairment and provisions;*
12. *Other;*
13. *Profit and loss before tax;*
14. *Tax expenses or income;*
15. *Net profit/loss;*

Key risk indicators by country:

16. *RoE – return of equity;*
17. *RoA – return of assets;*
18. *CIR – cost to income ratio;*

T02.02.2 Key performance indicators by country

(percentages)

Country [1) 2)] (Q4 2021)	Return on equity (RoE)	Return on assets (RoA)	Cost-to-income ratio (CIR)	Cost of risk (CoR) [4)]
Belgium	9.35%	0.64%	60.94%	0.15%
Bulgaria	C	C	C	C
Germany	4.09%	0.23%	72.71%	0.22%
Estonia	8.31%	0.98%	56.90%	0.00%
Ireland	5.88%	0.61%	72.82%	-0.01%
Greece	-20.15%	-1.51%	66.38%	5.26%
Spain	10.85%	0.68%	54.64%	1.02%
France	7.15%	0.45%	67.20%	0.41%
Croatia [3)]	-	-	-	-
Italy	5.08%	0.35%	68.01%	0.64%
Cyprus	0.97%	0.07%	74.06%	1.36%
Latvia	10.53%	1.14%	59.12%	C
Lithuania	10.68%	0.81%	47.40%	0.07%
Luxembourg	3.49%	0.26%	78.15%	0.07%
Malta	2.93%	0.21%	88.07%	0.34%
Netherlands	8.35%	0.49%	60.96%	0.09%
Austria	8.73%	0.70%	59.09%	0.41%
Portugal	4.13%	0.33%	48.68%	0.70%
Slovenia	10.48%	1.18%	62.59%	0.22%
Slovakia [3)]	-	-	-	-
Finland	9.34%	0.64%	51.31%	0.17%

Source: ECB.
Notes: Significant institutions at the highest level of consolidation for which common reporting (COREP) and financial reporting (FINREP) are available.
C: the value is suppressed for confidentiality reasons.
1) RoE and RoA are computed by dividing "net profit/loss" by, respectively, "equity" and "total assets" at the end of the corresponding reference period. The values of "net profit/loss", originally year-to-date, are annualised to increase the comparability of the ratios across quarters.
2) Returns figures may be based on different financial years. To increase consistency, if the end of the financial year is not 31 December, a linear projection of the figures has been made for each reporting period.
3) There are no significant institutions at the highest level of consolidation in Croatia and Slovakia.
4) As set out in the list of definitions of ITS data points, the numerator of the cost of risk indicator is adjustments in allowances for estimated loan losses during the relevant period (annualised). Those adjustments may be negative in certain circumstances.

Fig. 1. Key performance indicators by EU country

19. *COR – cost of risk*;

<u>*Non performing loans by country:*</u>

20. *Loans and advances;*
21. *Non-performing loans and advances*;
22. *Non-performing loans ratio %*.

In the middle of 2021 European parliament publish the Regulation 451/2021. This implementing regulation establishes uniform reporting formats and templates. The EU Regulation 451/2021, [11] replaces the Regulation 680/2014, [12] (Fig. 1).

The following table shows the original data used in the study, from the European Central Bank's report - Supervisory Banking Statistics for the fourth quarter of 2021, on the quality of assets by EU Member States, [10].

There are small differences in the reporting figures. They are in the section profit and loss figures by country in the following position – in the report for 2020, they are Net trading income and Exchange differences, net. In the report for 2021 this position are separate in Net gains or losses on financial assets and liabilities held for trading, Net gains or losses on financial assets and liabilities at fair value through profit and loss and Net gains or losses from hedge accounting (Fig. 2).

T04.02.2 Non-performing loans and advances by country
(EUR billions; percentages)

Country (Q4 2021)	Loans and advances [2]	Non-performing loans and advances	Non-performing loans ratio
Belgium	478.39	6.38	1.33%
Bulgaria	C	C	C
Germany	2,951.17	30.63	1.04%
Estonia	33.29	0.27	0.82%
Ireland	415.87	10.26	2.47%
Greece	215.85	15.19	7.04%
Spain	2,705.72	82.43	3.05%
France	5,831.57	110.91	1.90%
Croatia [1]	-	-	-
Italy	1,958.54	61.21	3.13%
Cyprus	37.46	1.49	3.97%
Latvia	C	C	1.14%
Lithuania	28.79	0.20	0.68%
Luxembourg	163.16	0.95	0.58%
Malta	18.20	0.55	3.03%
Netherlands	2,005.66	28.90	1.44%
Austria	521.58	9.65	1.85%
Portugal	170.57	6.49	3.81%
Slovenia	26.84	0.60	2.22%
Slovakia [1]	-	-	-
Finland	550.66	6.91	1.25%
Total	18,138.38	373.68	2.06%

Source: ECB.
Notes: Significant institutions at the highest level of consolidation for which common reporting (COREP) and financial reporting (FINREP) are available.
C: the value is suppressed for confidentiality reasons.
1) There are no significant institutions at the highest level of consolidation in Croatia and Slovakia.
2) Loans and advances in the asset quality tables are displayed at gross carrying amount. In line with FINREP: i) held for trading exposures are excluded, ii) cash balances at central banks and other demand deposits are included.

Fig. 2. Asset quality, by country for 2021

The indicators analyzed for the fourth quarter of 2021 are the same as the indicators analyzed from the Supervisory Banking Statistics report for the fourth quarter of 2020, [10]. There is a difference in the indicators in the Report for 2020, where the position "Net revenues from trade" exists. As a result of changed reporting in 2021, it has been replaced with the more detailed breakdown of these revenues, namely: "Net gains or losses on financial assets and liabilities held for trading", "Net gains or losses on financial assets and liabilities accounted for at fair value in profit or loss" and "Net gains or losses from hedge accounting".

In [8], the conclusions reached in the previous report are confirmed. The end data is different, consecutive financial years are analyzed, but the results obtained using the InterCriteria Analysis method show the same positive and negative consonance.

3 Experimental results

The results that receive after using the multicriteria decision-making method - InterCriteria Analysis are presented in the follow tables – Table 1 and Table 2.

From the data presented in them, the following dependencies can be observed between the indicators.

Table 1. Values of μ

μ	1	2	3	4	5	6	7	8	9	10	11	12	13	14	15	16	17	18	19	20	21	22
1	1	0,8571	0,8214	0,5714	0,5357	0,7857	0,75	0,9643	0,0714	0,9286	0,1786	0,8214	0,9643	0,0357	0,9643	0,6786	0,6071	0,5	0,5714	0,9643	0,9286	0,5
2	0,8571	1	0,8929	0,5714	0,5357	0,7143	0,8214	0,8929	0,0714	0,8571	0,25	0,75	0,8214	0,1786	0,8214	0,5357	0,4643	0,6429	0,5	0,8929	0,8571	0,5
3	0,8214	0,8929	1	0,4643	0,6429	0,6071	0,7857	0,8571	0,1786	0,8214	0,2143	0,7143	0,7857	0,2143	0,7857	0,5	0,4286	0,6071	0,5357	0,8571	0,8214	0,5357
4	0,5714	0,5714	0,4643	1	0,1786	0,6429	0,3929	0,5357	0,4286	0,5	0,5357	0,3929	0,5357	0,4643	0,5357	0,5357	0,4643	0,5	0,4643	0,6071	0,5	0,3571
5	0,5357	0,5357	0,6429	0,1786	1	0,3929	0,6429	0,5	0,5357	0,5357	0,5	0,5714	0,5714	0,4286	0,5714	0,5	0,5714	0,5357	0,3929	0,5714	0,5357	0,5357
6	0,7857	0,7143	0,6071	0,6429	0,3929	1	0,6071	0,75	0,2143	0,7143	0,3929	0,75	0,75	0,25	0,75	0,6786	0,6071	0,4286	0,5714	0,75	0,7143	0,5
7	0,75	0,8214	0,7857	0,3929	0,6429	0,6071	1	0,7857	0,1786	0,8214	0,2857	0,7857	0,7857	0,2143	0,7857	0,5714	0,5	0,6071	0,4643	0,7143	0,8214	0,5357
8	0,9643	0,8929	0,8571	0,5357	0,5	0,75	0,7857	1	0,0357	0,9643	0,1429	0,8571	0,9286	0,0714	0,9286	0,6429	0,5714	0,5357	0,6071	0,9286	0,9643	0,5357
9	0,0714	0,0714	0,1786	0,4286	0,5357	0,2143	0,1786	0,0357	1	0,0714	0,8214	0,1786	0,1071	0,8929	0,1071	0,3929	0,4643	0,4286	0,3929	0,1071	0,0714	0,5
10	0,9286	0,8571	0,8214	0,5	0,5357	0,7143	0,8214	0,9643	0,0714	1	0,1071	0,8929	0,8929	0,1071	0,8929	0,6786	0,6071	0,5	0,6429	0,8929	1	0,5714
11	0,1786	0,25	0,2143	0,5357	0,5	0,3929	0,2857	0,1429	0,8214	0,1071	1	0,2143	0,2143	0,7857	0,2143	0,3571	0,4286	0,6071	0,2143	0,2143	0,1071	0,3214
12	0,8214	0,75	0,7143	0,3929	0,5714	0,75	0,7857	0,8571	0,1786	0,8929	0,2143	1	0,8571	0,1429	0,8571	0,7857	0,7143	0,3929	0,6786	0,7857	0,8929	0,6786
13	0,9643	0,8214	0,7857	0,5357	0,5714	0,75	0,7857	0,9286	0,1071	0,8929	0,2143	0,8571	1	0	1	0,7143	0,6429	0,4643	0,5357	0,9286	0,8929	0,5357
14	0,0357	0,1786	0,2143	0,4643	0,4286	0,25	0,2143	0,0714	0,8929	0,1071	0,7857	0,1429	0	1	0	0,2857	0,3571	0,5357	0,4286	0,0714	0,1071	0,4643
15	0,9643	0,8214	0,7857	0,5357	0,5714	0,75	0,7857	0,9286	0,1071	0,8929	0,2143	0,8571	1	0	1	0,7143	0,6429	0,4643	0,5357	0,9286	0,8929	0,5357
16	0,6786	0,5357	0,5	0,5357	0,5	0,6786	0,5714	0,6429	0,3929	0,6786	0,3571	0,7857	0,7143	0,2857	0,7143	1	0,9286	0,1786	0,6429	0,6429	0,6786	0,6071
17	0,6071	0,4643	0,4286	0,4643	0,5714	0,6071	0,5	0,5714	0,4643	0,6071	0,4286	0,7143	0,6429	0,3571	0,6429	0,9286	1	0,25	0,5714	0,5714	0,6071	0,5357
18	0,5	0,6429	0,6071	0,5	0,5357	0,4286	0,6071	0,5357	0,4286	0,5	0,6071	0,3929	0,4643	0,5357	0,4643	0,1786	0,25	1	0,2143	0,5357	0,5	0,2143
19	0,5714	0,5	0,5357	0,4643	0,3929	0,5714	0,4643	0,6071	0,3929	0,6429	0,2143	0,6786	0,5357	0,4286	0,5357	0,6429	0,5714	0,2143	1	0,5357	0,6429	0,8571
20	0,9643	0,8929	0,8571	0,6071	0,5714	0,75	0,7143	0,9286	0,1071	0,8929	0,2143	0,7857	0,9286	0,0714	0,9286	0,6429	0,5714	0,5357	0,5357	1	0,8929	0,4643
21	0,9286	0,8571	0,8214	0,5	0,5357	0,7143	0,8214	0,9643	0,0714	1	0,1071	0,8929	0,8929	0,1071	0,8929	0,6786	0,6071	0,5	0,6429	0,8929	1	0,5714
22	0,5	0,5	0,5357	0,3571	0,5357	0,5	0,5357	0,5357	0,5	0,5714	0,3214	0,6786	0,5357	0,4643	0,5357	0,6071	0,5357	0,2143	0,8571	0,4643	0,5714	1

Table 2. Values of ν

ν	1	2	3	4	5	6	7	8	9	10	11	12	13	14	15	16	17	18	19	20	21	22
1	0	0,1429	0,1786	0,4286	0,4643	0,2143	0,25	0,0357	0,9286	0,0714	0,8214	0,1786	0,0357	0,9643	0,0357	0,3214	0,3929	0,5	0,3929	0,0357	0,0714	0,5
2	0,1429	0	0,1071	0,4286	0,4643	0,2857	0,1786	0,1071	0,9286	0,1429	0,75	0,25	0,1786	0,8214	0,1786	0,4643	0,5357	0,3571	0,4643	0,1071	0,1429	0,5
3	0,1786	0,1071	0	0,5357	0,3571	0,3929	0,2143	0,1429	0,8214	0,1786	0,7857	0,2857	0,2143	0,7857	0,2143	0,5	0,5714	0,3929	0,4286	0,1429	0,1786	0,4643
4	0,4286	0,4286	0,5357	0	0,8214	0,3571	0,6071	0,4643	0,5714	0,5	0,4643	0,6071	0,4643	0,5357	0,4643	0,4643	0,5357	0,5	0,5	0,3929	0,5	0,6429
5	0,4643	0,4643	0,3571	0,8214	0	0,6071	0,3571	0,5	0,4643	0,4643	0,5	0,4286	0,4286	0,5714	0,4286	0,5	0,4286	0,4643	0,5714	0,4286	0,4643	0,4643
6	0,2143	0,2857	0,3929	0,3571	0,6071	0	0,3929	0,25	0,7857	0,2857	0,6071	0,25	0,25	0,75	0,25	0,3214	0,3929	0,5714	0,3929	0,25	0,2857	0,5
7	0,25	0,1786	0,2143	0,6071	0,3571	0,3929	0	0,2143	0,8214	0,1786	0,7143	0,2143	0,2143	0,7857	0,2143	0,4286	0,5	0,3929	0,5	0,2857	0,1786	0,4643
8	0,0357	0,1071	0,1429	0,4643	0,5	0,25	0,2143	0	0,9643	0,0357	0,8571	0,1429	0,0714	0,9286	0,0714	0,3571	0,4286	0,4643	0,3571	0,0714	0,0357	0,4643
9	0,9286	0,9286	0,8214	0,5714	0,4643	0,7857	0,8214	0,9643	0	0,9286	0,1786	0,8214	0,8929	0,1071	0,8929	0,6071	0,5357	0,5714	0,5714	0,8929	0,9286	0,5
10	0,0714	0,1429	0,1786	0,5	0,4643	0,2857	0,1786	0,0357	0,9286	0	0,8929	0,1071	0,1071	0,8929	0,1071	0,3214	0,3929	0,5	0,3214	0,1071	0	0,4286
11	0,8214	0,75	0,7857	0,4643	0,5	0,6071	0,7143	0,8571	0,1786	0,8929	0	0,7857	0,7857	0,2143	0,7857	0,6429	0,5714	0,3929	0,75	0,7857	0,8929	0,6786
12	0,1786	0,25	0,2857	0,6071	0,4286	0,25	0,2143	0,1429	0,8214	0,1071	0,7857	0	0,1429	0,8571	0,1429	0,2143	0,2857	0,6071	0,2857	0,2143	0,1071	0,3214
13	0,0357	0,1786	0,2143	0,4643	0,4286	0,25	0,2143	0,0714	0,8929	0,1071	0,7857	0,1429	0	1	0	0,2857	0,3571	0,5357	0,4286	0,0714	0,1071	0,4643
14	0,9643	0,8214	0,7857	0,5357	0,5714	0,75	0,7857	0,9286	0,1071	0,8929	0,2143	0,8571	1	0	1	0,7143	0,6429	0,4643	0,5357	0,9286	0,8929	0,5357
15	0,0357	0,1786	0,2143	0,4643	0,4286	0,25	0,2143	0,0714	0,8929	0,1071	0,7857	0,1429	0	1	0	0,2857	0,3571	0,5357	0,4286	0,0714	0,1071	0,4643
16	0,3214	0,4643	0,5	0,4643	0,5	0,3214	0,4286	0,3571	0,6071	0,3214	0,6429	0,2143	0,2857	0,7143	0,2857	0	0,0714	0,8214	0,3214	0,3571	0,3214	0,3929
17	0,3929	0,5357	0,5714	0,5357	0,4286	0,3929	0,5	0,4286	0,5357	0,3929	0,5714	0,2857	0,3571	0,6429	0,3571	0,0714	0	0,75	0,3929	0,4286	0,3929	0,4643
18	0,5	0,3571	0,3929	0,5	0,4643	0,5714	0,3929	0,4643	0,5714	0,5	0,3929	0,6071	0,5357	0,4643	0,5357	0,8214	0,75	0	0,75	0,4643	0,5	0,7857
19	0,3929	0,4643	0,4286	0,5	0,5714	0,3929	0,5	0,3571	0,5714	0,3214	0,75	0,2857	0,4286	0,5357	0,4286	0,3214	0,3929	0,75	0	0,4286	0,3214	0,1071
20	0,0357	0,1071	0,1429	0,3929	0,4286	0,25	0,2857	0,0714	0,8929	0,1071	0,7857	0,2143	0,0714	0,9286	0,0714	0,3571	0,4286	0,4643	0,4286	0	0,1071	0,5357
21	0,0714	0,1429	0,1786	0,5	0,4643	0,2857	0,1786	0,0357	0,9286	0	0,8929	0,1071	0,1071	0,8929	0,1071	0,3214	0,3929	0,5	0,3214	0,1071	0	0,4286
22	0,5	0,5	0,4643	0,6429	0,4643	0,5	0,4643	0,4643	0,5	0,4286	0,6786	0,3214	0,4643	0,5357	0,4643	0,3929	0,4643	0,7857	0,1071	0,5357	0,4286	0

The indicators/criteria in strong positive consonance are:

1. *Net interest income* and 8. *Operating income;*

1. *Net interest income* and 13. *Profit and loss before tax are;*

1. *Net interest income* and 15. *Net profit/loss;*

1. *Net interest income* and 20. *Loans and advances;*

8. *Operating income* and 10. *Net income before impairment, provision and taxes;*

8. *Operating income* and 21. *Non-performing loans and advances;*

10. *Net income before impairment, provision and taxes* and 21. *Non-performing loans and advances;*

13. *Profit and loss before tax* and 15. *Net profit/loss.*

The indicators/criteria in positive consonance are:

1. *Net interest income* and 10. *Net income before impairment, provision and taxes;*

1. *Net interest income* and 21. *Non-performing loans and advances;*

2. *Net fee and commission income* and 3. *Net gains or losses on financial assets and liabilities held for trading;*

2. *Net fee and commission income* and 8. *Operating income;*

2. *Net fee and commission income* and 20. *Loans and advances;*

8. *Operating income* and 13. *Profit and loss before tax;*

8. *Operating income* and 15. *Net profit/loss;*

8. *Operating income* and 20. *Loans and advances;*

9. *Administrative expenses and depreciation* and 14. *Tax expenses or income;*

10. *Net income before impairment, provision and taxes* and 13. *Profit and loss before tax;*

10. *Net income before impairment, provision and taxes* and 15. *Net profit/loss;*

10. *Net income before impairment, provision and taxes* and 20. *Loans and advances;*

13. *Profit and loss before tax* and 20. *Loans and advances;*

13. *Profit and loss before tax* and 21. *Non-performing loans and advances;*

15. *Net profit/loss* and 20. *Loans and advances;*

15. *Net profit/loss* and 21. *Non-performing loans and advances;*

16. *RoE – return of equity* and 17. *RoA – return of assets;*

20. *Loans and advances* and 21. *Non-performing loans and advances.*

The indicators/criteria in strong dissonance are:

4. *Net gains or losses on financial assets and liabilities at fair value through profit and loss* and 14. *Tax expenses or income;*

4. *Net gains or losses on financial assets and liabilities at fair value through profit and loss* and 17. *RoA – return of assets;*

4. *Net gains or losses on financial assets and liabilities at fair value through profit and loss* and 19. *COR – cost of risk;*

7. *Net other operating income* and 19. *COR – cost of risk;*

9. *Administrative expenses and depreciation* and 17. *RoA – return of assets;*

13. *Profit and loss before tax* and 18. *CIR – cost to income ratio*;

14. *Tax expenses or income* and 22. *Non-performing loans ratio %;*

15. *Net profit/loss* and 18. *CIR – cost to income ratio*;

20. *Loans and advances* and 22. *Non-performing loans ratio %.*

The indicators/criteria in negative consonance are:

1. *Net interest income* and 9. *Administrative expenses and depreciation;*

2. *Net fee and commission income* and 9. *Administrative expenses and depreciation;*

8. *Operating income* and 14. *Tax expenses or income;*

9. *Administrative expenses and depreciation* and 10. *Net income before impairment, provision and taxes;*

9. *Administrative expenses and depreciation* and 13. *Profit or loss before taxes;*

9. *Administrative expenses and depreciation* and 15. *Net profit/loss;*

9. *Administrative expenses and depreciation* and 20. *Loans and advances;*

9. *Administrative expenses and depreciation* and 21. *Non-performing loans and advances;*

10. *Net income before impairment, provision and taxes* and 11. *Impairment and provisions;*

10. *Net income before impairment, provision and taxes* and 14. *Tax expenses or income;*

11. *Impairment and provisions* and 21. *Non-performing loans;*

14. *Tax expenses or income* and 20. *Loans and advances;*

14. *Tax expenses or income* and 21. *Non-performing loans and advances.*

The indicators/criteria in strong negative consonance are:

1. *Net interest income* and 14. *Tax expenses or income;*

8. *Operating income* and 9. *Administrative expenses and depreciation;*

13. *Profit and loss before tax* and 14. *Tax expenses or income;*

14. *Tax expenses or income* and 15. *Net profit/loss.*

The indicators/criteria, which are independent:

1. *Net interest income* and 18. *CIR – cost to income ratio*;

1. *Net interest income* and 22. *Non-performing loans ratio %*;

2. *Net fee and commission income* and 19. *COR – cost of risk;*

2. *Net fee and commission income* and 22. *Non-performing loans ratio %*;

3. *Net gains or losses on financial assets and liabilities held for trading* and 16. *RoE – return of equity*;

4. *Net gains or losses on financial assets and liabilities at fair value through profit and loss* and 10. *Net income before impairment, provision and taxes;*

4. *Net gains or losses on financial assets and liabilities at fair value through profit and loss* and 18. *CIR – cost to income ratio*;

4. *Net gains or losses on financial assets and liabilities at fair value through profit and loss* and 21. *Non-performing loans and advances*;

5. *Net gains or losses from hedge accounting* and 8. *Operating income;*

5. *Net gains or losses from hedge accounting* and 11. *Impairment and provisions;*

5. *Net gains or losses from hedge accounting* and 16. *RoE – return of equity*;

6. *Exchange differences, net* and 22. *Non-performing loans ratio %*;

7. *Net other operating income* and 17. *RoA – return of assets;*

9. *Administrative expenses and depreciation* and 22. *Non-performing loans ratio %*;

10. *Net income before impairment, provision and taxes* and 18. *CIR – cost to income ratio*;

18. *CIR – cost to income ratio* and 21. *Non-performing loans and advances.*

The results of the analysis show that there is a strong dependence between indicator/criterion 1. *Net interest income* and indicator/criterion 8. *Operating income*, indicator/criterion 13. *Profit or loss before taxes*, indicator/criterion 15. *Net profit or loss* and indicator/criterion 20. *Loans and advances.* The main reason for this is the fact that net interest income is the main operating income and forms a large part of the financial result for banks. Loans are the main interest-bearing assets and bring income to the bank in the form of interest, fees and commissions.

Another strongly pronounced dependence is observed in indicator/criterion 8. *Operating income* and indicator/criterion 10. *Net income, before impairment, provisions and taxes*, due to the fact that operating income is the main income forming the net income before taxes, charges and impairment. Such a strong dependence also exists between indicator/criterion 8. *Operating income* and indicator/criterion 21. *Non-performing loans and advances.* There is also a strong dependence between indicator/criteria 13. *Profit or loss before taxes* and indicator/criterion 15. *Net profit or loss.* This is due to the fact that the first criterion is the main formative element of the bank's profit.

When we analyze the data from the table, we again see dependence between the 16. *RoE – return of equity* and 17. *RoA – return of assets* ratios. This is because they are the main indicators of how profitable a bank is.

There is an insignificant dependence between indicator/criterion 1. *Net interest income* and indicator/criterion 14. *Tax income or expense*s because one criterion is income and the other is an expense for the bank. The same conclusion can be drawn for indicator/criterion 8. *Operating income* and indicator/criterion 9. *Administrative expenses and impairment*.

The reason for the weak positive dependence between indicator/criterion 16. *RoA - return on assets ratio* and indicator/criterion 18. *COR – cost of risk ratio* lies in the economic nature and the way these criteria are quantified. The *RoA* is calculated as the ratio of income from assets to their size, and the *COR* is the ratio of impairment and the level of loans.

4 Conclusion

The analyzed data from the table show a pronounced dependence between the criteria that are part of the Income Statement and their essence, as economic importance is among the main reasons for the relationship between them. Also between criteria from the Income Statement and assets from the Institution's Balance Sheet, as a consequence of these assets bringing a corresponding income for the banks. This is due to both their economic nature and the way in which they are calculated.

The purpose of this paper is to analyze banking processes using modern methods from the field of intelligent systems. Original results have been achieved related to the research of modern paradigms in the field of intelligent systems, using analytical and experimental models.

Acknowledgements. This paper was partially supported by the Bulgarian National Science Fund, Grant agreement No. KP-06-N22/1, project "Theoretical research and applications of the InterCriteria Analysis" and partially supported by the Bulgarian Ministry of Education and Science under the National Research Program "Smart crop production", Grant agreement No. D01–65/19.03.2021, approved by Decision of the Ministry Council №866/26.11.2020.

References

1. Atanassov, K., Szmidt, E., Kacprzyk, J.: On intuitionistic fuzzy pairs. In: Notes on Intuitionistic Fuzzy Sets, vol. 19, No. 3, pp. 1–13 (2013). ISSN: 1310–4926
2. Atanassov, K.: Index Matrices: Towards an Augmented Matrix Calculus, Studies in Computational Intelligence, vol. 573. Springer, Switzerland (2014). ISBN: 978-3319109442
3. Atanassov, K., Mavrov, D., Atanassova, V.: InterCriteria decision making: a new approach for multicriteria decision making, based on index matrices and intuitionistic fuzzy sets. In: Proceedings of the 12th International Workshop on Intuitionistic Fuzzy Sets and Generalized Nets, Warsaw, Poland (2013)
4. Atanassova, V., Mavrov, D., Doukovska, L., Atanassov, K.: Discussion on the threshold values in the InterCriteria decision making approach. Notes Intuitionistic Fuzzy Sets **20**(2), 94–99 (2014). ISSN: 1310–4926
5. Europe's role in the global financial system, Speech by Luis de Guindos, Vice-President of the ECB, at the SUERF/De Nederlandsche Bank Conference on Forging a new future between the UK and the EU. https://www.ecb.europa.eu/press/
6. Eurostat, Statistic explained, Enlargement countries - economic developments. https://ec.europa.eu/eurostat/statistics-explained/
7. InterCriteria Research Portal. http://www.intercriteria.net/publications
8. Schwab, K.: The Global Competitiveness Report 2019. World Economic Forum, Geneva (2019). ISBN: 978-2-940631-02-5
9. EU recommendation 2003/361
10. Supervisory Banking Statistics, Fourth quarter 2020 (2020). ISSN 2467–4303
11. Methodological note for the publication of aggregated Supervisory Banking Statistics, fourth quarter 2021 (2021). https://www.bankingsupervision.europa.eu/banking/statistics/shared/pdf/ssm.methodologicalnote_supervisorybankingstatistics202204.en.pdf
12. Regulation (EU) No 680/2014 of the European Parliament. https://www.en-standard.eu/
13. Regulation (EU) No 575/2013 of the European Parliament. https://www.en-standard.eu/

Industrial and Business Applications

Generalized Net Models of Traffic Quality Evaluation of a Service Stage

Velin Andonov(✉), Stoyan Poryazov, and Emiliya Saranova

Institute of Mathematics and Informatics, Bulgarian Academy of Sciences, Acad. G. Bonchev Str., Block 8, 1113 Sofia, Bulgaria
{velin_andonov,stoyan,emiliya}@math.bas.bg

Abstract. In the present paper, Generalized Net (GN) models of a service phase and a service stage in an overall telecommunication system are proposed. The proposed models are aimed at traffic quality evaluation. The values of some traffic quality indicators are obtained in the form of characteristics of tokens of the GNs. The proposed GN models are an important step towards the monitoring, prediction and management of the Quality of Service (QoS) in overall telecommunication systems. The GN approach described in the paper can be applied to all service systems and, in particular, to the next generation smart service networks.

Keywords: Generalized net · Service network · Quality of service

1 Introduction

In recent years, Generalized Nets (GNs, see [4]) have been successfully used in the modeling of service systems and for Quality of Service (QoS) evaluation of service systems. In [2], GN models of the causal structure of a queuing system in an overall telecommunication system are constructed. They are based on conceptual models of causal decomposition of the traffic in the queuing system. In [3], a GN model of a serial composition of services with intuitionistic fuzzy estimations of uncertainty is proposed. It is the first GN model of a composition of services which is constructed for the evaluation of the QoS of a comprise service as a composition of the QoS of the embedded services.

The present paper is a continuation of the work of the authors on the modeling and estimation of QoS in telecommunication systems using the apparatus of the GNs theory. Two GN models of a service phase and a service stage of an overall telecommunication system are constructed on the basis of the conceptual models of a service phase and a service stage described in [8]. As a result of the functioning of the GNs, the values of some of the traffic quality indicators defined

The work of Velin Andonov, Stoyan Poryazov and Emiliya Saranova is supported by the research project “Perspective Methods for Quality Prediction in the Next Generation Smart Informational Service Networks” (KP-06-N52/2) financed by the Bulgarian National Science Fund.

K. T. Atanassov et al. (Eds.): BOS/SOR/IWIFSGN 2022, LNNS 793, pp. 141–155, 2023.
https://doi.org/10.1007/978-3-031-45069-3_13

in [8] are evaluated through the characteristic functions of particular places of the GNs. The GN approach allows the values of other traffic quality indicators to be obtained by modifying the characteristic functions of the places without changing the other parameters of the nets.

The two proposed GN models are the first step towards the construction of a GN model of overall network efficiency evaluation. They are also important step in the study of quality composition in overall telecommunication systems. In recent years, the problem of quality composition has been a hot topic of research [5,10].

The rest of the paper is organized as follows. In Sect. 2, the basic notions which are used in the paper are introduced. In Sect. 3, a GN model of traffic quality evaluation of a service phase is described. In Sect. 4, a GN model of traffic quality evaluation of a service stage is described. Finally, some conclusions are drawn and a direction for future research is outlined.

2 Preliminaries

2.1 Base Virtual Service Devices and Their Parameters

In the conceptual models of service systems in general, and telecommunication systems in particular, the notion of base virtual service device plays a fundamental role. Every base virtual device x has the following parameters: intensity of the flow of requests (Fx), probability of directing the flow of requests towards the device (Px), service time in the device (Tx), traffic intensity (Yx, measured in Erlang) and device capacity (N_x). Various types of base virtual devices are used (see [7]). Their graphical representation is shown in Fig. 1.

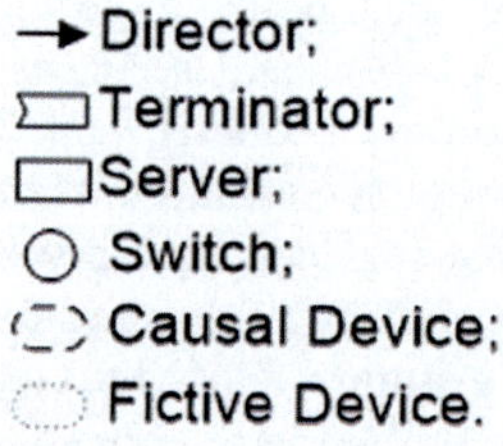

Fig. 1. Base virtual device types.

Each type of base virtual device has a specific function. These functions are described in detail in [9].

An important step in the modeling of service systems through GNs is done in [1] where GN representations of the base virtual service devices are proposed. These representations are used in the construction of the GN models in the present paper.

2.2 Conceptual Model of a Service Phase

The notion of a service phase is explained in [8]. It is based on the ITU-T definition of a service (see [6]). Since it is important for the present work, we give that definition below.

Definition 1. *A service phase is a service presentation containing:*

- *One of the functions, realizing the service, which is considered indivisible.*
- *All modeled reasons for ending/finishing of this function, i.e. the causal structure of the function.*
- *Hypothetic characteristics related to the causal structure of the function (a well-known example of a hypothetic characteristic is the offered traffic concept).*

A conceptual model of a service phase is proposed in [8]. It is shown in Fig. 2.

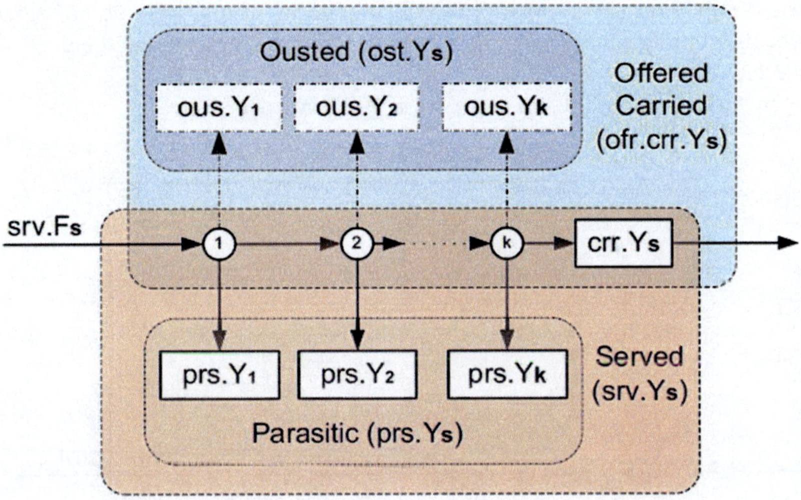

Fig. 2. Conceptual model of a service phase (see [8]).

The service phase is represented by $k + 1$ base virtual causal devices. Each of them represents a reason for ending of the service phase. Apart from the base service devices in Fig. 2 there are also comprise virtual service devices which have base virtual devices embedded in them. For instance, the comprise device of the ousted traffic which encompasses all k ousted base devices. In the graphical representation above special qualifiers are used to characterize the traffic. These are: $crr.$, $srv.$, $ous.$, $ofr.crr.$ and $prs.$. They are abbreviations of carried, served, ousted, offered carried and parasitic, respectively. For the definitions of these traffic characterizations see [8].

Many quality indicators of a service phase are defined in [8]. Here, we shall use the *Service efficiency indicator* defined as

$$I_4 = \frac{crr.Ys}{srv.Ys} = 1 - \frac{prs.Ys}{srv.Ys}. \quad (1)$$

2.3 Conceptual Model of a Service Stage

The service stage concept is defined in [8]. Since this notion is important for the present work, we shall give the definition below.

Definition 2. *Service stage is a service presentation containing:*

- *One service phase, realizing one function of the service.*
- *All auxiliary service phases that directly support this function realization but are not part of the realized function itself.*

Auxiliary service phases are for example buffer, entry, etc. Their performance depends directly on the service phase. The concept of a service stage allows for the telecommunication system to be divided into subservices which enhances the modeling.

A conceptual model of a service stage g is shown in Fig. 3.

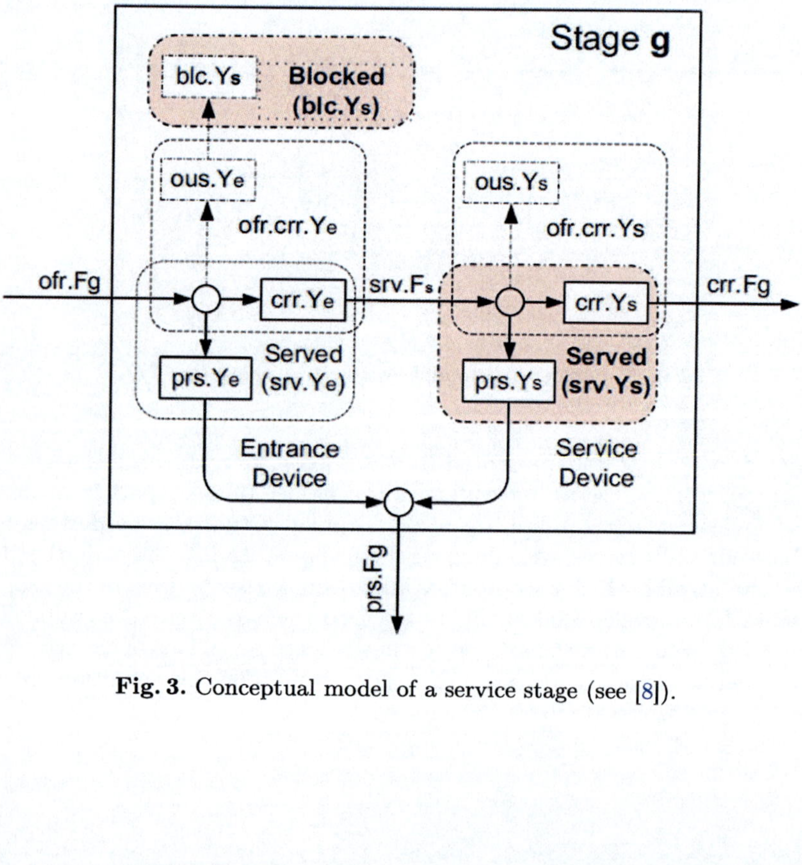

Fig. 3. Conceptual model of a service stage (see [8]).

The service stage consists of Entrance and Service phases.

Among the many traffic quality indicators of a service stage defined in [8], in the present paper we shall use the Carried effectiveness indicator defined as:

$$I_9 = \frac{eff.crr.Y}{crr.Y}. \tag{2}$$

3 A Generalized Net Model of Traffic Quality Evaluation of a Service Phase

In this section, a GN model of a service phase is constructed on the basis of the conceptual model shown in Fig. 2. Its graphical representation is shown in Fig. 4.

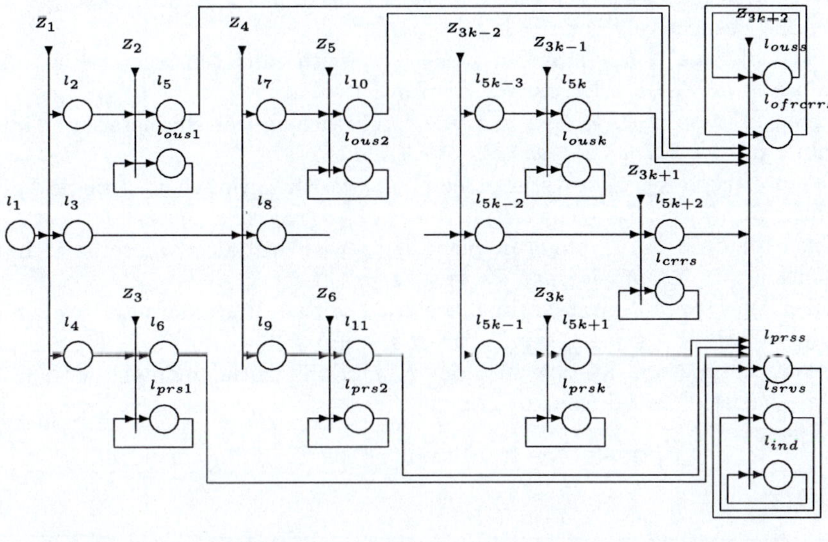

Fig. 4. A GN model of traffic quality evaluation of a service phase.

The GN is a reduced one (see [4]) and consists of $3k + 2$ transitions, $7k + 8$ places and $2k + 7$ types of tokens. The transitions represent the following functions:

- Transitions $Z_2, Z_5, \cdots, Z_{3k-1}$ represent the function of the k ousted devices.
- Transitions $Z_3, Z_6, \cdots, Z_{3k}$ represent the function of the k parasitic devices.
- Transition Z_{3k+1} represents the function of the carried device.
- Transition Z_{3k+2} represents the function of the comprise ousted, offered carried, parasitic and served devices.

A special notation for the places of the GN is used. The places which correspond to virtual service device, and where the tokens obtain as characteristics the values of the parameters of the corresponding device are labeled l_x where 'x' is the name of the device and the symbol '.' is omitted. The place l_{ind} is where the evaluation of the values of the quality indicators is performed.

The $2k+7$ tokens are denoted by $\alpha, \alpha_1, \cdots, \alpha_{2k+6}$. In the initial moment of the functioning of the GN:

- Token of type α enters place l_1 with initial characteristic "formula for generating the $srv.Fs$ flow of requests".
- Tokens of types $\alpha_1, \alpha_3, \cdots, \alpha_{2k-1}$ stay respectively in places $l_{ous1}, l_{ous2}, \cdots, l_{ous2k-1}$ with initial characteristic "initial values of $ous.Y_i, ous.F_i, ous.P_i, ous.T_i, ous.N_i$" respectively, for $i = 1, 2, \cdots, k$.
- Tokens of types $\alpha_2, \alpha_4, \cdots, \alpha_{2k}$ stay in places $l_{prs1}, l_{prs2}, \cdots, l_{prsk}$ respectively with initial characteristic "initial values of $prs.Y_i, prs.F_i, prs.P_i, prs.T_i, prs.N_i$" respectively, for $i = 1, 2, \cdots, k$.
- Token of type α_{2k+1} stays in place l_{crrs} with initial characteristic "initial values of $crr.Ys, crr.Fs, crr.Ps, crr.Ts, crr.Ns$".
- Token of type α_{2k+2} stays in place l_{ouss} with initial characteristic "initial values of $ous.Ys, ous.Fs, ous.Ps, ous.Ts, ous.Ns$".
- Token of type α_{2k+3} stays in place $l_{ofrcrrs}$ with initial characteristic "initial values of $ofr.crr.Ys, ofr.crr.Fs, ofr.crr.Ps, ofr.crr.Ts, ofr.crr.Ns$".
- Token of type α_{2k+4} stays in place l_{prss} with initial characteristic "initial values of $prs.Ys, prs.Fs, prs.Ps, prs.Ts, prs.Ns$".
- Token of type α_{2k+5} stays in place l_{srvs} with initial characteristic "initial values of $srv.Ys, srv.Fs, srv.Ps, prs.Ts, srv.Ns$".
- Token of type α_{2k+6} stays in place l_{ind} in the initial moment with initial characteristic "initial value of I_4".

A formal description of the transitions of the GN follows below.

$$Z_1 = \langle \{l_1\}, \{l_2, l_3, l_4\}, r_1 \rangle,$$

where

$$r_1 = \begin{array}{c|ccc} & l_2 & l_3 & l_4 \\ \hline l_1 & true & true & true \end{array}.$$

The token of type α in place l_1 splits into three identical tokens which enter respectively places l_2, l_3, l_4 without obtaining new characteristics.

$$Z_2 = \langle \{l_2, l_{ous1}\}, \{l_5, l_{ous1}\}, r_2 \rangle,$$

where

$$r_2 = \begin{array}{c|cc} & l_5 & l_{ous1} \\ \hline l_2 & true & true \\ l_{ous1} & false & true \end{array}.$$

The token of type α in place l_2 splits into two identical tokens one of which enters place l_{ous1} where it merges with the token of type α_1. The other one enters place l_5 where it obtains the characteristic "current values of $ous.Y_1, ous.F_1, ous.P_1, ous.T_1, ous.N_1$". The token α_1 in place l_{ous1} obtains the characteristic "current values of $ous.Y_1, ous.F_1, ous.P_1, ous.T_1, ous.N_1$".

$$Z_3 = \langle \{l_4, l_{prs1}\}, \{l_6, l_{prs1}\}, r_3 \rangle,$$

where

$$r_3 = \begin{array}{c|cc} & l_5 & l_{prs1} \\ \hline l_4 & true & true \\ l_{prs1} & false & true \end{array}.$$

The token of type α in place l_4 splits into two identical tokens one of which enters place l_{prs1} where it merges with the token of type α_2. The other one enters place l_6 where it obtains the characteristic "current values of $prs.Y_1, prs.F_1, prs.P_1, prs.T_1, prs.N_1$". The token α_2 in place l_{prs1} obtains the characteristic "current values of $prs.Y_1, prs.F_1, prs.P_1, prs.T_1, prs.N_1$".

The formal description of transitions $Z_5, Z_8, \cdots, Z_{3k-1}$ is analogous to that of transition Z_2. The formal description of transitions $Z_6, Z_9, \cdots, Z_{3k}$ is analogous to that of transitions Z_3.

The next transition which has a different formal description is Z_{3k+1}.

$$Z_{3k+1} = \langle \{l_{5k-2}, l_{crrs}\}, \{l_{5k+2}, l_{crrs}\}, r_{3k+1} \rangle,$$

where

$$r_{3k+1} = \begin{array}{c|cc} & l_{5k+2} & l_{crrs} \\ \hline l_{5k-2} & true & true \\ l_{crrs} & false & true \end{array}.$$

The token of type α in place l_{5k-2} splits into two identical tokens one of which enters place l_{crrs} where it merges with the token of type α_{2k+1}. The other one enters place l_{5k+2} where it obtains the characteristic "current values of $crr.Ys, crr.Fs, crr.Ps, crr.Ns$". The token α_{2k+1} in place l_{crrs} obtains the characteristic "current values of $crr.Ys, crr.Fs, crr.Ps, crr.Ts, crr.Ns$".

$$Z_{3k+2} = \langle \{l_5, \cdots, l_{5k}, l_6, \cdots, l_{5k+1}, l_{5k+2}, l_{ouss}, l_{ofrcrrs}, l_{prss}, l_{srvs}, l_{ind}\},$$
$$\{l_{ouss}, l_{ofrcrrs}, l_{prss}, l_{srvs}, l_{ind}\}, r_{3k+2} \rangle,$$

where

$$r_{3k+2} = \begin{array}{c|ccccc} & l_{ouss} & l_{ofrcrrs} & l_{prss} & l_{srvs} & l_{ind} \\ \hline l_5 & true & true & false & false & false \\ \vdots & \vdots & \vdots & \vdots & \vdots & \vdots \\ l_{5k} & true & true & false & false & false \\ l_6 & false & false & true & true & false \\ \vdots & \vdots & \vdots & \vdots & \vdots & \vdots \\ l_{5k+1} & false & false & true & true & false \\ l_{5k+2} & false & true & false & true & false \\ l_{ouss} & true & false & false & false & false \\ l_{ofrcrrs} & false & true & false & false & false \\ l_{prss} & false & false & true & false & false \\ l_{srvs} & false & false & false & true & false \\ l_{ind} & false & false & false & false & true \end{array}.$$

The tokens in each of the places $l_5, l_{10}, \cdots, l_{5k}$ split into two identical tokens one of which enters l_{ouss} and the other one enters place $l_{ofrcrrs}$. All tokens entering place l_{ouss} merge with the token α_{2k+2} which obtains the characteristic "current values of $ous.Ys, ous.Fs, ous.Ps, ous.Ts, ous.Ns$". Token α in place l_{5k+2} splits into two identical tokens one of which enters place $l_{ofrcrrs}$. The other one enters place l_{srvs}. All tokens entering place $l_{ofrcrrs}$ merge with the token α_{2k+3} which obtains the characteristic "current values of $ofr.crr.Ys, ofr.crr.Fs, ofr.crr.Ps, ofr.crr.Ts, ofr.crr.Ns$". The tokens in each of the places $l_6, l_{11}, \cdots, l_{5k+1}$ split into two identical tokens one of which enters l_{prss} and the other one enters place l_{srvs}. All tokens entering place l_{prss} merge with the token α_{2k+4} which obtains the characteristic "current values of $prs.Ys, prs.Fs, prs.Ps, prs.Ts, prs.Ns$". All tokens entering place l_{srvs} merge with the token α_{2k+5} which obtains the characteristic "current values of $srv.Ys, srv.Fs, srv.Ps, srv.Ts, srv.Ns$". Token α_{2k+6} in place l_{ind} obtains the characteristic "current value of indicator I_4". The value of the indicator I_4 is evaluated using Eq. (1).

4 Generalized Net Model of a Service Stage

In this section, a GN model of a service stage is constructed on the basis of the conceptual model shown in Fig. 3. Its graphical representation is shown in Fig. 5.

The GN is a reduced one and consists of 14 transitions, 34 places and 15 types of tokens. The transitions represent the following functions:

- Transitions Z_2 represents the function of the ousted subdevice of the entrance device.
- Transitions Z_3 represents the function of the carried subdevice of the entrance device.
- Transition Z_4 represents the function of the parasitic subdevice of the entrance device.
- Transition Z_5 represents the function of the offered carried subdevice of the entrance device.

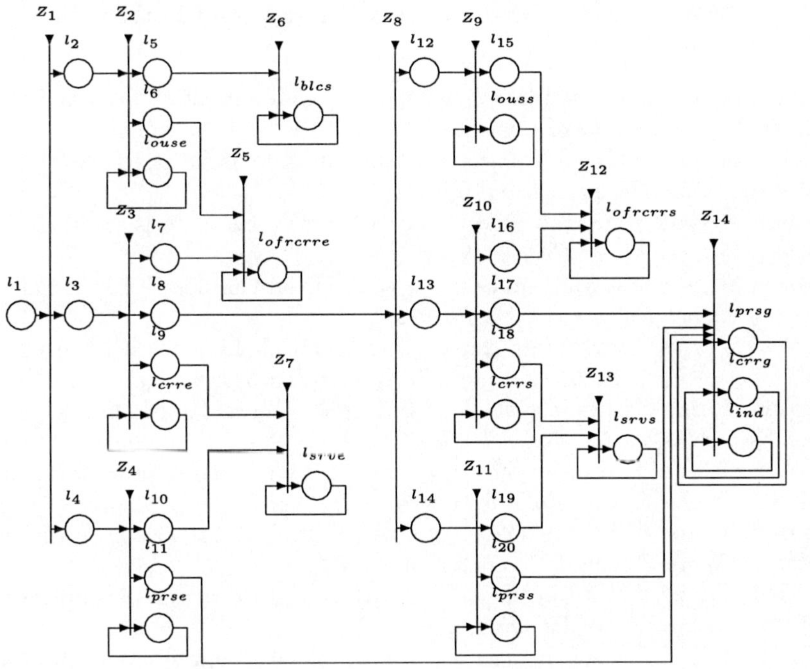

Fig. 5. Generalized net model of traffic quality evaluation of a service stage.

- Transition Z_6 represents the function of the blocked subdevice of the service device.
- Transition Z_7 represents the function of the served subdevice of the entrance device.
- Transitions Z_9 represents the function of the ousted subdevice of the service device.
- Transitions Z_{10} represents the function of the carried subdevice of the service device.
- Transition Z_{11} represents the function of the parasitic subdevice of the service device.
- Transition Z_{12} represents the function of the offered carried subdevice of the service device.
- Transition Z_{13} represents the function of the served subdevice of the service device.
- Transition Z_{14} represents the function of the parasitic and carried device of the service stage and evaluates the chosen traffic quality indicator.

A special notation for the places of the GN is used. The places which correspond to virtual service devices and where the tokens obtain as characteristics the values of the parameters of the corresponding device are labeled l_x where 'x' is the name of the device and the symbol '.' is omitted. The place l_{ind} is where the evaluation of the values of the quality indicators is performed.

The 15 tokens are denoted by lowercase Greek letters. In the initial moment of the functioning of the GN:

- Token of type α enters place l_1 with initial characteristic "formula for generating the $ofr.Fg$ flow of requests".
- Token of type β stays in place l_{ouse} with initial characteristic "initial values of $ous.Ye, ous.Fe, ous.Pe, ous.Te, ous.Ne$".
- Token of type γ stays in place l_{crre} with initial characteristic "initial values of $crr.Ye, crr.Fe, crr.Pe, crr.Te, crr.Ne$".
- Token of type δ stays in place l_{prse} with initial characteristic "initial values of $prs.Ye, prs.Fe, prs.Pe, prs.Te, prs.Ne$".
- Token of type ϵ stays in place $l_{ofrcrre}$ with initial characteristic "initial values of $ofr.crr.Ye, ofr.crr.Fe, ofr.crr.Pe, ofr.crr.Te, ofr.crr.Ne$".
- Token of type ζ stays in place l_{blcs} with initial characteristic "initial values of $blc.Ys, blc.Fs, blc.Ps, blc.Ts, blc.Ns$".
- Token of type η stays in place l_{srve} with initial characteristic "initial values of $srv.Ye, srv.Fe, srv.Pe, srv.Te, srv.Ne$".
- Token of type θ stays in place l_{ouss} with initial characteristic "initial values of $ous.Ys, ous.Fs, ous.Ps, ous.Ts, ous.Ns$".
- Token of type κ stays in place l_{crrs} with initial characteristic "initial values of $crr.Ys, crr.Fs, crr.Ps, crr.Ts, crr.Ns$".
- Token of type λ stays in place l_{prss} with initial characteristic "initial values of $prs.Ys, prs.Fs, prs.Ps, prs.Ts, prs.Ns$".
- Token of type μ stays in place $l_{ofrcrrs}$ with initial characteristic "initial values of $ofr.crr.Ys, ofr.crr.Fs, ofr.crr.Ps, ofr.crr.Ts, ofr.crr.Ns$".
- Token of type ν stays in place l_{srvs} with initial characteristic "initial values of $srv.Ys, srv.Fs, srv.Ps, srv.Ts, srv.Ns$".
- Token of type ξ stays in place l_{prsg} with initial characteristic "initial values of $prs.Yg, prs.Fg, prs.Pg, prs.Tg, prs.Ng$".
- Token of type o stays in place l_{crrg} with initial characteristic "initial values of $crr.Yg, crr.Fg, crr.Pg, crr.Tg, crr.Ng$".
- Token of type π stays in place l_{ind} with initial characteristic "initial value of I_9".

A formal description of the transitions of the GN follows below.

$$Z_1 = \langle \{l_1\}, \{l_2, l_3, l_4\}, r_1 \rangle,$$

where

$$r_1 = \begin{array}{c|ccc} & l_2 & l_3 & l_4 \\ \hline l_1 & true & true & true \end{array}.$$

Token of type α in place l_1 splits into three identical tokens which enter respectively places l_2, l_3, l_4 without obtaining new characteristics.

$$Z_2 = \langle \{l_2, l_{ouse}\}, \{l_5, l_6, l_{ouse}\}, r_2 \rangle,$$

where

$$r_2 = \begin{array}{c|ccc} & l_5 & l_6 & l_{ous1} \\ \hline l_2 & false & false & true \\ l_{ous1} & true & true & true \end{array}.$$

The token of type α in place l_2 enters place l_{ouse} where it merges with the β token. Token β in place l_{ouse} splits into three identical tokens. The first one remains in place l_{ouse} where it obtains the characteristic "current values of $ous.Ye, ous.Fe, ous.Pe, ous.Te, ous.Ne$". The other two enter places l_5 and l_6 respectively without obtaining a new characteristic.

$$Z_3 = \langle\{l_3, l_{crre}\}, \{l_7, l_8, l_9, l_{crre}\}, r_3\rangle,$$

where

$$r_3 = \begin{array}{c|cccc} & l_7 & l_8 & l_9 & l_{crre} \\ \hline l_3 & true & true & true & false \\ l_{crre} & false & false & false & true \end{array}.$$

The token of type α in place l_3 splits into three identical tokens which enter places l_7, l_8, l_9 respectively where they obtain the characteristic "current values of $crr.Ye, crr.Fe, crr.Pe, crr.Te, crr.Ne$". Token γ in place l_{crre} obtains the characteristic "current values of $crr.Ye, crr.Fe, crr.Pe, crr.Ne$".

$$Z_4 = \langle\{l_4, l_{prse}\}, \{l_{10}, l_{11}, l_{prse}\}, r_4\rangle,$$

where

$$r_4 = \begin{array}{c|ccc} & l_{10} & l_{11} & l_{prse} \\ \hline l_4 & false & false & true \\ l_{prse} & true & true & true \end{array}.$$

The token of type α in place l_4 enters place l_{prse} where it merges with the δ token. Token δ in place l_{prse} splits into three identical tokens. The first one remains in place l_{prse} where it obtains the characteristic "current values of $prs.Ye, prs.Fe, prs.Pe, prs.Te, prs.Ne$". The other two enter places l_{10} and l_{11} respectively without obtaining a new characteristic.

$$Z_5 = \langle\{l_6, l_7, l_{ofrcrre}\}, \{l_{ofrcrre}\}, r_5\rangle,$$

where

$$r_5 = \begin{array}{c|c} & l_{ofrcrre} \\ \hline l_6 & true \\ l_7 & true \\ l_{ofrcrre} & true \end{array}.$$

The tokens from places l_6 and l_7 enter place $l_{ofrcrre}$ where they merge with token ϵ. Token ϵ in place $l_{ofrcrre}$ obtains the characteristic "current values of $ofr.crr.Ye, ofr.crr.Fe, ofr.crr.Pe, ofr.crr.Te, ofr.crr.Ne$".

$$Z_6 = \langle\{l_5, l_{blce}\}, \{l_{blce}\}, r_6\rangle,$$

where

$$r_6 = \begin{array}{c|c} & l_{ofrcrre} \\ \hline l_5 & true \\ l_{blce} & true \end{array}.$$

The token from places l_5 enter place l_{blce} where it merges with token ζ. Token ζ in place l_{blce} obtains the characteristic "current values of $blc.Ye, blc.Fe, blc.Pe, blc.Te, blc.Ne$".

$$Z_7 = \langle \{l_9, l_{10}, l_{srve}\}, \{l_{srve}\}, r_7 \rangle,$$

where

$$r_7 = \begin{array}{c|c} & l_{srve} \\ \hline l_9 & true \\ l_{10} & true \\ l_{srve} & true \end{array}.$$

The tokens from places l_9 and l_{10} enter place l_{srve} where they merge with token η. Token η in place l_{srve} obtains the characteristic "current values of $srv.Ye, srv.Fe, srv.Pe, srv.Te, srv.Ne$".

$$Z_8 = \langle \{l_8\}, \{l_{12}, l_{13}, l_{14}\}, r_8 \rangle,$$

where

$$r_8 = \begin{array}{c|ccc} & l_{12} & l_{13} & l_{14} \\ \hline l_8 & true & true & true \end{array}.$$

The token of type α in place l_8 splits into three identical tokens which enter respectively places l_{12}, l_{13}, l_{14} without obtaining new characteristics.

$$Z_9 = \langle \{l_{12}, l_{ouss}\}, \{l_{15}, l_{ouss}\}, r_9 \rangle,$$

where

$$r_9 = \begin{array}{c|cc} & l_{15} & l_{ouss} \\ \hline l_{12} & true & false \\ l_{ouss} & true & true \end{array}.$$

The α token from places l_{12} enters place l_{15}. Token θ in place l_{ouss} splits into two tokens one of which remains in place l_{ouss} with a characteristic "current values of $ous.Ys, ous.Fs, ous.Ps, ous.Ts, ous.Ns$". The other one enters place l_{15} where it merges with the token coming from place l_{12} into a new α token which obtains the characteristic "current values of $ous.Ys, ous.Fs, ous.Ps, ous.Ts, ous.Ns$".

$$Z_{10} = \langle \{l_{13}, l_{crrs}\}, \{l_{16}, l_{17}, l_{18}, l_{crrs}\}, r_{10} \rangle,$$

where

$$r_{10} = \begin{array}{c|cccc} & l_{16} & l_{17} & l_{18} & l_{crrs} \\ \hline l_{13} & true & true & true & false \\ l_{crrs} & false & false & false & true \end{array}.$$

The token of type α in place l_{13} splits into three identical tokens which enter places l_{16}, l_{17}, l_{18} respectively where they obtain the characteristic "current values of $crr.Ys, crr.Fs, crr.Ps, crr.Ts, crr.Ns$". Token κ in place l_{crrs} obtains the characteristic "current values of $crr.Ys, crr.Fs, crr.Ps, crr.Ts, crr.Ns$".

$$Z_{11} = \langle \{l_{14}, l_{prss}\}, \{l_{19}, l_{20}, l_{prss}\}, r_{11} \rangle,$$

where

$$r_{11} = \begin{array}{c|ccc} & l_{19} & l_{20} & l_{prss} \\ \hline l_{14} & false & false & true \\ l_{prss} & true & true & true \end{array}.$$

The token of type α in place l_{14} enters place l_{prss} where it merges with the λ token. Token λ in place l_{prss} splits into three identical tokens. The first one remains in place l_{prss} where it obtains the characteristic "current values of $prs.Ys, prs.Fs, prs.Ps, prs.Ts, prs.Ns$". The other two enter places l_{19} and l_{20} respectively without obtaining a new characteristic.

$$Z_{12} = \langle \{l_{15}, l_{16}, l_{ofrcrrs}\}, \{l_{ofrcrrs}\}, r_{12} \rangle,$$

where

$$r_{12} = \begin{array}{c|c} & l_{ofrcrrs} \\ \hline l_{15} & true \\ l_{16} & true \\ l_{ofrcrrs} & true \end{array}.$$

The tokens from places l_{15} and l_{16} enter place $l_{ofrcrrs}$ where they merge with token μ. Token μ in place $l_{ofrcrrs}$ obtains the characteristic "current values of $ofr.crr.Ys, ofr.crr.Fs, ofr.crr.Ps, ofr.crr.Ns$".

$$Z_{13} = \langle \{l_{18}, l_{19}, l_{srvs}\}, \{l_{srvs}\}, r_{13} \rangle,$$

where

$$r_{13} = \begin{array}{c|c} & l_{srvs} \\ \hline l_{18} & true \\ l_{19} & true \\ l_{srvs} & true \end{array}.$$

The tokens from places l_{18} and l_{19} enter place l_{srvs} where they merge with token ν. Token ν in place l_{srvs} obtains the characteristic "current values of $srv.Ys, srv.Fs, srv.Ps, srv.Ts, srv.Ns$".

$$Z_{14} = \langle \{l_{11}, l_{17}, l_{20}, l_{prsg}, l_{crrg}, l_{ind}\}, \{l_{prsg}, l_{crrg}, l_{ind}\}, r_{14} \rangle,$$

where

$$r_{14} = \begin{array}{c|ccc} & l_{prsg} & l_{crrg} & l_{ind} \\ \hline l_{11} & true & false & false \\ l_{17} & false & true & false \\ l_{20} & true & false & false \\ l_{prsg} & true & false & false \\ l_{crrg} & false & true & false \\ l_{ind} & false & false & true \end{array}.$$

The tokens from places l_{11} and l_{20} enter place l_{prsg} where they merge with the token ξ. Token ξ in place l_{prsg} obtains the characteristic "current values of $prs.Yg, prs.Fg, prs.Pg, prs.Tg, prs.Ng$". The α token in place l_{17} enters place l_{crrg} where it merges with the token o. Token o in place l_{crrg} obtains the characteristic "current values of $crr.Yg, crr.Fg, crr.Pg, prs.Tg, crr.Ng$". Token π in place l_{ind} obtains the characteristic "current value of I_9". The value of I_9 is obtained from Eq. (2).

5 Conclusions

The present paper is a continuation of the research on the modeling of QoS in overall telecommunication systems with GNs. The GN model of traffic quality evaluation of a service phase and the GN model of traffic quality evaluation of a service stage described in the paper are important step towards the construction of a GN model of traffic quality evaluation of an overall telecommunication system including users. One of the advantages of the GNs approach is that the two GN models presented here can be easily included in a GN model of an overall telecommunication system using some of the operators defined over GNs. Such GN model of an overall telecommunication system can be used for QoS monitoring, prediction and management. Another advantage of the use of GNs is that different quality indicators can be evaluated by modifying the characteristic functions of some places of the net. The model can be also extended to predict the Quality of Experience (QoE) and would allow dynamic pricing policy execution depending on the load of the network.

The presented GN approach to the QoS evaluation can be applied to any service system in general and in particular to next generation smart service networks.

References

1. Andonov, V., Poryazov, S., Saranova, E.: Generalized net representations of elements of service systems theory. Adv. Stud. Contemp. Math. (Kyungshang) **29**(2), 179–189 (2019)
2. Andonov, V., Poryazov, S., Saranova, E.: Generalized net representations of the causal structure of a queuing system. In: Proceedings of the 2020 IEEE 10th International Conference on Intelligent Systems (IS), Varna, Bulgaria, pp. 80–86 (2020)
3. Andonov, V., Poryazov, S., Saranova, E.: Generalized net model of a serial composition of services with intuitionistic fuzzy estimations of uncertainty. In: Kahraman, C., Tolga, A.C., Cevik Onar, S., Cebi, S., Oztaysi, B., Sari, I.U. (eds.) INFUS 2022. LNNS, vol. 504, pp. 616–623. Springer, Cham (2022). https://doi.org/10.1007/978-3-031-09173-5_71
4. Atanassov, K.: On Generalized Nets Theory. Prof. M. Drinov Academic Publ. House, Sofia (2007)
5. De Gyvés Avila, S.: QoS Awareness and Adaptation in Service Composition. The University of Leeds, Leeds (2014)

6. ITU-T Recommendation E.800 (09/08): Definitions of terms related to quality of service
7. Poryazov, S., Saranova, E.: Models of Telecommunication Networks with Virtual Channel Switching and Applications. Prof. Marin Drinov Academic Publishing House (2012). (in Bulgarian)
8. Poryazov, S., Saranova, E., Ganchev, I.: Scalable traffic quality and system efficiency indicators towards overall telecommunication system's QoE management. In: Ganchev, I., van der Mei, R.D., van den Berg, H. (eds.) Autonomous Control for a Reliable Internet of Services. LNCS, vol. 10768, pp. 81–103. Springer, Cham (2018). https://doi.org/10.1007/978-3-319-90415-3_4
9. Poryazov, S., Andonov, V., Saranova, E.: Three intuitionistic fuzzy estimations of uncertainty in service compositions. In: Atanassov, K.T., et al. (eds.) IWIFSGN BOS/SOR 2020. LNNS, vol. 338, pp. 72–84. Springer, Cham (2022). https://doi.org/10.1007/978-3-030-95929-6_6
10. Strunk, A.: QoS-aware service composition: a survey. In: Proceedings of the 2010 Eighth IEEE European Conference on Web Services, Washington, DC, USA, 1–3 December 2010, pp. 67–74 (2010)

Generalized Net Model of Overall Network Efficiency Evaluation

Velin Andonov(✉), Stoyan Poryazov, and Emiliya Saranova

Institute of Mathematics and Informatics, Bulgarian Academy of Sciences, Acad. G. Bonchev Str., Block 8, 1113 Sofia, Bulgaria
{velin_andonov,stoyan,emiliya}@math.bas.bg

Abstract. In recent years, Generalized Nets (GNs) have been successfully used in the modeling of telecommunication systems. GN models of traffic quality evaluation of a service phase and a service stage in an overall telecommunication system have been developed. In the present paper, a GN model of overall network efficiency evaluation is described. The model is based on a conceptual model of the Quality of Service (QoS) contributions in an overall telecommunication system including users. It can be used for monitoring, prediction and management of the QoS in overall telecommunication systems. The GN approach described in the paper can be applied to all service systems and, in particular, to the next generation smart service networks.

Keywords: Generalized net Next generation network · Telecommunication system · Quality of service · Network efficiency

1 Introduction

The present paper is a continuation of the paper "Generalized net models of traffic quality evaluation of a service stage" by the same authors (see [2]). An overall telecommunication system including users is considered. In [7], scalable traffic quality and system efficiency indicators are proposed at a phase, stage and system level. The aim of the present research is to construct a Generalized Net (GN, see [3]) model of an overall telecommunication system including users based on the conceptual model of the schematic contributions to the Quality of Service (QoS) in an overall telecommunication system described in [7]. The GN model can be used for QoS monitoring, prediction and management.

The basic notions used in the paper are described in [2]. Here, we shall mention them briefly for completeness.

The work of Velin Andonov, Stoyan Poryazov and Emiliya Saranova is supported by the research project "Perspective Methods for Quality Prediction in the Next Generation Smart Informational Service Networks" (KP-06-N52/2) financed by the Bulgarian National Science Fund.

K. T. Atanassov et al. (Eds.): BOS/SOR/IWIFSGN 2022, LNNS 793, pp. 156–168, 2023.
https://doi.org/10.1007/978-3-031-45069-3_14

1.1 Base Virtual Service Devices and Their Parameters

The notion of a base virtual service device plays a central role in the conceptual models of service systems. The base virtual devices do not have other devices embedded in them. Let x denote an arbitrary base virtual device. Every such device has the following parameters: intensity of the flow of requests (Fx), probability of directing the flow of requests towards the device (Px), service time in the device (Tx), traffic intensity (Yx, measured in Erlang) and device capacity (Nx). In the conceptual models of telecommunication systems various types of base virtual devices are used (see [6]). Their graphical representation is shown in Fig. 1.

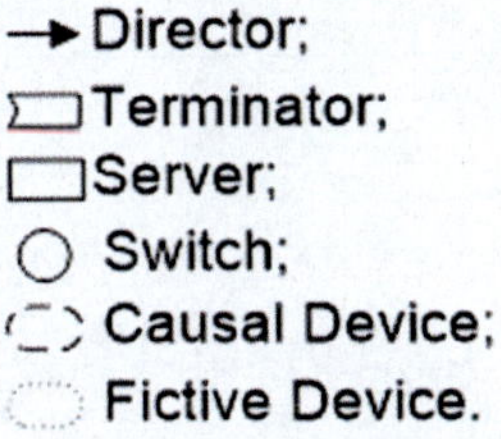

Fig. 1. Base virtual device types.

Every type of base virtual device has a specific function. For detailed description of these functions, the reader may refer to [8].

As a starting point in the construction of the GN models the conceptual models in terms of service systems theory are used. The construction of the GN models is greatly enhanced by the proposed in [1] GN representations of the base virtual service devices. These GN representations are also used in the present paper.

1.2 Conceptual Model of QoS Contributions in an Overall Telecommunication System Including Users

A conceptual model of QoS contributions in an overall telecommunication system including users is described in [7]. Its graphical representation is shown in Fig. 2.

In this conceptual model, the calling (denoted by A) and the called (denoted by B) terminals and users are included. The five main service stages of the telecommunication system are presented by virtual service devices. They are: A-terminal, Dialing (denoted by d), Switching (s), B-terminal Seizure and B-terminal. The users are represented in the model by the A-user and B-user stages.

A special notation for the parameters and for the virtual devices is developed based on the notion of a qualifier. For detailed definition of the qualifiers, the

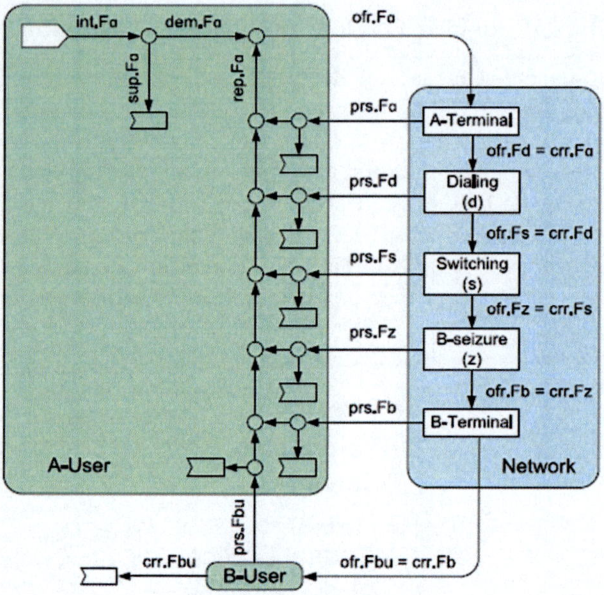

Fig. 2. Schematic contributions to QoS in an overall telecommunication system including users (see [7]).

reader may refer to [7]. The used qualifiers are abbreviations of traffic characterizations. The following qulifiers are used in the model: int. (from intent); dem. (demand); rep. (repeated); ofr. (offered); prs. (parasitic); crr. (carried).

1.3 Network Efficiency Indicators

In [7], many efficiency indicators are proposed on five levels: service phase, service stage, part of the network, overall network and overall telecommunication system level. Here we shall mention only some of the efficiency indicators on network and overall system level. The classic network efficiency indicators are the "Answer Seizure Ratio (ASR)", "Answer Bid Ratio (ABR)" and "Network Efficiency Ratio (NER)" (see [4]). These indicators do not take into account initiated but unsuccessful attempts and the effect of the repeated attempts. Instead, we shall use the indicators defined in [7] and listed below.

Indicator 18: Network efficiency indicator Ea on the A-terminal stage:

$$I_{18} = Ea = \frac{crr.Fa}{ofr.Fa}. \tag{1}$$

Indicator 22: Network efficiency indicator Eb on the B-terminal stage:

$$I_{22} = Eb = \frac{crr.Fb}{ofr.Fa}. \tag{2}$$

Indicator 24: System efficiency indicator Eu on the Ad sub-stage:

$$I_{24} = Eu = I_{11}I_{23} = \frac{dem.Fa}{ofr.Fa}\frac{crr.Fbu}{ofr.Fa}. \tag{3}$$

Indicator 25: System efficiency indicator Ei on the Ai sub-stage:

$$I_{25} = Ei = I_{10}I_{24} = \frac{dem.Fa}{int.Fa}\frac{dem.Fa}{ofr.Fa}\frac{crr.Fbu}{ofr.Fa}. \tag{4}$$

2 A Generalized Net Model of Overall Network Efficiency Evaluation

In this section, a GN model of overall network efficiency evaluation is described. The GN is a reduced one (see [3]). The GN consists of 18 transitions, 43 places and 21 types of tokens. The graphical representation of the GN is shown in Fig. 3.

A special notation is used for the places of the GN where tokens obtain as characteristic the values of the parameters of some service device. These places are denoted by l_x where x is the name of the corresponding virtual device and the symbol '.' is omitted. For example, l_N is the place where the tokens obtain as characteristic the values of the parameters of the comprise Network device. The function of each of the transitions is given below.

- Z_1 represents the function of the generator of the intent call attempts with intensity $int.Fa$.
- Z_2 represents the function of the generator of the suppressed intent call attempts with intensity $sup.Fa$.
- Z_3 represents the function of the comprise A device.
- Z_4 represents the intensity of the call attempts trying to occupy A-terminals.
- Z_5 represents the function of the A-Terminal device.
- Z_6 represents the intensity of all parasitic call attempts in the A-terminals.
- Z_7 represents the function of the Dialing device.
- Z_8 represents the intensity of all parasitic call attempts in the Dialing stage.
- Z_9 represents the function of the Switching stage (the s device).
- Z_{10} represents the intensity of all parasitic call attempts in the s device.
- Z_{11} represents the function of the B-seizure device.
- Z_{12} represents the intensity of all parasitic call attempts in the z device.
- Z_{13} represents the function of the B-Terminal device.
- Z_{14} represents the intensity of all parasitic call attempts of the B-terminal device.
- Z_{15} represents the function of the comprise Network device.
- Z_{16} represents the function of the B-User device.
- Z_{17} represents the intensity of the offered call attempts to the B-User device.
- Z_{18} represents the intensity of the carried flow of call attempts of the B-User device.

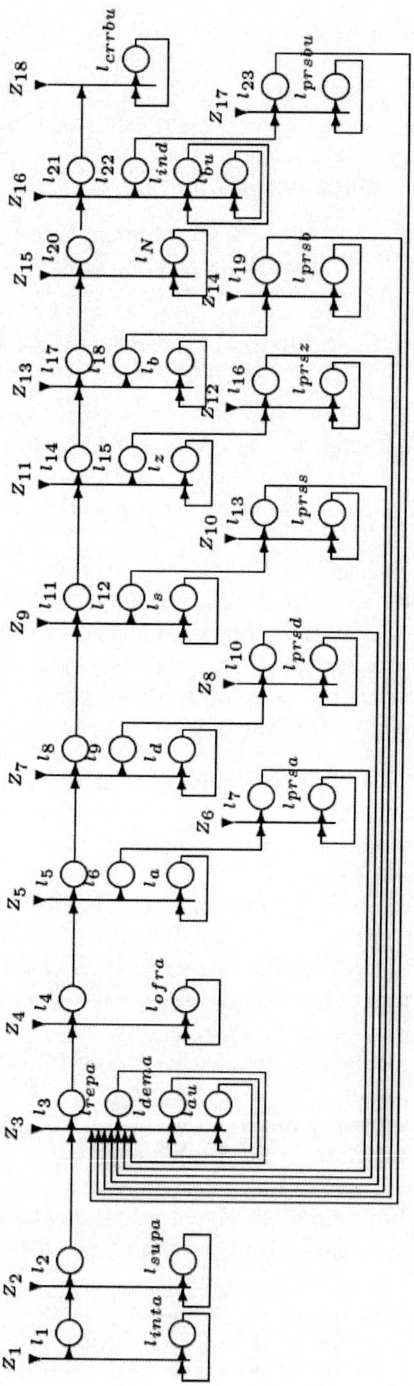

Fig. 3. Graphical representation of a GN of overall network efficiency evaluation.

The 21 types of tokens used in the model are denoted by lowercase Greek letters. In the initial moment of functioning of the GN the tokens are placed in some of the places of the GN as described below.

- Token of type α stays in place l_{inta} with initial characteristic "initial values of $int.Fa, int.Ya, int.Ta$".
- Token of type β stays in place l_{supa} with initial characteristic "initial values of $sup.Fa, sup.Ya, sup.Ta, sup.Pa$".
- Token of type γ stays in place l_{repa} with initial characteristic "initial values of $rep.Fa, rep.Ya, rep.Ta$".
- Token of type δ stays in place l_{dema} with initial characteristic "initial values of $dem.Fa, dem.Ya, dem.Ta, dem.Pa$".
- Token of type ϵ stays in place l_{au} with initial characteristic "initial values of Fau, Yau, Tau, Pau".
- Token of type ζ stays in place l_{ofra} with initial characteristic "initial values of $ofr.Fa, ofr.Ya, ofr.Ta, ofr.Pa$".
- Token of type η stays in place l_a with initial characteristic "initial values of Fa, Ya, Ta, Pa".
- Token of type θ stays in place l_{prsa} with initial characteristic "initial values of $prs.Fa, prs.Ya, prs.Ta, prs.Pa$".
- Token of type κ stays in place l_d with initial characteristic "initial values of Fd, Yd, Td, Pd".
- Token of type λ stays in place l_{prsd} with initial characteristic "initial values of $prs.Fd, prs.Yd, prs.Td, prs.Pd$".
- Token of type μ stays in place l_s with initial characteristic "initial values of Fs, Ys, Ts, Ps".
- Token of type ν stays in place l_{prss} with initial characteristic "initial values of $prs.Fs, prs.Ys, prs.Ts, prs.Ps$".
- Token of type ξ stays in place l_z with initial characteristic "initial values of Fz, Yz, Tz, Pz".
- Token of type o stays in place l_{prsz} with initial characteristic "initial values of $prs.Fz, prs.Yz, prs.Tz, prs.Pz$".
- Token of type π stays in place l_b with initial characteristic "initial values of Fb, Yb, Tb, Pb".
- Token of type ρ stays in place l_{prsb} with initial characteristic "initial values of $prs.Fb, prs.Yb, prs.Tb, prs.Pb$".
- Token of type σ stays in place l_N with initial characteristic "initial values of F_N, Y_N, T_N, P_N".
- Token of type τ stays in place l_{ind} with initial characteristic "initial values of $I_{18}, I_{22}, I_{24}, I_{25}$".
- Token of type υ stays in place l_{bu} with initial characteristic "initial values of Fbu, Ybu, Tbu, Pbu".
- Token of type ϕ stays in place l_{prsbu} with initial characteristic "initial values of $prs.Fbu, prs.Ybu, prs.Tbu, prs.Pbu$".
- Token of type χ stays in place l_{crrbu} with initial characteristic "initial values of $crr.Fbu, crr.Ybu, crr.Tbu, crr.Pbu$".

The formal description of the transitions of the GN is given below.

$$Z_1 = \langle \{l_{inta}\}, \{l_1, l_{inta}\}, r_1 \rangle,$$

where

$$r_1 = \begin{array}{c|cc} & l_1 & l_{inta} \\ \hline l_{inta} & true & true \end{array}.$$

The token of type α in place l_{inta} splits into two identical tokens one of which enters place l_1 with characteristic "current values of $int.Fa, int.Ya, int.Ta$". The other one remains in place l_{inta} with characteristic "current values of $int.Fa, int.Ya, int.Ta$".

$$Z_2 = \langle \{l_1, l_{supa}\}, \{l_2, l_{supa}\}, r_2 \rangle,$$

where

$$r_2 = \begin{array}{c|cc} & l_2 & l_{supa} \\ \hline l_1 & true & true \\ l_{supa} & false & true \end{array}.$$

The token of type α in place l_1 splits into two identical tokens one of which enters place l_{supa} where it merges with the β token.The other one enters place l_2 without obtaining a new characteristic. Token β in place l_{supa} obtains the characteristic "current values of $sup.Fa$, $sup.Ya$, $sup.Ta$, $sup.Pa$".

$$Z_3 = \langle \{l_2, l_{repa}, l_7, l_{10}, l_{13}, l_{16}, l_{19}, l_{23}, l_{dema}, l_{au}\}, \{l_3, l_{repa}, l_{dema}, l_{au}\}, r_3 \rangle,$$

where

$$r_3 = \begin{array}{c|cccc} & l_3 & l_{repa} & l_{dema} & l_{au} \\ \hline l_2 & true & false & true & false \\ l_{repa} & true & true & false & false \\ l_7 & false & true & false & false \\ l_{10} & false & true & false & false \\ l_{13} & false & true & false & false \\ l_{16} & false & true & false & false \\ l_{19} & false & true & false & false \\ l_{23} & false & true & false & false \\ l_{dema} & true & false & true & false \\ l_{au} & false & false & false & true \end{array}.$$

The α token in place l_2 splits into two identical tokens one of which enters place l_3 while the other enters place l_{dema} where it merges with the δ token. Token δ in place l_{dema} splits into two tokens one of which remains in place l_{dema} and obtains the characteristic "current values of $dem.Fa$, $dem.Ya$, $dem.Ta$, $dem.Pa$". The other one enters place l_3. The tokens from places $l_7, l_{10}, l_{13}, l_{16}, l_{19}, l_{23}$ enter place l_{repa} where they merge with the γ token. Token γ obtains the characteristic "current values of $rep.Fa$, $rep.Ya$, $rep.Ta$". Token ϵ in place l_{au} obtains the characteristic "current values of Fau, Yau, Tau, Pau". All tokens entering l_3 merge into a new α token without obtaining a new characteristic.

$$Z_4 = \langle \{l_3, l_{ofra}\}, \{l_4, l_{ofra}\}, r_4 \rangle,$$

where

$$r_4 = \begin{array}{c|cc} & l_4 & l_{ofra} \\ \hline l_3 & true & true \\ l_{ofra} & false & true \end{array}.$$

The token of type α in place l_3 splits into two identical tokens one of which enters place l_{ofra} where it merges with the ζ token. The other one enters place l_4 without obtaining a new characteristic. Token ζ in place l_{ofra} obtains the characteristic "current values of $ofr.Fa$, $ofr.Ya, ofr.Ta, ofr.Pa$".

$$Z_5 = \langle \{l_4, l_a\}, \{l_5, l_6, l_a\}, r_5 \rangle,$$

where

$$r_5 = \begin{array}{c|ccc} & l_5 & l_6 & l_a \\ \hline l_4 & true & false & true \\ l_a & false & true & true \end{array}.$$

The token of type α in place l_4 splits into two identical tokens one of which enters place l_a where it merges with the η token. The other one enters place l_5 without obtaining a new characteristic. Token η in place l_a splits into two tokens. One of them remains in place l_a where it obtains the characteristic "current values of Fa, Ya, Ta, Pa". The other one enters place l_6 without obtaining a new characteristic.

$$Z_6 = \langle \{l_6, l_{prsa}\}, \{l_7, l_{prsa}\}, r_6 \rangle,$$

where

$$r_6 = \begin{array}{c|cc} & l_7 & l_{prsa} \\ \hline l_6 & true & true \\ l_{prsa} & false & true \end{array}.$$

The token of type ζ in place l_6 splits into two identical tokens one of which enters place l_{prsa} where it merges with the θ token. The other one enters place l_7 without obtaining a new characteristic. Token θ in place l_{prsa} obtains the characteristic "current values of $prs.Fa, prs.Ya, prs.Ta, prs.Pa$".

$$Z_7 = \langle \{l_5, l_d\}, \{l_8, l_9, l_d\}, r_7 \rangle,$$

where

$$r_7 = \begin{array}{c|ccc} & l_8 & l_9 & l_d \\ \hline l_5 & true & false & true \\ l_d & false & true & true \end{array}.$$

The token of type α in place l_5 splits into two identical tokens one of which enters place l_d where it merges with the κ token. The other one enters place l_8 without obtaining a new characteristic. Token κ in place l_d splits into two tokens. One of them remains in place l_d where it obtains the characteristic "current values of Fd, Yd, Td, Pd". The other one enters place l_9 without obtaining a new characteristic.

$$Z_8 = \langle \{l_9, l_{prsd}\}, \{l_{10}, l_{prsd}\}, r_8 \rangle,$$

where

$$r_8 = \begin{array}{c|cc} & l_{10} & l_{prsd} \\ \hline l_9 & true & true \\ l_{prsd} & false & true \end{array}.$$

The token of type κ in place l_9 splits into two identical tokens one of which enters place l_{prsd} where it merges with the λ token. The other one enters place l_{10} without obtaining a new characteristic. Token λ in place l_{prsd} obtains the characteristic "current values of $prs.Fd, prs.Yd, prs.Td, prs.Pd$ ".

$$Z_9 = \langle \{l_8, l_s\}, \{l_{11}, l_{12}, l_s\}, r_9 \rangle,$$

where

$$r_9 = \begin{array}{c|ccc} & l_{11} & l_{12} & l_s \\ \hline l_8 & true & false & true \\ l_s & false & true & true \end{array}.$$

The token of type α in place l_8 splits into two identical tokens one of which enters place l_s where it merges with the μ token. The other one enters place l_{11} without obtaining a new characteristic. Token μ in place l_s splits into two tokens. One of them remains in place l_s where it obtains the characteristic "current values of Fs, Ys, Ts, Ps". The other one enters place l_{12} without obtaining a new characteristic.

$$Z_{10} = \langle \{l_{12}, l_{prss}\}, \{l_{13}, l_{prss}\}, r_{10} \rangle,$$

where

$$r_{10} = \begin{array}{c|cc} & l_{13} & l_{prss} \\ \hline l_{12} & true & true \\ l_{prss} & false & true \end{array}.$$

The token of type μ in place l_{12} splits into two identical tokens one of which enters place l_{prss} where it merges with the ν token. The other one enters place l_{13} without obtaining a new characteristic. Token ν in place l_{prss} obtains the characteristic "current values of $prs.Fs, prs.Ys, prs.Ts, prs.Ps$".

$$Z_{11} = \langle \{l_{11}, l_z\}, \{l_{14}, l_{15}, l_z\}, r_{11} \rangle,$$

where

$$r_{11} = \begin{array}{c|ccc} & l_{14} & l_{15} & l_z \\ \hline l_{11} & true & false & true \\ l_z & false & true & true \end{array}.$$

The token of type α in place l_{11} splits into two identical tokens one of which enters place l_z where it merges with the ξ token. The other one enters place l_{14} without obtaining a new characteristic. Token ξ in place l_z splits into two tokens.

One of them remains in place l_z where it obtains the characteristic "current values of Fz, Yz, Tz, Pz". The other one enters place l_{15} without obtaining a new characteristic.

$$Z_{12} = \langle \{l_{15}, l_{prsz}\}, \{l_{16}, l_{prsz}\}, r_{12} \rangle,$$

where

$$r_{12} = \begin{array}{c|cc} & l_{16} & l_{prsz} \\ \hline l_{15} & true & true \\ l_{prsz} & false & true \end{array}.$$

The token of type ξ in place l_{15} splits into two identical tokens one of which enters place l_{prsz} where it merges with the o token. The other one enters place l_{16} without obtaining a new characteristic. Token o in place l_{prsz} obtains the characteristic "current values of $prs.Fz, prs.Yz, prs.Tz, prs.Pz$".

$$Z_{13} = \langle \{l_{14}, l_b\}, \{l_{17}, l_{18}, l_b\}, r_{13} \rangle,$$

where

$$r_{13} = \begin{array}{c|ccc} & l_{17} & l_{18} & l_b \\ \hline l_{14} & true & false & true \\ l_b & false & true & true \end{array}.$$

The token of type α in place l_{14} splits into two identical tokens one of which enters place l_b where it merges with the π token. The other one enters place l_{17} without obtaining a new characteristic. Token π in place l_b splits into two tokens. One of them remains in place l_b where it obtains the characteristic "current values of Fb, Yb, Tb, Pb". The other one enters place l_{18} without obtaining a new characteristic.

$$Z_{14} = \langle \{l_{18}, l_{prsb}\}, \{l_{19}, l_{prsb}\}, r_{14} \rangle,$$

where

$$r_{14} = \begin{array}{c|cc} & l_{19} & l_{prsb} \\ \hline l_{18} & true & true \\ l_{prsb} & false & true \end{array}.$$

The token of type π in place l_{18} splits into two identical tokens one of which enters place l_{prsb} where it merges with the ρ token. The other one enters place l_{19} without obtaining a new characteristic. Token ρ in place l_{prsb} obtains the characteristic "current values of $prs.Fb, prs.Yb, prs.Tb, prs.Pb$".

$$Z_{15} = \langle \{l_{17}, l_N\}, \{l_{20}, l_N\}, r_{15} \rangle,$$

where

$$r_{15} = \begin{array}{c|cc} & l_{20} & l_N \\ \hline l_{17} & true & true \\ l_N & false & true \end{array}.$$

The token of type α in place l_{17} splits into two identical tokens one of which enters place l_N where it merges with the σ token. The other one enters place

l_{20} without obtaining a new characteristic. Token σ in place l_N obtains the characteristic "current values of Y_N, F_N, T_N, P_N".

$$Z_{16} = \langle \{l_{20}, l_{ind}, l_{bu}\}, \{l_{21}, l_{22}, l_{ind}, l_{bu}\}, r_{16} \rangle,$$

where

$$r_{16} = \begin{array}{c|cccc} & l_{21} & l_{22} & l_{ind} & l_{bu} \\ \hline l_{20} & true & false & true & true \\ l_{ind} & false & false & true & false \\ l_{bu} & false & true & false & true \end{array}.$$

The token of type α in place l_{20} splits into three identical tokens one of which enters place l_{bu} where it merges with the v token. The second one enters place l_{21} without obtaining a new characteristic. The third one enters place l_{ind} where it merges with the τ token. Token τ in place l_{ind} obtains the characteristic "current values of $I_{18}, I_{22}, I_{24}, I_{25}$". Token v in place l_{bu} splits into two tokens. One of them remains in place l_{bu} where it obtains the characteristic "current values of Fbu, Ybu, Tbu, Pbu". The other one enters place l_{22} without obtaining a new characteristic.

$$Z_{17} = \langle \{l_{22}, l_{prsbu}\}, \{l_{23}, l_{prsbu}\}, r_{17} \rangle,$$

where

$$r_{17} = \begin{array}{c|cc} & l_{23} & l_{prsbu} \\ \hline l_{22} & true & true \\ l_{prsbu} & false & true \end{array}.$$

The token of type v in place l_{22} splits into two identical tokens one of which enters place l_{prsbu} where it merges with the ϕ token. The other one enters place l_{23} without obtaining a new characteristic. Token ϕ in place l_{prsbu} obtains the characteristic "current values of $prs.Fbu, prs.Ybu, prs.Tbu, prs.Pbu$".

$$Z_{18} = \langle \{l_{21}, l_{crrbu}\}, \{l_{crrbu}\}, r_{18} \rangle,$$

where

$$r_{18} = \begin{array}{c|c} & l_{20} \\ \hline l_{21} & true \\ l_{crrbu} & true \end{array}.$$

The token of type α in place l_{21} enters place l_{crrbu} where it merges with the χ token. Token χ in place l_{crrbu} obtains the characteristic "current values of $crr.Fbu, crr.Ybu, crr.Tbu, crr.Pbu$".

3 Conclusions

In the present paper, for the first time a GN model of overall network efficiency evaluation is described. The values of two network efficiency indicators are evaluated during the functioning of the GN: network efficiency indicator on the A terminal stage (Ea) and network efficiency indicator on the B terminal stage

(Eb). One of the advantages of the GN approach is that the values of other network efficiency indicators such as network efficiency indicator on the B-terminal seizure stage, network efficiency indicator on the Dialing stage, etc. can be evaluated by changing the characteristic function of the GN related to the place where the indicators values are assigned to the tokens. This can be done without changing the other components of the net and the graphical representation.

The proposed GN model is a model of an overall telecommunication system including users. This allows for the values of efficiency indicators on overall system level to be evaluated. In the present model, the values of the system efficiency indicator on the Ad sub-stage (considering the demand call attempts) and the system efficiency indicator on the Ai sub-stage (considering the intent call attempts) are also evaluated.

The GN model can be used to evaluate network cost/quality ratios indicators such as mean cost/quality ratio, instantaneous cost/quality ratio, etc. It can be used for monitoring, prediction and management of the Quality of Service (QoS) in an overall telecommunication system and for establishment and utilization of dynamic pricing policies based on the network load. The proposed GN approach to quality evaluation can be applied to any service system and in particular to next generation informational service networks where the dynamic pricing is considered as an additional dimension of the call admission control process in order to efficiently and effectively control the use of the network resources (see [5,9]).

References

1. Andonov, V., Poryazov, S., Saranova, E.: Generalized net representations of elements of service systems theory. Adv. Stud. Contemp. Math. (Kyungshang) **29**(2), 179–189 (2019)
2. Andonov, V., Poryazov, S., Saranova, E.: Generalized net models of traffic quality evaluation of a service stage. In: Proceedings of the Twentieth International Workshop on Intuitionistic Fuzzy Sets and Generalized Nets, Warsaw, Poland, 14 October 2022. (in press)
3. Atanassov, K.: On Generalized Nets Theory. Prof. M. Drinov Academic Publ. House, Sofia (2007)
4. ITU-T Rec. E.425 (03/2002): Network management—internal automatic observations
5. Hou, J., Yang, J., Papavassiliou, S.: Integration of pricing with call admission control to meet QoS requirements in cellular networks. IEEE Trans. Parallel Distrib. Syst. **13**(9), 898–910 (2002)
6. Poryazov, S., Saranova, E.: Models of Telecommunication Networks with Virtual Channel Switching and Applications. Prof. Marin Drinov Academic Publishing House (2012). (in Bulgarian)
7. Poryazov, S., Saranova, E., Ganchev, I.: Scalable traffic quality and system efficiency indicators towards overall telecommunication system's QoE management. In: Ganchev, I., van der Mei, R.D., van den Berg, H. (eds.) Autonomous Control for a Reliable Internet of Services. LNCS, vol. 10768, pp. 81–103. Springer, Cham (2018). https://doi.org/10.1007/978-3-319-90415-3_4

8. Poryazov, S., Andonov, V., Saranova, E.: Three intuitionistic fuzzy estimations of uncertainty in service compositions. In: Atanassov, K.T., et al. (eds.) IWIFSGN BOS/SOR 2020. LNNS, vol. 338, pp. 72–84. Springer, Cham (2022). https://doi.org/10.1007/978-3-030-95929-6_6
9. Radonjić, V., Aćimović-Raspopović, V.: Dynamic pricing models in next generation networks. In: 2011 19th Telecommunications Forum (TELFOR) Proceedings of Papers, Belgrade, Serbia, pp. 174–181 (2011)

Generalized Net Model of the General Claim Process – Cassation Proceedings before the Supreme Court of Cassation

Hristo Blidov and Lyubka Doukovska(✉)

Intelligent Systems Department, Institute of Information and Communication Technologies, Bulgarian Academy of Sciences, Acad. Georgi Bonchev Str., 1113 Sofia, Bulgaria
{hristo.blidov,lyubka.doukovska}@iict.bas.bg

Abstract. In the present paper an application of the apparatus of Generalized nets is proposed for modeling the cassation proceedings before the Supreme Court of Cassation. The advantages of using such model are very different, some of them are expressed in helping to detect and eliminate unnecessary complications in the process itself, leading to its simplification and optimization, presenting a complex process simpler and more understandable, for scientific and teaching needs etc.

Keywords: Intelligent systems · Generalized Net · General claim process

1 Introduction

Cassation Appeal - General Characteristics: The cassation appeal in Bulgarian procedural law is that it is regulated as regular, but not always possible due to the assessment of admissibility according to Art. 280 of the Civil Code, the third instance for review of the decisions of the Appeal courts, [8].

Another feature of the cassation proceedings consists in the expressly listed in Art. 280, para. 1 of the Civil Code strict grounds for allowing the cassation appeal. This legislative decision is one of the most contested in Bulgarian procedural law. For the parties to the case, the cassation instance has the character of exclusivity, in the sense of limited accessibility, [8].

Decisions Subject to Cassation Appeal. Grounds for Allowing the Cassation Appeal: The cassation proceeding is a specific court proceeding, since in order to develop the case it goes through 2 different phases. The first phase is related to the admission to consideration of the case on its merits. Only after the case has been admitted for consideration, the Supreme Court of Cassation owes a ruling on the legal dispute on the merits. The Supreme Court of Cassation is competent to consider the case.

According to Art. 280, para. 1 of the Code of Civil Procedure, appellate decisions in which the court ruled on a substantive or procedural issue are subject to cassation appeal before the Supreme Court, which is:

K. T. Atanassov et al. (Eds.): BOS/SOR/IWIFSGN 2022, LNNS 793, pp. 169–178, 2023.
https://doi.org/10.1007/978-3-031-45069-3_15

1. decided in contradiction with the mandatory practice of the Supreme Court of Cassation and the Supreme Court in interpretative decisions and rulings, as well as in contradiction with the practice of the Supreme Court of Cassation;
2. decided in contradiction with acts of the Constitutional Court of the Republic of Bulgaria or the Court of the European Union;
3. important for the correct application of the law, as well as for the development of the law.

Regardless of the above prerequisites, the appellate decision is admissible for cassation appeal in case of probable nullity or inadmissibility, as well as in case of obvious incorrectness.

Not subject to cassation appeal: decisions on appellate cases with a claim cost of up to BGN 5,000 - for civil cases, and up to BGN 20,000 - for commercial cases; decisions on appellate cases on claims for alimony, matrimonial claims, proceedings for distribution of the use of jointly owned property, requests for change of name, etc.; the decisions on some appellate cases on labor disputes, etc., [9].

Grounds of Cassation. Examination of the Cassation Appeal on the Merits: These grounds for admission are different from the grounds for a cassation appeal within the meaning of Art. 281 of the Civil Code, which are related to vices of the appellate decision. This is related to the verification of the merits of the complaint. Void, inadmissible or incorrect decision (as in appeal), [6].

The cassation appeal, regulated in the Civil, falls under the control-overturning type of judicial appeal of judicial acts - the purpose is control and verification of the appealed decisions on the grounds specified in the law, referred to in Art. 281 of the Civil Code. If at least 1 of them is present, the Supreme Court is obliged to accept that the complaint is well-founded and to proceed in view of the type of vice, [7].

Grounds

1. All cases where the decision is void;
2. Inadmissible decisions;
3. Incorrect decision - violations of substantive law; of the procedural rules; unreasonableness of the decision.

Examination of the Cassation Appeal on the Merits

1. The party that appeals to the Supreme Court of Cassation with a cassation appeal is called the "appellant", and the other party - the defendant in the cassation appeal. The cassation appeal is submitted within a one-month preclusion period from the delivery of the decision to the party. The submission is made through the court that issued the appeal decision. In terms of content, the complaint must comply with the requirements specified in Art. 284, par. 1 of the Civil Code requirements, but since the cassation appeal as a control-cancellation involves to a lesser extent the official functions of the Supreme Court, non-compliance with these requirements makes it inadmissible.

The Appeal court, not the Supreme Court of Cassation, has the obligation to conduct an ex officio examination regarding the possible presence of deficiencies in the appeal. If the appellate court finds reparable deficiencies, it gives the assessor a seven-day period to remove them, and if they are not removed, it returns the appeal with an order that is subject to appeal.

The Appeal court checks the regularity of the complaint and if it does not meet the established legal requirements, it informs the party to remove the admitted irregularities within one week.

If the appeal is regular, the Appeal court sends it together with the exchanged documents and the case to the Supreme Court of Cassation.

The appeal is returned by the appellate court when: 1. it is filed after the expiration of the appeal period; 2. The admitted irregularities are not remedied in time; 3. The appellate decision is not subject to cassation appeal. The Appeal court issues a return order, which in turn can be appealed by private appeal to the Supreme Court of Cassation.

2. After accepting the appeal, the appellate court sends a copy of it to the defendant and gives him a one-month deadline to respond. The defendant may also file a counter-appeal in cassation within the response period. In this case, in addition to being a defendant in the cassation appeal, he also acquires the status of assessor in the counter-appeal in cassation. The cross-appeal in cassation is subject to the same review by the appellate court as the original appeal.

3. After the appeal and the answer are sent to the Supreme Court of Cassation, it checks the admissibility of the appeal in a closed session and pronounces on this with a ruling - in the event that the Supreme Court of Cassation finds the cassation appeal inadmissible, it is barring, because it deprives the party of its right to cassation proceedings.

4. If the Supreme Court of Cassation accepts the appeal as admissible and admits it for examination on the merits, the Supreme Court of Cassation proceeds to summon the parties.

Cassation appeals (if there is a counter appeal) are considered by a three-member panel of the Supreme Court of Cassation in an open session. The Supreme Court monitors ex officio the grounds related to the nullity or inadmissibility of the appellate decision. In relation to the correctness of the judicial act, the court rules only on the grounds established in the appeal.

Cassation Decision. Return of the Case to the Appeal Court and Powers of the Supreme Court of Cassation in Re-appeal

1. With its decision, the Supreme Court of Cassation may uphold the decision of the Appeal court or annul it in whole or in part.
2. If it finds that the decision is void or inadmissible - the Supreme Court of Cassation will act as the Appeal court in this case - if it is void or terminate the case or if it is not subject to termination, return it for a new consideration. If he finds that the decision is inadmissible - he invalidates it and terminates the case or sends it to the competent court if it is inadmissible due to wrong jurisdiction.
3. The decision is annulled as incorrect when the substantive law is violated or significant violations of the judicial procedure rules are committed or the decision is unfounded. The court returns the case for a new examination by another panel of the appellate court only if it is necessary to repeat or carry out new judicial actions.

In this paper, Generalized Nets (GN) and Index Matrices (IM) are used for analysis purposes, [1–5]. In paper [10] is presented the development of court proceedings in civil disputes or the so-called "general claim process" from the initiation of the case to its conclusion by the court of first instance through the use of temporal intuitionist fuzzy pairs.

2 Generalized Net Model of the General Claim Process

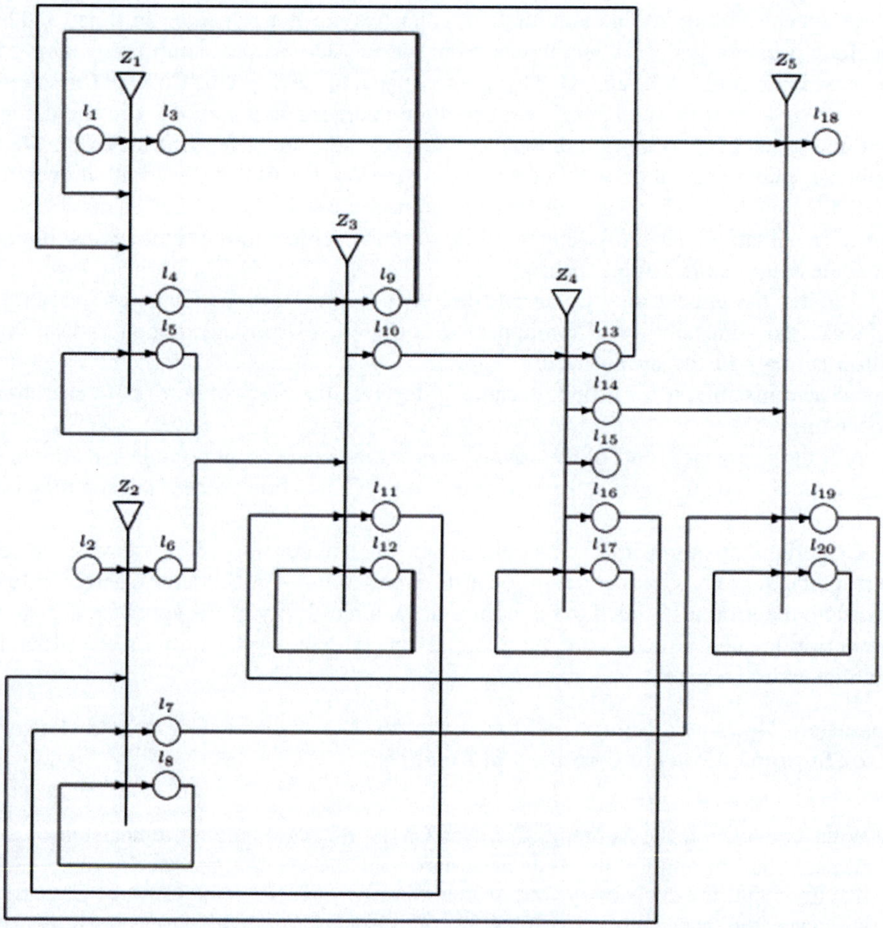

Fig. 1. GN-model

The GN-model (see Fig. 1) contains 5 transitions, 20 places and 4 types of tokens that have the following sence:

Token "G"– "*Appeal Court*" – this token permanently stays in place l_{12} with characteristic:

"trial chamber of the Appeal Court"

Token "C"– *"Appellant"*,, - is possible that in some moment token "C" enters place l_1 with characteristic:

"cassation appeal"

Token "D"– *"Respondent party"* - It is possible that in some moment token "D" enters place l_2 with characteristic:

"cassation appeal"

Token "H"– *"Supreme Court of Cassation"* - this token permanently stays in place l_{17} with characteristic:

"trial chamber of the Supreme Court of Cassation"

The GN transitions have the following form:

First transition:

$$Z_1 = \{\langle l_1, l_5, l_9, l_{13}\}, \{l_3, l_4, l_5\},$$

	l_3	l_4	l_5	
l_1	f	f	t	
l_5	$W_{5,3}$	$W_{5,4}$	$W_{5,5}$	$\rangle$.
l_9	f	f	t	
l_{13}	f	f	t	

Where:

- $W_{5,3} =$ *"the cassation appeal applicant should appear before the Supreme Court of Cassation";*
- $W_{5,4} =$ *"the cassation appeal applicant files a cassation appeal or an answer of the counter cassation appeal";*
- $W_{5,5} =$ *"the cassation appeal applicant is not ready with the answer of the counter cassation appeal";*

Characteristics:

$l_3 =$ *"without new characteristic";*
$l_4 =$ *"content of the cassation appeal or content of the answer of the counter cassation appeal";*
$l_5 =$ *"without new characteristic".*

When token "C" enters place l_1 on the next time moment it enters l_5 without new characteristic.

When predicate $W_{5,3}$ is true than token "C" splits to 2 tokens – the original token "C" and token "C1" that enter place l_3 without a new characteristic.

When predicate $W_{5,4}$ is true than token "C" splits to 2 tokens – the original token "C" and token "C2" that enter place l_4 with characteristic:

"content of the cassation appeal or content of the answer of the counter cassation appeal"

When predicate $W_{5,5}$ is true than token "C" continues to stay in place l_5 without a new characteristic.

Second transition:

$$Z_2 = \langle \{ l_2, l_8, l_{11}, l_{16} \}, \{ l_6, l_7, l_8 \},$$

	l_6	l_7	l_8
l_2	f	f	t
l_8	$W_{8,6}$	$W_{8,7}$	$W_{8,8}$
l_{11}	f	f	t
l_{16}	f	f	t

$\rangle$.

Where:

- $W_{8,6}$ = *"the respondent party files an answer of the cassation appeal or a counter cassation appeal"*;
- $W_{8,7}$ = *"the respondent party should appear before the Supreme Court of Cassation"*;
- $W_{8,8}$ = *"the respondent party is not ready with the answer of the cassation appeal"*.

Characteristics:

l_6 = *"content of the answer of the cassation appeal or content of the counter cassation appeal"*.
l_7 = *"without new characteristic"*;
l_8 = *"without new characteristic"*.

When token "D" enters place l_2 on the next time moment it enters l_8 without a new characteristic.

When predicate $W_{8,6}$ is true than token "D" splits to 2 tokens – the original token "D" and token "D2" that enter place l_6 with characteristic:

"content of the answer of the cassation appeal or content of the counter cassation appeal"

When predicate $W_{8,7}$ is true than token "D" splits to 2 tokens – the original token "D" and token "D1" that enter place l_7 without a new characteristic.

When predicate $W_{8,8}$ is true than token "D" continues to stay in place l_8 without a new characteristic.

Transition 3:

$$Z_3 = \langle \{ l_4, l_6, l_{12}, l_{19}, l_{11} \}, \{ l_9, l_{10}, l_{11}, l_{12} \},$$

	l_6	l_7	l_8	l_9
l_4	f	f	t	t
l_6	f	f	f	t
l_{12}	$W_{12,9}$	$W_{12,10}$	$W_{12,11}$	$W_{12,12}$
l_{19}	f	f	t	t

$\rangle$.

Where:

- $W_{12,9}$ = *"the Appeal Court has instructions to the cassation appeal applicant or there is still no answer of the cassation appeal or counter cassation appeal from the respondent party"*;
- $W_{12,10}$ = *"the documents are ready to be sent to the Supreme Court of Cassation"*;
- $W_{12,11}$ = *"the Appeal Court has instructions to the respondent party or there is still no cassation appeal or answer of the counter cassation appeal from the cassation appeal applicant"*;
- $W_{12,12}$ = *"the work of the Appeal Court continues"*.

Characteristics:

l_9 = *"content of the answer of the cassation appeal or of the counter cassation appeal or instructions of the Appeal Court to the cassation appeal applicant"*;

l_{10} = *"content of the documents sent to the Supreme Court of Cassation"*;

l_{11} = *"content of the cassation appeal or of the answer of counter cassation appeal or instructions of the Appeal Court to the respondent party"*;

l_{12} = *"without new characteristic"*.

Each one of the tokens "C2"and "D2"enters place l_{12} and unites with token "G" without a new characteristic.

When predicate $W_{12,9}$ is true token "G" splits to 2 tokens – the original token "G" and token "G1" that enter place l_9 with characteristic:

"content of the answer of the cassation appeal or of the counter cassation appeal or instructions of the Appeal Court to the cassation appeal applicant"

When predicate $W_{12,11}$ is true token "G" splits to 2 tokens – the original token "G" and token "G2" that enter place l_{11} with characteristic:

"content of the cassation appeal or of the answer of counter cassation appeal or instructions of the Appeal Court to the respondent party"

When predicate $W_{12,10}$ is true token "G" splits to 2 tokens – the original token "G" and token "G3" that enter place l_{10} with characteristic:

"content of the documents sent to the Supreme Court of Cassation"

Transition 4:

$$Z_4 = \langle \{ l_{10}, l_{17} \}, \{ l_{13}, l_{14}, l_{15}, l_{16}, l_{17} \},$$

	l_{13}	l_{14}	l_{15}	l_{16}	l_{17}	
l_{10}	f	f	f	f	t	$\rangle$.
l_{17}	$W_{17,13}$	$W_{17,14}$	$W_{17,15}$	$W_{17,16}$	$W_{17,16}$	

Where:

- $W_{17,13}$ = *"the Supreme Court of Cassation has instructions to the documents of the cassation appeal applicant or sets a date for an open court hearing";*
- $W_{17,14}$ = *"the Supreme Court of Cassation allows the cassation appeal to be examined on the merits and sets a date for an open court hearing";*
- $W_{17,15}$ = *"the Supreme Court of Cassation does not allow the cassation appeal to be examined on the merits and terminates the court case";*
- $W_{17,16}$ = *"the Supreme Court of Cassation has instructions to the documents of the respondent or sets a date for an open court hearing";*
- $W_{17,17}$ = *"the work of the Supreme Court of Cassation continues".*

Characteristics:

l_{13} = *"content of the instructions to the documents of the cassation appeal applicant or concrete date"*;

l_{14} = *"a ruling of the Supreme Court of Cassation has been issued for admissibility on the merits of the cassation appeal and a date for an open court hearing";*

l_{15} = *"a ruling of the Supreme Court of Cassation has been issued for inadmissibility on the merits of the cassation appeal and termination of the court case"*;

l_{16} = *"content of the instructions to the documents of the respondent party or concrete date"*;

l_{17} = *"without new characteristic".*

Token "G3" enters place l_{17} and unites with token "H" without a new characteristic.

When predicate $W_{17,13}$ is true token "H" splits to 2 tokens – the original token "H" and token "H1" that enter place l_{13} with characteristic:

"content of the instructions to the documents of the cassation appeal applicant or concrete date"

When predicate $W_{17,16}$ is true token "H" splits to 2 tokens – the original token "H" and token "H2" that enter place l_1 with characteristic:

"content of the instructions to the documents of the respondent party or concrete date"

When predicate $W_{17,14}$ is true token "H" enters place l_{14} with characteristic:

"a ruling of the Supreme Court of Cassation has been issued for admissibility on the merits of the cassation appeal and a date for an open court hearing"

When predicate $W_{17,15}$ is true token "H" enters place l_{15} with characteristic:

"a ruling of the Supreme Court of Cassation has been issued for inadmissibility on the merits of the cassation appeal and termination of the court case"

Transition 5:

$$Z_5 = \langle \{ l_3, l_7, l_{14}, l_{20} \}, \{ l_{18}, l_{19}, l_{20} \},$$

	l_{18}	l_{19}	L_{20}
l_3	f	f	t
l_7	f	f	t
l_{14}	f	f	t
l_{20}	$W_{20,18}$	$W_{20,19}$	$W_{20,20}$

$\rangle$.

Where:

- W20,18 = *"the Supreme Court of Cassation is ready with the final decision on the case";*
- W20,19 = *"the Supreme Court of Cassation returns the case to the lower court";*
- W20,20 = *"the work of the Supreme Court of Cassation continues";*

Characteristics:

l_{18} = *"content of the final court decision"*;
l_{19} = *"legal grounds for reurning the case to a lower court"*;
l_{20} = *"without new characteristic"*;

Token "H" enters place l_{20} without a new characteristic.
When predicate $W_{20,18}$ is true token "H" enters place l_{18} with characteristic

"content of the final court decision"

When predicate $W_{20,19}$ is true token "H" enters place l_{19} with characteristic

"legal grounds for reurning the case to a lower court".

3 Conclusion

The paper presents in details the different steps and options of development of the cassation proceedings before the Supreme Court of Cassation. An in-depth analysis of the process shows that it is more complicated that it can be, so there are opportunities to facilitate and simplify the procedure, which can be achieved through legislative changes.

The apparatus of generalized nets can be used further for the description of the whole general claim process which will be very useful for simplification and optimization the complicate process. The time limit for hearing and resolving the case may be shortened, which would satisfy the public interest.

Acknowledgements. This work was partially supported by the Bulgarian National Science Fund, Grant agreement No. KP-06-N36, BG PLANTNET "Establishment of National Information Network GenBank – Plant genetic resources" and partially supported by the Bulgarian Ministry of Education and Science under the National Research Program "Smart crop production", Grant agreement No. D01-65/19.03.2021, approved by Decision of the Ministry Council №866/26.11.2020.

References

1. Alexieva, J., Choy, E., Koycheva, E.: Review and bibliography on generalized nets theory and applications. In: Choy, E., Krawczak, M., Shannon, A., Szmidt, E. (eds.) A Survey of Generalized Nets, Raffles KvB Monograph, no. 10, pp. 207–301 (2007)
2. Atanassov, K.: Generalized Nets. World Scientific, Singapore, New Jersey, London (1991)
3. Atanassov, K.: On Generalized Nets Theory, Prof. Marin Drinov Academic Publishing House, Sofia (2007)
4. Atanassov, K., Generalized index matrices, Comptes rendus de l'Academie Bulgare des Sciences, vol. 40, no. 11, pp. 15–18 (1987). ISSN: 1310-1331
5. Atanassov, K.: Index Matrices: Towards An Augmented Matrix Calculus. Springer, Cham (2014). https://doi.org/10.1007/978-3-319-10945-9
6. Code of Civil Procedure, The State Gazette, № 59, 20.07.2007 in force from 01.03.2008 (in Bulgarian)
7. Code of Civil Procedure, The State Gazette, № 12, 08.02.1952 in force from 08.02.1952 (in Bulgarian)
8. Civil Procedure Code. https://lex.bg/laws/ldoc/2135558368
9. Stalev, J.: Bulgarian civil procedural law, p. 1448, First Edition, Tenth Edition (2020). ISBN: 9789542831532
10. Blidov, H., Doukovska, L., Atanassov, K.: Generalized net model of the first phase of the general claim process. In: Proceedings of the 10-th International Conference on Intelligent Systems - IS'20, Varna, Bulgaria, IEEE Xplore, pp. 626–629 (2020). ISBN: 978-1-7281-5456-5, ISSN: 1541-1672, https://doi.org/10.1109/IS48319.2020.9200126

Generalized Net Model of the General Claim Process – Annulment Proceedings Before the Supreme Court of Cassation

Hristo Blidov and Lyubka Doukovska(✉)

Intelligent Systems Department, Institute of Information and Communication Technologies, Bulgarian Academy of Sciences, Acad. Georgi Bonchev Str., 1113 Sofia, Bulgaria
{hristo.blidov,lyubka.doukovska}@iict.bas.bg

Abstract. The present work is dedicated to the application of innovative, intelligent methods for the analysis of processes in the administration of justice and, in particular, the application of the Generalized Networks (GN) apparatus for modeling the annulment proceedings before the Supreme Court of Cassation is proposed. The annulment proceedings of an effective court decision, as part of the general claim process, represent a complex process in its essence and nature, which is often difficult to perceive and understand. The use and application of the GN model in the specific case enables the proceedings before the Supreme Court of Cassation to be analyzed in more detail. At the same time, shortcomings are discovered that could be avoided in the future, which would lead to the improvement of the judicial procedure itself.

Keywords: Intelligent systems · Generalized Net · General claim process

1 Introduction

Annulment of Decisions that have Entered into Legal Force. Annulment Proceedings

The proceedings for annulment of an effective court decision under art. 303 of the Civil Code constitute a remedy against defective judgments entered into force, [6]. The procedure is not an appeal because it takes place after the decision has entered into force and the trial is over. Therefore, it is not a phase, a stage of the general claim process, but an independent, extrajudicial proceeding.

It is a means of protection against incorrect decisions, and when the incorrectness consists in a discrepancy between the decision and the actual legal situation and is due to the comprehensively listed in Art. 303 reasons, [7].

Therefore, the annulment is related to the principle of establishing the truth and aims to ensure it even in relation to decisions that have entered into force, by removing their force of res judicata and requiring a retrial of the case in order to replace the incorrect decision with a new, correct decision. In addition, here, as in the cassation appeal, the Supreme Court of Cassation is competent to examine and rule on the raised dispute.

K. T. Atanassov et al. (Eds.): BOS/SOR/IWIFSGN 2022, LNNS 793, pp. 179–188, 2023.
https://doi.org/10.1007/978-3-031-45069-3_16

Based on the above, the proceedings for annulment of an effective court decision can be defined as an independent judicial proceeding for extrajudicial control and annulment of decisions that have entered into force, when they are incorrect, due to any of the reasons expressly stated in Art.303 of the Civil Code. Only non-appealable and res judicata decisions are subject to annulment.

According to established judicial practice, when there is another order to attack the judicial act or to protect the right, the annulment is inadmissible.

In all cases, when the Supreme Court of Cassation finds that judicial acts are not subject to annulment (divorce decision, for example), it should leave the applications for annulment without consideration, [8].

Legitimate to initiate proceedings for annulment is above all the "interested party". Such is the person bound by a decision unfavorable to him, which is incorrect due to the presence of a vice, expressly mentioned in Art. 303 affecting this person (for example, he was irregularly summoned).

The cassation appeal in Bulgarian procedural law is that it is regulated as regular, but not always possible due to the assessment of admissibility according to Art. 280 of the Civil Code, the third instance for review of the decisions of the Appeal courts.

Legal Grounds for Annulment

1. Art. 303, par. 1, item 1 of the Civil Code - new circumstances or new written evidence of essential importance for the case are discovered, which could not have been known at the time of its decision or which the party could not obtain in a timely manner. Here we are talking about the incompleteness of the factual and evidentiary material, which is due to an objective impossibility;
2. Art. 303, par. 1, item 2 of the Civil Code - in accordance with the due process of law, the falsity of a document, of the testimony of a witness, of the conclusion of an expert, on which the decision is based, or a criminal act of the party, of its representative, of a member of the court or of an agent in connection with the resolution of the case;
3. Art. 303, par. 1, item 3 of the Civil Code - the decision is based on a decree of a court or other state institution, which was subsequently revoked;
4. Art. 303, par. 1, item 4 of the Civil Code - between the same parties, for the same request and on the same grounds, another effective decision that contradicts it was passed before it;
5. Art. 303, par. 1, item 5 of the Civil Code - the party, as a result of a violation of the relevant rules, was deprived of the opportunity to participate in the case or was not properly represented, or when he could not appear in person or through a representative due to special unforeseen circumstances that were not could overcome;
6. Art. 303, par. 1, item 6 of the Civil Code - the party in violation of the relevant rules was or, accordingly, was not represented by a person under Art. 29 of the Civil Code (special procedural representation);
7. Art. 303, par. 1, item 7 of the Civil Code - The European Court of Human Rights has established a violation of the Convention for the Protection of Human Rights and Fundamental Freedoms or of the protocols thereto and a new trial of the case is necessary in order to remove the consequences of the violation.

Proceedings for Annulment of an Effective Court Decision
Competent to consider the application for annulment is always and only the Supreme Court of Cassation, [9].

Referral to the Supreme Court of Cassation is made at the request of the interested party. The application for annulment is filed through the court of first instance and must meet the requirements that apply to the appeal and contain a precise and reasoned statement of the grounds for annulment.

If the application does not meet these requirements, the party is sent a notice to remove them within a week.

If the irregularities of the application for cancellation are not removed within the deadline, the same will be returned to the applicant.

A transcript shall be attached to the application, which shall be served on the opposing party – the defendant who can give an answer within a week of receiving the transcript.

1. The Supreme Court of Cassation rules on the admissibility of the application in closed session by checking whether the contested act is subject to annulment under Art. 303, if it originates from an authorized person, as well as whether the application was submitted in time.
2. Essentially, i.e. on the merits of the application, the SC decides in an open session.

The Supreme Court of Cassation either rejects the request or respects it.

1. When canceling the challenged decision, the Supreme Court of Cassation never decides on the merits, but returns the case for consideration by another panel of the relevant court. In the annulment decision, the Supreme Court of Cassation indicates where to start the consideration of the case.
2. In the event that, between the same parties, for the same request and on the same grounds, another valid decision that contradicts it was passed before it, the court shall cancel the incorrect decision.

In the presented paper, Generalized Nets (GN) and Index Matrices (IM) are used for analysis purposes, [1–5]. In paper [10] is presented the development of court proceedings in civil disputes or the so-called "general claim process" from the initiation of the case to its conclusion by the court of first instance through the use of temporal intuitionist fuzzy pairs.

2 Generalized Net Model of the General Claim Process

The GN-model (see Fig. 1) contains 6 transitions, 23 places and 5 types of tokens that have the following meaning:

Token "A"– "*Archive*"– permanently stay in place l_{13} with characteristic:

"*file of the entire court proceedings*"

Token "E"– "*Court of first instance*" - permanently stays in place l_8 with characteristic:

"*trial chamber of the* Court of first instance"

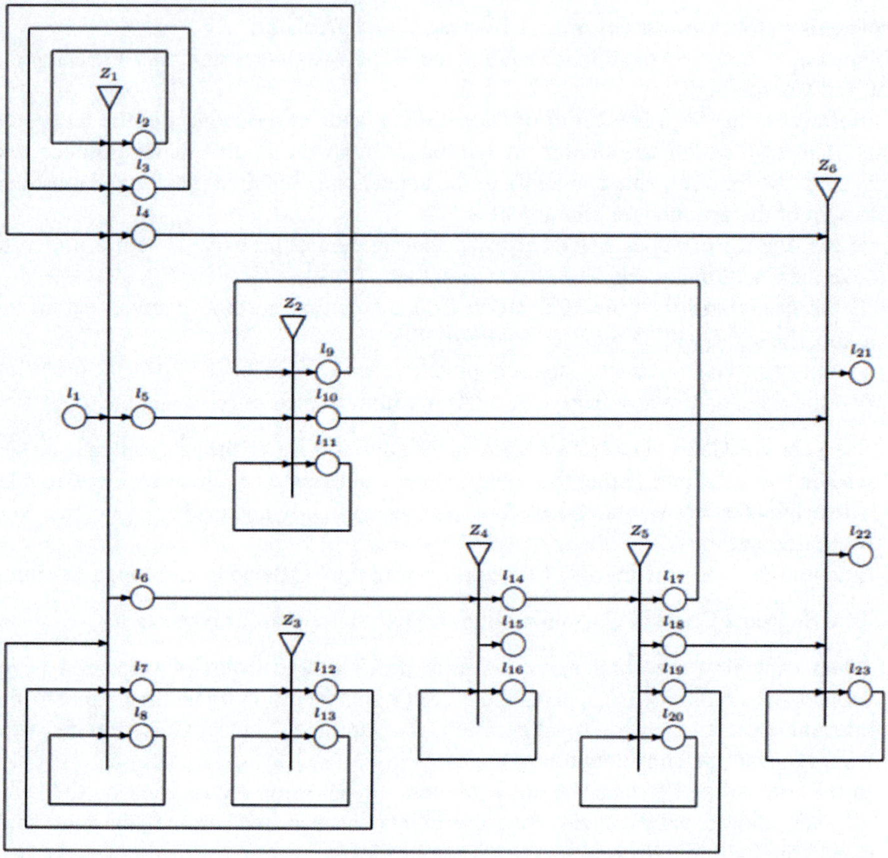

Fig. 1. GN-model

Token "C"– "*Claimant*" - token "C" enters place l_1 with characteristic:

"name and full data of the claimant, application for annulment"

Token "D"– "*Defendant*" – stays in place l_{11} with characteristic:

"name and full data of the claimant, response of the application for annulment"

Token "B"– "*Supreme Court of Cassation*" this token permanently stays in place l_{17} with characteristic:

"trial chamber of the Supreme Court of Cassation"

The GN transitions have the following form:

First transition:

$$Z_1 = \{\langle\, l_1, l_2, l_8, l_9, l_{12}, l_{19}\},\{\, l_2, l_3, l_4, l_5, l_6, l_7, l_8\},$$

	l_2	l_3	l_4	l_5	l_6	l_7	l_8
l_1	t	f	f	f	f	f	f
l_2	$W_{2,2}$	$W_{2,3}$	$W_{2,4}$	$W_{2,5}$	$W_{2,5}$	$W_{2,7}$	$W_{2,8}$
l_8	$W_{8,2}$	$W_{8,3}$	f	$W_{8,5}$	$W_{8,6}$	$W_{8,7}$	t
l_9	f	f	f	f	f	f	t
l_{12}	f	f	f	f	f	f	t
l_{19}	f	f	f	f	f	f	t

$\rangle$.

Where:

- $W_{2,2}$ = *"the applicant waits the results of the check-ups from the Court of first instance; disposition of the Court of first instance or calling before the Supreme Court of Cassation"*;
- $W_{2,3}$ = *"the Court of first instance has considered that the application for annulment is irregular and terminates the case"*;
- $W_{2,4}$ = *"the applicant shall appear at an open court hearing before the Supreme Court of Cassation"*;
- $W_{2,8}$ = *"the applicant has complied the instructions of the Court of first instance"*;
- $W_{8,2}$ = *"there is an information to be received from the applicant"*;
- $W_{8,5}$ = *"there is an information to be received from the defendant"*;
- $W_{8,6}$ = *"the documents are ready to be sent to the Supreme Court of Cassation"*;
- $W_{8,7}$ = *"an information from the archive is necessary"*;

Token "C" from place l_1 splits to two tokens – the same token "C" that enters place l_2 and token "C1" that enters place l_8 and unites with token "E" with characteristic:

"the documents of the applicant have been filed in the Court of first instance"

When predicate $W_{2,3}$ is true, token "C" from place l_2 enters place l_3 with characteristic:

"the Court of first instance has considered that the application for annulment is irregular and terminates the case"

When predicate $W_{2,3}$ is true, token "C" from place l_2 enters place l_4 without any characteristic.

When predicate $W_{2,8}$ is true, token "C" splits to two tokens – the same token "C" that continues to stay in place l_2 and token "C2" that enters place l_8 and unites with token "E" that obtains characteristic:

"documents of the applicant in compliance with the instructions of the Court of first instance"

Tokens "D1" from place l_9, "A1" from l_1, "B2" from l_{19} enter place l_9 and unites with token "E" that obtains characteristic:

"new documents related with the annulment proceedings"

When predicate $W_{8,2}$ is true token "E" splits to two tokens– the same token "E" that continues to stay in place l_8 and token "E1" that enters place l_2 and unites with token "C" that obtains characteristic:

"instructions of the Court of first instance to the applicant or the response of the application for annulment which are sent to the applicant by the court"

When predicate $W_{8,5}$ is true token "E" splits to two tokens– the same token "E" that continues to stay in place l_8 and token "E2" that enters place l_5 and unites with token "D" that obtains characteristic:

"the *application for annulment which the Court of first instance is sending to the defendant"*

When predicate $W_{8,6}$ is true token "E" splits to two tokens– the same token "E" that continues to stay in place l_8 and token "E3" that enters place l_8 with characteristic:

"the case file is sent to the Supreme Court of Cassation"

When predicate $W_{8,7}$ is true Token "E" splits to two tokens– the same token "E" that continues to stay in place l_8 and token "E4" that enters place l_7 with characteristic:

"request for information from the court archive"

Second transition:

$$Z_2 = \langle \{ l_5, l_{11}, l_{17} \}, \{ l_9, l_{10}, l_{11} \},$$

$$\begin{array}{c|ccc} & l_9 & l_{10} & l_{11} \\ \hline l_1 & f & f & t \\ l_{11} & W_{11,9} & W_{11,10} & W_{11,11} \\ l_{17} & f & f & t \end{array} \rangle.$$

Where:

- $W_{11,9}$ = *"the defendant prepares a response to the application for annulment";*
- $W_{11,10}$ = *"the defendant shall before the Supreme Court of Cassation";*
- $W_{11,11}$ = *"the defendant awaits a summons from the Supreme Court of Cassation".*

Token "E2" from place l_2 enters place l_{11} and unites with token "D" that obtains characteristic:

"application for annulment sent to the defendant"

When predicate $W_{11,9}$ is true token "D" splits to two tokens– the same token "D" that continues to stay in place l_{11} and token "D1" that enters place l_9 with characteristic:

"response of the defendant to the application for annulment"

When predicate $W_{11,10}$ is true token "D" enters place l_{11} with characteristic:

"appearing of the defendant s before the Supreme Court of Cassation"

Transition 3:

$$Z_3 = \langle \{ l_7, l_{13}\}, \{ l_{12}, l_{13}\},$$

	l_{12}	l_{13}	
l_7	t	t	$\rangle$.
l_{13}	t	f	

Token "E4" from place l_7 enters place l_{13} and unites with token "A" without a new characteristic.

On the next time moment token "A" splits to two tokens – the same token "A" that continues ot stay in place l_{13} and token "A1" that enters place l_{12} with characteristic:

"requested information from the archive"

Transition 4:

$$Z_4 = \langle \{ l_{10}, l_{17}\}, \{l_{14}, l_{15}, l_{16} \},$$

	l_{14}	l_{15}	l_{16}	
l_6	f	f	t	$\rangle$.
l_{16}	$W_{16,14}$	$W_{16,15}$	t	

Where:

- $W_{16,14}$ = *"the application for annulment is admissible according to the Supreme Court of Cassation";*
- $W_{16,15}$ = *"the application for annulment is inadmissible according to the Supreme Court of Cassation and the case is terminated".*

Token "E3" from place l_6 enters place l_{16} and unites with token "B" without a new characteristic.

When predicate $W_{16,14}$ is true token "B" enters place l_{14} without any characteristic.

Transition 5:

$$Z_5 = \langle \{ l_{14}, l_{20} \}, \{ l_{17}, l_{18}, l_{19}, l_{20} \},$$

	l_{17}	l_{18}	l_{19}	l_{20}
l_{14}	f	f	f	t
l_{20}	$W_{20,17}$	$W_{20,18}$	$W_{20,19}$	t

$\rangle$.

Where:

- $W_{20,17}$ = *"a date for an open court hearing is set";*
- $W_{20,18}$ = *"waiting for the open court hearing to begin";*
- $W_{20,19}$ = *"a date for an open court hearings is set";*

Token "B" from l_{14} enters place l_{20} with a characteristic:

"*order of the Supreme Court of Cassation to summon the parties*"

When predicate $W_{20,17}$ and $W_{20,19}$ are true, token "B" splits to three tokens – the same token "B" that stays in place l_{20} and token "B1" and "B2" that enter place l_{17} and l_{19} with equal characteristics:

"date for an open court hearings"

When predicate $W_{20,18}$ is true, token "B" enters place l_{18} with characteristic:

"*waiting the beginning of the open court hearing*"

Transition 6:

$$Z_6 = \langle \{ l_4, l_{10}, l_{18}, l_{20}, \}, \{ l_{21}, l_{22}, l_{23} \},$$

	l_{21}	l_{22}	l_{23}
l_4	f	f	t
l_{10}	f	f	t
l_{18}	f	f	t
l_{23}	$W_{23,21}$	$W_{23,22}$	t

$\rangle$.

Where:

- $W_{23,21}$ = *"the case is sent to the competent court according the Supreme Court of Cassation";*
- $W_{23,22}$ = *"the Supreme Court of Cassation considers which court decision is correct and should be left into legal force";*

Token "C" from place l_4, token "D" from place l_{10} and token "B" from place l_{18} enter place l_{23} without any characteristic.

When predicate $W_{23,21}$ is true, the three tokens enter place l_{21} with characteristic:

"*sending the case to the competent court according the Supreme Court of Cassation*"

When predicate $W_{23,22}$ is true, the three tokens enter place l_{22} with characteristic:

"*consideration which court decision is correct and should be left into legal force according the Supreme Court of Cassation considers*".

3 Conclusion

The present paper examines in detail the proceedings for annulment of an effective court decision before the Supreme Court of Cassation as part of the general claim process. The application of the GN apparatus demonstrates the presence of unnecessary complications and repetitions in the proceedings in question, which is detrimental to the public interest. Also the GN apparatus gives clear indications for the optimization and simplification of the described procedure. However, this is possible through the adoption of the necessary changes in the legislation.

The obtained results are applicable to the solution of a wider range of tasks related to the analysis of the processes in the administration of justice. This could be a direction for future research that will lead to the enrichment of the researched scientific field.

Acknowledgements. This work was partially supported by the Bulgarian National Science Fund, Grant agreement No. KP-06-N36, BG PLANTNET "Establishment of National Information Network GenBank – Plant genetic resources" and partially supported by the Bulgarian Ministry of Education and Science under the National Research Program "Smart crop production", Grant agreement No. D01-65/19.03.2021, approved by Decision of the Ministry Council №866/26.11.2020.

References

1. Alexieva, J., Choy, E., Koycheva, E.: Review and bibliography on generalized nets theory and applications. In: Choy, E., Krawczak, M., Shannon, A., Szmidt, E. (eds.) A Survey of Generalized Nets, Raffles KvB Monograph, no. 10, pp. 207–301 (2007)
2. Atanassov, K.: Generalized Nets. World Scientific, Singapore, New Jersey, London (1991)
3. Atanassov, K.: On Generalized Nets Theory, Prof. Marin Drinov Academic Publishing House, Sofia (2007)
4. Atanassov, K.: Generalized index matrices, Comptes rendus de l'Academie Bulgare des Sciences, vol. 40, no. 11, pp. 15–18 (1987). ISSN: 1310-1331
5. Atanassov, K.: Index Matrices: Towards an Augmented Matrix Calculus. Springer, Cham (2014). https://doi.org/10.1007/978-3-319-10945-9
6. Code of Civil Procedure, The State Gazette, №59, 20.07.2007 in force from 01.03.2008 (in Bulgarian)

7. Code of Civil Procedure, The State Gazette, №12, 08.02.1952 in force from 08.02.1952 (in Bulgarian)
8. Civil Procedure Code. https://lex.bg/laws/ldoc/2135558368
9. Stalev, J.: Bulgarian civil procedural law, p. 1448, First Edition, Tenth Edition (2020). ISBN: 9789542831532
10. Blidov, H., Doukovska, L., Atanassov, K.: Generalized net model of the first phase of the general claim process. In: Proceedings of the 10-th International Conference on Intelligent Systems - IS'20, Varna, Bulgaria, IEEE Xplore, pp. 626–629 (2020). ISBN: 978-1-7281-5456-5, ISSN: 1541-1672, https://doi.org/10.1109/IS48319.2020.9200126

Medical and Health care Applications

Selected Artificial Intelligence Technologies in the Practice of the Clinician and Researcher in Physiotherapy

Dariusz Mikołajewski[1(✉)] and Emilia Mikołajewska[2]

[1] Kazimierz Wielki University, Chodkiewicza 30, 85-064 Bydgoszcz, Poland
dmikolaj@ukw.edu.pl

[2] Ludwik Rydygier Collegium Medicum in Bydgoszcz, Nicolaus Copernicus University in Toruń, Jagiellońska 13/15, 85-067 Bydgoszcz, Poland

Abstract. Clinical decision support systems based on artificial intelligence (AI) can help specialists improve the accuracy of diagnostic decisions, and thus increase the effectiveness of therapy, shorten hospitalization and absenteeism, and improve the quality of life related to health. The complexity of the assessment of the functional state in physiotherapy and its susceptibility to the influence of a number of disorders requires precise methods of analysis. The aim of this study was to determine the scope of application of adaptive computational intelligence systems based on fuzzy logic to analyze the results of physiotherapy, taking into account previous research in this area and own research. Results to date, including this study, confirm the potential of artificial intelligence, including fuzzy logic, to improve various areas of physiotherapy. The aforementioned approach may lead to a breakthrough in physiotherapy service delivery through the application of AI. Fuzzy logic as a tool for describing uncertainty and linguistically described parameters can contribute particularly much here. To date, the objectification of physiotherapy outcomes is not fully developed, and these areas of physiotherapy can provide scope for effective applications of both fuzzy logic and ordered fuzzy numbers.

Keywords: Artificial Intelligence · Fuzzy Logic · Physiotherapy · Clinical Applications

1 Introduction

Clinical decision support systems based on artificial intelligence (AI) can help specialists improve the accuracy of diagnostic decisions, and thus increase the effectiveness of therapy, shorten hospitalization and absenteeism, and improve the quality of life related to health. The complexity of the assessment of the functional state in physiotherapy and its susceptibility to the influence of a number of disorders requires precise methods of analysis [1, 2]. It is also worth paying attention to AI-based preventive medicine, especially in the area of preventing deficits in the musculoskeletal system. It is much cheaper and faster than the subsequent use of physiotherapy, not to mention surgical procedures [3].

K. T. Atanassov et al. (Eds.): BOS/SOR/IWIFSGN 2022, LNNS 793, pp. 191–199, 2023.
https://doi.org/10.1007/978-3-031-45069-3_17

Narrative review of publications from six major bibliographic databases according to the agreed inclusion criteria coupled with a critical analysis of the current and future uses of AI in physiotherapy. Six major bibliographic databases were reviewed using the specified keywords in the English language. Both number and percentage of publications concerning AI applications is very low compared to the other areas of the medical and health sciences as far as clinical areas (Table 1). It can be seen that the number of publications on the use of AI in physiotherapy and nursing is much lower than the number of publications in other areas of medicine. In addition, the percentage of publications on the use of AI in physiotherapy and nursing is also lower.

Table 1. Results of literature review.

Medical specialty	General number of publications	Number of publications concerning AI applications	Percentage [%]
Physiotherapy	219 935	187	0.09
Rehabilitation	743 711	1 299	0.17
Cardiology	468 110	1 297	0.28
Neurology	557 527	1 457	0.26
Geriatry	206 579	332	0.16
Nursing	847 391	788	0.09

The increase in the number of publications is slow, it only became visible from 2017 (Fig. 1).

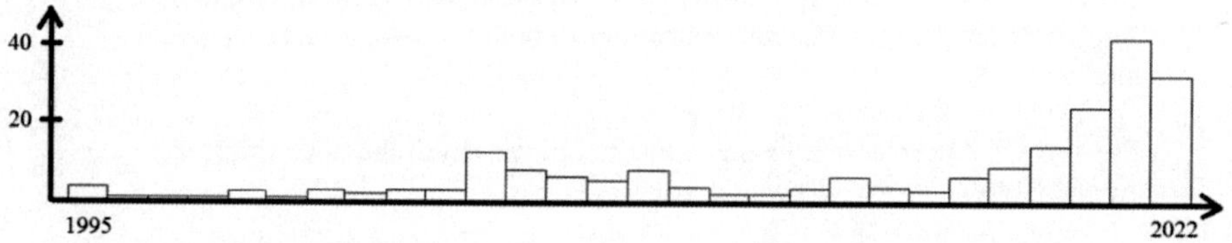

Fig. 1. Number of publications concerning applications of AI in physiotherapy.

AI in physiotherapy most often refers to following parts of the body:

- lower limb: 19,
- upper limb: 10,
- trunk: 9,
- head: 6.

AI in physiotherapy is most often used in the following diseases:

- cardiovascular diseases: 19,
- gait disorders: 19,

- stroke/cerebrovascular accident (CVA): 11,
- orthopaedic injuries: 11,
- spinal cord injury (SCI): 10,
- pulmonary diseases: 4,
- traumatic brain injury (TBI): 0.

Artificial neural networks (ANN) were the most frequently used AI paradigm in physiotherapy, while the fuzzy logic studies were the least numerous. The aim of this study was to determine the scope of application of adaptive computational intelligence systems based on fuzzy logic to analyze the results of physiotherapy, taking into account previous research in this area and own research.

The conclusions from the research to date are as follows: although progress has been made in the number of studies and publications, there has been little effort in helping medical professionals to manage physiotherapy, e.g. to predict the patient's condition and the duration of recovery to full or best possible functional capacity. Therefore, further research should focus on predictive applications of AI rather than analyzing research results [1–3].

We will present the applications of fuzzy logic in physiotherapy on the example of the following three case studies based on the own research. We also built more complex commercial preventive medicine systems based on measurement of various biomedical parameters and diet, but they cannot be described here.

1.1 Case Study 1: Fuzzy-Based Gait Analysis in Physiotherapy

The first area in which we applied fuzzy logic is gait analysis. Changes in the patients' health status (described by gait parameters) were reflected in the results of fuzzy analysis, which increased the possibility of clinical gait evaluation towards the use of mHealth solutions [4]. This study provided simple tools for screening in the primary care level to quickly detect general trends expressed as a single number. If this number is in the range of 51–100, the patient's gait is normal, while a patient's gait of a score of 1–50 requires further, more detailed diagnosis. It is based on the aggregation of the scores of six gait parameters. Possible early and accurate identification and initial assessment of gait disorders is not only a prerequisite for the rapid implementation of specific therapy, but also a useful tool for assessing the effectiveness of rehabilitation at various stages of the physiotherapy of the gait function. The research was carried out on various groups of patients [4–9], improving the obtained computational tool.

1.2 Case Study 2: Fuzzy-Based Quality of Life Analysis in Physiotherapy

The second area in which we applied fuzzy logic is life quality analysis. For post-stroke patients undergoing neurological physiotherapy, a health-related quality of life (HRQoL) assessment was developed based on a two-level hierarchical fuzzy system [9]. It is based on the aggregation of the scores of five clinimetric scales and reports the score as a single value in the range [0; 1]. The higher value (i.e. closer to 1) is better. Such a semi-automatic fuzzy quality of life tool is useful for a complementary assessment of the patient's functioning, quality of life and therapy motivation, as well as for the cyclical evaluation of the physiotherapy process. This stimulates taking into account the patient's goals and preferences in the therapy process, increasing his involvement in the therapeutic process, which translates into its effectiveness (the so-called therapeutic alliance).

This tool was successfully studied in a group of patients after stroke, because it is one of the main causes of long-term disability, and effective treatment of post-stroke changes requires the involvement of the patient and his family. In post-stroke patients, fast, accurate, reliable, and reproducible assessment of HRQoL is essential. Although there are many tools for this purpose, none of them meets all the requirements. Computational tools based on fuzzy logic can introduce a breakthrough here, as they can aggregate results from different clinimetric scales.

1.3 Case Study 3: Fuzzy-Based Physical Influence of Occupational Stress and Burnout Analysis in Physiotherapy

The third area in which we applied fuzzy logic is assessment of long term stress and burnout reflected i.a. in musculoskeletal changes [10, 11]. We compared two professional groups: physiotherapists and IT specialists as groups burdened with different types of physical involvement and stress at work. In our research, we searched for new, new and more reliable computational models for work-related stress and burnout. Aggregation ranged from one (with three subscales) to four clinimetric scales. The results of the clinimetric tests are converted in the model into a universal percentage scale while maintaining the characteristics of the guidelines underlying these scales.

This allowed for the development of three different fuzzy models, a comparative study and the selection of the model best suited to the studied groups and the method of their assessment. Fuzzy analysis allowed for the optimal selection of the parameters of the computational model for the study group with different characteristics of occupational stress factors. The study shows that the computational algorithm can optimize the selection of models or their parameters faster, more accurately and efficiently, becoming an important auxiliary tool for a diagnostician.

2 Materials and Methods

2.1 Material

For the purposes of this article, our own research was conducted on the basis of archival data of 50 healthy walkers and 50 patients after stroke. The study involved 50 men and 50 women, aged 37–75 (mean 57, SD = 7.24). Stroke survivors were up to 3 years after stroke.

The selection method in both groups was convenience sample based on archival data sets.

2.2 Methods

Standardized spatial-temporal gait parameters were used for the analysis. They were calculated from the gait parameters (walking speed, cadence and stride length) measured for each patient during the recording of the 10 m walk test and the patients' anthropometric measurements (leg length). Next Open Source Tracker Video Analysis and Modeling Tool software version 5.1.5. was used to generate time series that were exported to Matlab for multifractal time series analysis. The MATLAB R2022b software was used to build the models.

In order to determine time sequence variability reflecting the unevenness of the steps fractal dimension (D, range 1–2) was calculated using the formula:

$$a = 1/SD \tag{1}$$

$$D = (logA)/log\,(1/s) \tag{2}$$

where: a - number of elements obtained as a result of scaling the object, D - fractal dimension, self-similarity dimension, s - scaling factor.

In order to determine the possibility of changing the trend (e.g. in gait reeducation-process) Hurst Exponent (in the range 0–1) is calculated using the formula:

$$SD = aH \tag{3}$$

where: SD - standard deviation, a - length of time series.

The following interpretation of the H value was adopted:

- H in the range of 0–0.5: a series of high volatility with frequent changes of direction in short-term trends,
- H = 0.5: random sign, equal probability of changing and maintaining the trend,
- H in the range of 0.5–1: disorderly course, with a higher probability of maintaining the current trend.

The fuzzy parameter (in the range 0–100, where healthy as dynamic norm are described in the range 51–100) was computed using the original algorithm described earlier.

Although ANN, especially are capable of making an automated qualitative and quantitative gait assessment, they may be more difficult to use and match to a specific patient population. Even simple multilayer perceptron (MLP) can be used here.

The results of the computational analysis were compared with the results of the traditional spatio-temporal gait parameters analysis using the CGA Gait Analyzer program developed by prof. Chris Kirtley.

The Statistica 13 software was used for statistical analyzes.

3 Results

The results of the computational analysis for the studied group are shown in Table 2 (for the fuzzy-based analysis) and Table 3 (for the fractal analysis).

The experience showed that the combination of the fuzzy model and multifractal analysis (fractal dimension, Hurst index) allows for a better description of the gait compared to traditional spatio-temporal gait parameters.

Table 2. Results of fuzzy-based analysis.

	Mean	SD	Min	Q1	Median	Q3	Max
Fractal dimension							
Healthy people	76	16	51	65	74	83	100
Post stroke patients	31	6	12	26	33	39	46

Table 3. Results of multifractal analysis.

	Mean	SD	Min	Q1	Median	Q3	Max
Fractal dimension							
Healthy people	1.14	0.17	1.01	1.05	1.11	1.14	1.17
Post stroke patients	1.34	0.33	1.11	1.25	1.31	1.38	1.45
Hurst Exponent							
Healthy people	0.22	0.06	0.12	0.15	0.21	0.27	0.34
Post stroke patients	0.31	0.07	0.14	0.19	0.26	0.31	0.39

4 Discussion

Artificial intelligence manifests the potential to improve healthcare through new strategies for healthcare delivery, automation and decision support, prognostication from patient data and facilitating patient engagement [12–14].

The development of computational models reduces the effort needed to carry out even complex calculations or simulations and promotes their clinical use. Such tools (even simplified ones) can provide better imaging of physiological phenomena in physiotherapy and the entire spectrum of possible disorders. This is also due to the massive use of new tools in clinical conditions and in tests on patients who have not been subject to them in laboratories so far. Moreover, simplified procedures and IT support enable significant time savings, and at the same time the accuracy and repeatability greater than classic observation methods.

This fits into both the paradigm of preventive medicine (medicine for healthy people) and personalized medicine, focused on the patient and their goals and health-related quality of life. This is when treatment plans are adjusted almost in real time based on health records (patient history, history, test results) and analysis of data from sensors worn by the patient or placed in devices for diagnosis, therapy and care [13]. This is fostered by the emergence of low-cost, relatively accurate and reliable sensors and analytic tools, including in telerehabilitation. Hence, the development of AI-based approaches for the automated assessment of a patient's performance, as well as his or her progress towards returning to full or maximum achievable functional capacity for that patient [13]. This provides an important semi-automatic or automatic complement to traditional functional diagnosis by medical professionals, including in the therapy management of patients participating in home rehabilitation. The work by Liao et al. [13] grouped the aforementioned AI-based solutions for movement assessment into three main categories: discrete movement assessment, rule-based approaches and template-based approaches. This approach still develops taking into consideration advances both in the area of computational analysis [15–18], and devices for rehabilitation purposes [19, 20], including IoT based [20, 21].

The challenge is still the representation of data, features, motion segmentation and functions. The main limitation is the still small number of studies on large numbers of patients and, consequently, certified ready-to-use systems. While there are collections of historical data in hospitals, access to them is not easy, and the data itself is often not digitized in a way that allows it to be used consistently at the national level, e.g. for teaching and testing medical AI systems. Companies from the MedTech area even offer free software in exchange for access to data. Moreover, the implementation of a new software product takes longer due to the need to meet the requirements of MDR and ISO 13485. We have developed the first Polish ISO 13485 automation system in the area of medical devices AlexQMS (Alex Quality Management System).

Urgency of the aforementioned issue is also due to the fact that the population of elderly people and people with disabilities is growing due to injuries and chronic diseases of civilization (stroke, diseases of cardiopulmonary system, etc.), the epidemic of which has been stopped but not reversed. The lack of medical specialists, especially in smaller centers, with the inability to educate them quickly, is a serious challenge for the current health care system, hence the emphasis on computational tools.

Physiotherapy devices and software are necessary to maintain or improve both functioning and independence of patients with motor deficits. This allows to better meet the mobility expectations of such people. Modern robotic solutions, self-powered IoT sensor networks, and optimized machine learning algorithms are likely to play a leading role in providing support services to unmet global physiotherapy needs.

Fuzzy logic as a tool for describing uncertainty and linguistically described parameters can contribute particularly much here. To date, the objectivisation of physiotherapy outcomes is not fully developed, and these areas of physiotherapy can provide scope for effective applications of both fuzzy logic and ordered fuzzy numbers.

The field of application of fuzzy logic in health sciences is still expanding, from macroscale (e.g. assessment of aggregation and uncertainty) [22] to microscale (e.g. assessment of proteins) [23, 24]. The next step belongs to big data systems, due to the high availability of data sets, to biologically reliable neural networks [25, 26]. We hope that their rapid development will also be reflected in their application in physiotherapy, mainly due to many not fully understood mechanisms and dependencies. The rapid development of artificial intelligence-based tools for physiotherapists will ensure better effectiveness of rehabilitation methods.

Unfortunately, the potential of the data in our possession is currently not fully exploited: large amounts of data from laboratory tests result in attempts of reliable analysis by methods other than computational based on relatively limited data sets.

5 Conclusions

Results to date, including this study, confirm the potential of artificial intelligence, including fuzzy logic, to improve various areas of physiotherapy. This includes the automation of tasks requiring data analysis, classification and prediction for clinical decision-making. This is particularly true for therapy delivered using stationary rehabilitation robots and exoskeletons, as well as telerehabilitation, where data for AI systems can be extracted automatically.

The aforementioned approach may lead to a breakthrough in physiotherapy service delivery through the application of AI.

The results of the present study should not only contribute to the development of research on AI applications in clinical practice in physiotherapy and the undertaking of interdisciplinary research. They should also encourage physiotherapists to expand their knowledge and experience of therapy options based on new technologies. On the other hand, this will allow information systems to be prepared in line with clinicians' expectations. Regardless, the ability to analyze clinical data (including AI-based) is already an important part of professional development in physiotherapy.

References

1. Ozdemir, F., Ari, A., Kilcik, M.H., Hanbay, D., Sahin, I.: Prediction of neuropathy, neuropathic pain and kinesiophobia in patients with type 2 diabetes and design of computerized clinical decision support systems by using artificial intelligence. Med. Hypotheses **143**, 110070 (2020)
2. Caldas, R., Fadel, T., Buarque, F., Markert, B.: Adaptive predictive systems applied to gait analysis: a systematic review. Gait Posture **77**, 75–82 (2020)
3. Huang, C.C., Liu, H.M., Huang, C.L.: Intelligent scheduling of execution for customized physical fitness and healthcare system. Technol. Health Care **24**(Suppl 1), S385–S392 (2015)
4. Mikołajewska, E., Prokopowicz, P., Mikołajewski, D.: Computational gait analysis using fuzzy logic for everyday clinical purposes - preliminary findings. Bio-Algorithms Med-Syst. **13**(1), 37–42 (2017)
5. Prokopowicz, P., Mikołajewski, D., Tyburek, K., Mikołajewska, E.: Computational gait analysis for post-stroke rehabilitation purposes using fuzzy numbers, fractal dimension and neural networks. Bull. Pol. Acad. Sci. Tech. Sci. **68**(2), 191–198 (2020)
6. Mikołajewska, E.: Associations between results of post-stroke NDT-Bobath rehabilitation in gait parameters, ADL and hand functions. Adv. Clin. Exp. Med. **22**(5), 731–738 (2013)
7. Mikołajewska, E., Prokopowicz, P., Mikolajewski, D.: Computational gait analysis using fuzzy logic for everyday clinical purposes-preliminary findings. Bio-Algorithms Med-Syst. **13**(1), 37–42 (2017)
8. Mikołajewski, D., Mikołajewska, E., Sangho, B.: Fraktalna analiza i predykcja zmian parametrów chodu. Studia i Materiały Informatyki Stosowanej **13**(2), 21–25 (2021)
9. Prokopowicz, P., Mikołajewski, D., Mikołajewska, E., Kotlarz, P.: Fuzzy system as an assessment tool for analysis of the health-related quality of life for the people after stroke. In: Proceedings of the 16th International Conference on Artificial Intelligence and Soft Computing (ICAISC), vol. 10245, pp. 710–721 (2017)
10. Prokopowicz, P., Mikołajewski, D.: Fuzzy approach to computational classification of burnout—preliminary findings. Appl. Sci. **12**, 3767 (2022)
11. Mikołajewski, D., Prokopowicz, P.: Effect of COVID-19 on selected characteristics of life satisfaction reflected in a fuzzy model. Appl. Sci. **12**, 7376 (2022)

12. Xiao, X., Fang, Y., Xiao, X., Xu, J., Chen, J.: Machine-learning-aided self-powered assistive physical therapy devices. ACS Nano **15**(12), 18633–18646 (2021)
13. Liao, Y., Vakanski, A., Xian, M., Paul, D., Baker, R.: A review of computational approaches for evaluation of rehabilitation exercises. Comput. Biol. Med. **119**, 103687 (2020)
14. Tack, C.: Artificial intelligence and machine learning | applications in musculoskeletal physiotherapy. Musculoskelet. Sci. Pract. **39**, 164–169 (2019)
15. Galas, K.: Effectiveness of blink classification using selected neural networks. Studia i Materiały Informatyki Stosowanej **13**(1), 11–16 (2021)
16. Czerniak, J.M., Zarzycki, H.: Artificial Acari Optimization as a new strategy for global optimization of multimodal functions. J. Comput. Sci. **22**, 209–227 (2017)
17. Czerniak, J.M., Śmigielski, G., Ewald, D., Paprzycki, M., Dobrosielski, W.: New proposed implementation of ABC method to optimization of water capsule flight. In: Proceedings of 2015 Federated Conference on Computer Science and Information Systems (FedCSIS), pp. 489–493 (2015)
18. Czerniak, J.M., Dobrosielski, W.T., Apiecionek, Ł., Ewald, D., Paprzycki, M.: Practical application of OFN arithmetics in a crisis control center monitoring. In: Fidanova, S. (eds.) Recent Advances in Computational Optimization. SCI, vol. 655, pp. 51–64. Springer, Cham (2016). https://doi.org/10.1007/978-3-319-40132-4_4
19. Piszcz, A.: BCI in VR: an immersive way to make the brain-computer interface more efficient. Studia i Materiały Informatyki Stosowanej **13**(1), 11–16 (2021)
20. Molnár, J., et al.: Weather station IoT educational model using cloud services. J. Univ. Comput. Sci. **26**(11), 1495–1512 (2020)
21. Rojek, I., Macko, M., Mikołajewski, D., Saga, M., Burczyński, T.: Modern methods in the field of machine modelling and simulation as a research and practical issue related to industry 4.0. Bull. Pol. Acad. Sci. Tech. Sci. **69**(2), e136717 (2021)
22. Pękala, B., Dyczkowski, K., Grzegorzewski, P., Bentkowska, U.: Inclusion and similarity measures for interval-valued fuzzy sets based on aggregation and uncertainty assessment. Inf. Sci. **547**, 1182–1200 (2021)
23. Mrozek, D., Malysiak-Mrozek, B., Kozielski, S.: Alignment of protein structure energy patterns represented as sequences of fuzzy numbers. In: Proceedings of NAFIPS 2009 - 2009 Annual Meeting of the North American Fuzzy Information Processing Society, pp. 1–6 (2009)
24. Martínez-Fernández, S., et al.: Continuously assessing and improving software quality with software analytics tools: a case study. IEEE Access **7**, 68219–68239 (2019)
25. Barbierato, E., Gribaudo, M., Iacono, M.: Modeling and evaluating the effects of big data storage resource allocation in global scale cloud architectures. Int. J. Data Warehous. Min. **12**, 2 (2016)
26. Wojcik, G.M., Kaminski, W.A.: Self-organised criticality as a function of connections' number in the model of the rat somatosensory cortex. In: Proceedings of the 8th International Conference on Computational Science 2008; Computational Science - ICCS 2008, PT 1 5101, pp. 620–629 (2008)
27. Ważny, M., Wójcik, G.M.: Shifting spatial attention - numerical model of posner experiment. Neurocomputing **135**, 139–144 (2014)

A Generalized Net Model of Some Nephrological Diseases

Martin Lubich[1], Elenko Popov[2], Radostina Georgieva[2], Dmitrii Dmitrenko[2], Borislav Bojkov[2], Chavdar Slavov[2], Ludmila Todorova[3](✉), Vassia Atanassova[3], Peter Vassilev[3], and Krassimir Atanassov[3]

[1] DKC "Sofiamed", 16 Boul. G. M. Dimitrov, 1797 Sofia, Bulgaria
martinlub11@abv.bg

[2] UMBAL "Tsaritsa Yoanna – ISUL", Department of Urology and Andrology, Medical University-Sofia, 8, Byalo more Str., 1527 Sofia, Bulgaria

[3] Department of Bioinformatics and Mathematical Modelling, Institute of Biophysics and Biomedical Engineering, Bulgarian Academy of Sciences, Acad. G. Bonchev Str., Bl. 105, 1113 Sofia, Bulgaria
lpt@biomed.bas.bg, krat@bas.bg

Abstract. This paper describes a generalized net model of the process of diagnostic work-up of the basic nephrological diseases and conditions: AKI, oliguria, proteinuria, hematuria, and its'implementation in everyday clinical practice for education, differential diagnosis and multidisciplinary collaboration.

Keywords: Generalized net · Nephrological disease

1 Introduction

In the recent years, an increasing number of people suffering from Chronic Kidney Disease (CKD) has been observed [7], which leads to an increased frequency of reaching the End-Stage Renal Disease (ESRD) [12] and subsequently the need for long-term renal replacement treatment in the form of Hemodialysis Treatment (HT) or Kidney Transplantation (KT) [4]. One of the important reasons for the development of CKD, in some countries (Great Britain and USA up to 15–18%) are glomeru-lonephritic kidney diseases. CKD is associated with increased cardiovascular risk and therefore appears as a socially significant disease. In essence, glomerulonephritis is a heterogeneous group of diseases, with different etiology, pathogenesis, clinical course and outcome [13]. It mainly affects people in the young and active age [6]. GN occurs in the form of several syndromes [5]. These are:

K. T. Atanassov et al. (Eds.): BOS/SOR/IWIFSGN 2022, LNNS 793, pp. 200–206, 2023.
https://doi.org/10.1007/978-3-031-45069-3_18

1. Asymptomatic urinary abnormalities - the presence of protein in the urine, established with qualitative and quantitative research methods.
2. Acute nephritis - occurring with proteinuria and hematuria, renal function disorders, olig- uria.
3. Nephrotic syndrome with hematuria, proteinuria, decreased serum albumin, dyslipidemia, increased tendency to thrombosis. Almost always these symptoms are accompanied by Arterial Hypertension (AH) [9]. These clinical syndromes and laboratory changes are expressed to a different degree in different types of glomerulonephritis and only by their appearance, the exact diagnosis cannot be made and timely and adequate treatment cannot be started [11]. Therefore, it is of fundamental importance to carry out a Kidney Biopsy (KB) in a timely manner, which has the task of establishing a final histological diagnosis and, according to its result, starting adequate treatment [8]. Of all these abnormal findings, only hematuria could originated from pathological processes outside the glomerulus [10]. Hematuria could be a symptom of systematic coagulopathy - prerenal hematuria (diagnosed by aberrations in coagulation tests) or could be a symptom or urological disease - infection, stone or malignancy of the kidney, ureter, bladder, prostate (diagnosed by planar imaging (CT/MRI), ultrasound and PSA test) [14]. This differential diagnosis should be performed before the hypothesis of CKD could be accepted and verified by a biopsy.

This paper presents the possibilities of Generalized Net (GN, [1–3]) models for systematizing the approach for timely diagnosis of patients who are indicated for the diagnosis of Glomerulonephritis.

2 A Generalized Net Model

The definition of a GN and the algorithm of its functioning are given in [2,3]. The present GN model (see Fig. 1) contains 9 transitions, 23 places and tokens $\pi_1, \pi_2, \ldots,$ where π_i represents the i-th (the current) patient. For brevity, below, we will omit index i.

The current token π enters the GN through place l_1 with the initial characteristic

"name, parameters(age, sex, etc.), supposition for a nephrological disease".

$$Z_1 = \left\langle \{l_1\}, \{l_2\}, \begin{array}{c|c} & l_2 \\ \hline l_1 & true \end{array} \right\rangle.$$

In place l_2, token π obtains the characteristic

"results of the laboratory tests".

$$Z_2 = \langle \{l_2, l_9, l_{14}, l_{15}, l_{17}, l_{21}\}, \{l_3, l_4, l_5, l_6, l_7, l_8\},$$

	l_3	l_4	l_5	l_6	l_7	l_8
l_2	$W_{2,3}$	$W_{2,4}$	$W_{2,5}$	$false$	$W_{2,7}$	$W_{2,8}$
l_{14}	$W_{14,3}$	$W_{14,4}$	$W_{14,5}$	$W_{14,6}$	$W_{14,7}$	$W_{14,8}$
l_{15}	$W_{15,3}$	$W_{15,4}$	$W_{15,5}$	$W_{15,6}$	$W_{15,7}$	$W_{15,8}$
l_{17}	$W_{17,3}$	$W_{17,4}$	$W_{17,5}$	$W_{17,6}$	$W_{17,7}$	$W_{17,8}$
l_{21}	$W_{21,3}$	$W_{21,4}$	$W_{21,5}$	$W_{21,6}$	$W_{21,7}$	$W_{21,8}$

$\rangle$,

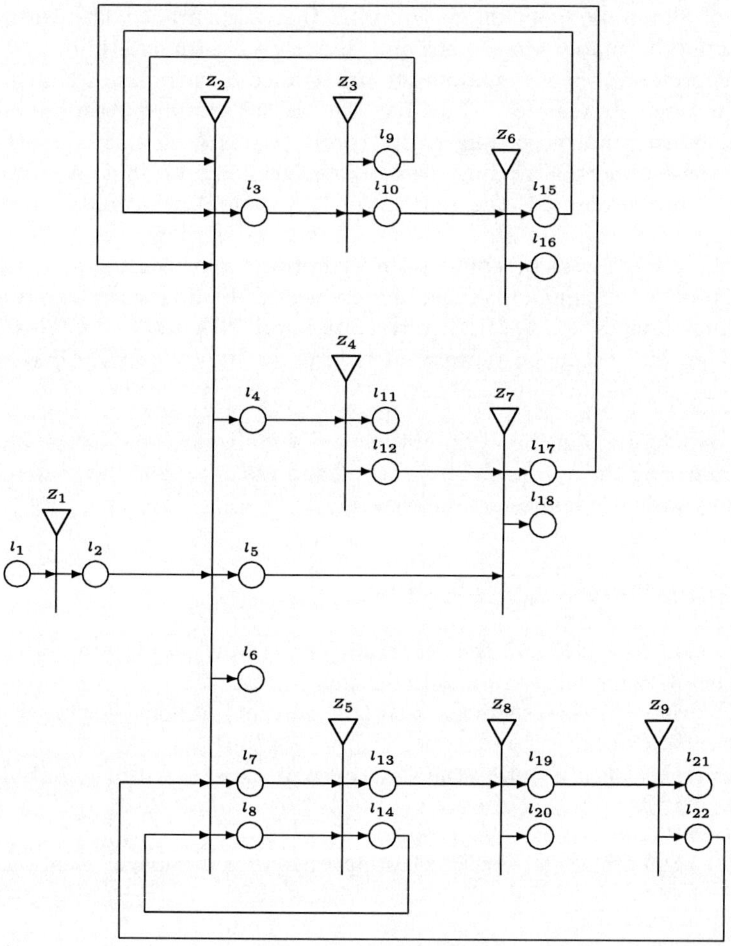

Fig. 1. A GN model

where

$W_{2,3} = W_{9,3} = W_{14,3} = W_{15,3} = W_{17,3} = W_{21,3} =$ "the laboratory results direct to a proteinuria",

$W_{2,4} = W_{9,3} = W_{14,4} = W_{15,4} = W_{17,4} = W_{21,4} =$ "the laboratory results show that a coagulation test is necessary",
$W_{2,5} = W_{9,5} = W_{14,5} = W_{15,5} = W_{17,5} = W_{21,5} =$ "the laboratory results direct to a hematuria",
$W_{2,7} = W_{9,7} = W_{14,7} = W_{15,7} = W_{17,7} = W_{21,7} =$ "the laboratory results show that a reduce volume of 24 h. dioresis is necessary",
$W_{2,8} = W_{9,8} = W_{14,8} = W_{15,8} = W_{17,8} = W_{21,8} =$ "the laboratory results show a high level of the blood pressure",
$W_{9,6} = W_{14,6} = W_{15,6} = W_{17,6} = W_{21,6} =$"the results of the repeated tests show positive progress".

The token π obtains a characteristic

"24 h. urine protein test and glucose intolerance tests are necessary"

in place l_3,

"results of the coagulation test"

in place l_4,

"result of the urine analysis"

in place l_5,

"all results of the test(s) are normal"

in place l_6,

"results of the 24 h. dioresis"

in place l_7,

"results of the blood pressure test"

in place l_8.

$$Z_3 = \left\langle \{l_3\}, \{l_9, l_{10}\}, \begin{array}{c|cc} & l_9 & l_{10} \\ \hline l_3 & W_{3,9} & W_{3,10} \end{array} \right\rangle,$$

where

$W_{3,10} =$ "the result of the urine protein test is higher than 3.5 g/day" OR "the glucose intolerance test is positive",
$W_{3,9} = \neg W_{3,10}$,

where $\neg P$ is the negation of predicate P.

The token π obtains a characteristic

"biopsy is necessary"

in place l_9,

"the observations must be repeat after 3-6 months"

in place l_{10}.

$$Z_4 = \left\langle \{l_4\}, \{l_{11}, l_{12}\}, \begin{array}{c|cc} & l_{11} & l_{12} \\ \hline l_4 & W_{4,11} & W_{4,12} \end{array} \right\rangle,$$

where

$W_{4,11}$ = "the coagulation test is pathological",
$W_{4,12} = \neg W_{4,11}$.

The token π obtains a characteristic

"a treatment of the coagulation disorder is necessary"

in place l_{11}, and

"a continuation of the assesment of hematuria is necessary"

in place l_{12}.

$$Z_5 = \left\langle \{l_7, l_8\}, \{l_{13}, l_{14}\}, \begin{array}{c|cc} & l_{13} & l_{14} \\ \hline l_7 & W_{7,13} & W_{7,14} \\ l_8 & W_{8,13} & W_{8,14} \end{array} \right\rangle,$$

where

$W_{7,13} = W_{8,13}$ = "the result is higher than the norm",
$W_{7,14} = W_{8,14} = \neg W_{6,13}$.

The token π obtains a characteristic

"a biopsy is necessary"

in place l_{13},

"the observation must be repeated immediately"

in place l_{14}.

$$Z_6 = \left\langle \{l_{10}\}, \{l_{15}, l_{16}\}, \begin{array}{c|cc} & l_{15} & l_{16} \\ \hline l_{10} & W_{10,15} & W_{10,16} \end{array} \right\rangle,$$

where

$W_{10,15}$ = "the observation must be repeated",
$W_{10,16}$ = "the treatment for a proteinuria is necessary".

The current token π does not obtain any characteristic in place l_{15} and it obtains the characteristic

"a treatment of the disease causing proteinuria"

in place l_{16}.

$$Z_7 = \left\langle \{l_5, l_{12}\}, \{l_{17}, l_{18}\}, \begin{array}{c|cc} & l_{11} & l_{12} \\ \hline l_4 & W_{4,11} & W_{4,12} \end{array} \right\rangle,$$

where

$W_{5,18}$ = "the result of the coagulation test is abnormal",
$W_{5,17} = \neg W_{5,18}$.

The token π obtains a characteristic

"the observations must be repeat after 3-6 months; the problem is a systematic and it is not originally from the urinary system"

in place l_{17}, and

"a treatment of haematuria with a glumerular origin"

in place l_{18}.

$$Z_8 = \left\langle \{l_{13}\}, \{l_{19}, l_{20}\}, \begin{array}{c|cc} & l_{19} & l_{20} \\ \hline l_{13} & W_{13,19} & W_{13,20} \end{array} \right\rangle,$$

where

$W_{13,20}$ = "the results of biopsy are positive for glomerular/nephrological disease",
$W_{13,19} = \neg W_{13,20}$.

The token π obtains a characteristic

"PSA-test and MRI of the prostate are necessary"

in place l_{19},

"a treatment of the glomerular/nephrological. disease"

in place l_{1720}.

$$Z_9 = \left\langle \{l_{19}\}, \{l_{21}, l_{22}\}, \begin{array}{c|cc} & l_{21} & l_{22} \\ \hline l_{19} & W_{19,21} & W_{19,22} \end{array} \right\rangle,$$

where

$W_{19,21}$ = "the PSA-test or the MRI are abnormal",
$W_{19,22} = \neg W_{19,21}$.

The token π obtains a characteristic

"treatment of a urological disease is necessary"

in place l_{21} and

"the observation must be repeated periodically"

in place l_{22}.

3 Conclusion

The proposed GN model for assessment of patients presenting with urine abnormalities, arterial hypertension or reduced urine production can be used for education, testing and differential diagnosis in this clinical scenario, as well as for establishing a multidisciplinary collaboration between several medical specialties.

Acknowledgements. The authors are grateful for the support provided by the Bulgarian National Science Fund under Grant Ref. No. KP-06-N43/7/30.11.2020 "Creating a prognostic model predicting life expectancy in prostate cancer patients and providing better quality of life after definitive surgical treatment".

References

1. Alexieva, J., Choy, E., Koycheva, E.: Review and bibliography on generalized nets theory and applications. In: Choy, E., Krawczak, M., Shannon, A., Szmidt, E. (eds.) A Survey of Generalized Nets Raffles, KvB Monograph No. 10, pp. 207–301 (2007)
2. Atanassov, K.: Generalized Nets. World Scientific, Singapore (1991)
3. Atanassov, K.: On Generalized Nets Theory. "Prof. M. Drinov" Academic Publishing House, Sofia (2007)
4. Barrett, K.E., Barman, S., Boitano, S., Brooks, H.: Ganong's Review of Medical Physiology, 25th edn. McGraw-Hill, New York (2016)
5. Dzhambazova, E.: Human Physiology. Climent Ohridski University Publishing House, Soifa (2015). (in Bulgarian)
6. Lv, J.-C., Zhang, L.-X.: Prevalence and disease burden of chronic kidney disease. In: Liu, B.-C., Lan, H.-Y., Lv, L.-L. (eds.) Renal Fibrosis: Mechanisms and Therapies. AEMB, vol. 1165, pp. 3–15. Springer, Singapore (2019). https://doi.org/10.1007/978-981-13-8871-2_1
7. O'Hare, A.M., Choi, A.I., Bertenthal, D., et al.: Age affects outcomes in chronic kidney disease. J. Am. Soc. Nephrol. **18**(10), 2758–2765 (2007). https://doi.org/10.1681/ASN.2007040422. Epub 12 Sept 2007. PMID: D17855638
8. Pesce, F., et al.: Glomerulonephritis in AKI: from pathogenesis to therapeutic intervention. Front. Med. **7**, 983 (2021)
9. Piryov, B., Vassilev, V., Belchev, C.: Human Anatomy and Physiology. Mnemozina, Sofia (1995). (in Bulgarian)
10. Saha, M.K., Massicotte-Azarniouch, D., Reynolds, M., et al.: Glomerular hematuria and the utility of urine microscopy: a review. Am. J. Kidney Dis. **80**(3), 383–392 (2022). https://doi.org/10.1053/j.ajkd.2022.02.022
11. Sethi, S., Fervenza, F.C.: Standardized classification and reporting of glomerulonephritis. Nephrol. Dial. Transplant. **34**(2), 193–199 (2019)
12. Steddon, S., et al.: Clinical syndromes - proteinuria. In: Oxford Handbook of Nephrology and Hypertension, 2nd edn. Oxford Medical Handbooks, Oxford (2014)
13. Taal, M.W., Glenn, M.G., Marsden, Ph.A., et al.: Laboratory assessment of kidney disease: glomerular filtration rate, urinalysis and proteinuria. Brenner & Rectors, The Kidney, Saunders Elsevier, Philadelphia, PA (9th edn.), pp. 868–896
14. Willis, G.C., Tewelde, S.Z.: The approach to the patient with hematuria. Emerg. Med. Clin. **37**(4), 755–769 (2019)

A Generalized Net Model of Acute Respiratory Distress Syndrome

Diana Petkova[1] and Krassimir Atanassov[1,2](✉)

[1] Institute of Biophysics and Biomedical Engineering, Bulgarian Academy of Sciences, Acad. G. Bonchev Str., Bl. 105, 1113 Sofia, Bulgaria
diana@biomat.bas.bg, krat@bas.bg

[2] Intelligent Systems Laboratory, Assen Zlatarov University, 8010 Burgas, Bulgaria

Abstract. A generalized net model of acute respiratory distress syndrome is described. This net model could be used for investigation of blood parameters after HCL administration, after inhalation of liposomes for improvement of lung functions as well as for training medical students to take decisions in case of ARDS pathology after HCl - inhalation or other pathological models.

Keywords: acute respiratory distress syndrome · Generalized net · Liposomes

1 Introduction

Acute respiratory distress syndrome (ARDS) is a syndrome of acute respiratory failure, caused by direct or indirect pulmonary insults such as sepsis due to nonpulmonary sources, trauma, aspiration of gastric content and in rare causes by pancreatitis and drug reactions [13,23,24]. This pathology very often results in high mortality of patients. ARDS is characterized by hypoxemia, decreased lung compliance, pulmonary edema [9,22]. The main problem in this disease which leads to lung disfunctions is the changes in endogenous surfactant system.

One of the most common etiology of ARDS is the gastric acid aspiration which causes changes in the pulmonary epithelium and flux of edema fluid into the alveolar space. Thus alveolar surfactant is inactivated and atelectasis occurs [12,17].

A lot of numerous clinical trials have been used for their efficacy for ARDS. One of them was application of different exogenous surfactants but the reported results for surfactant-based therapeutics to date [2,3,18,20] have been inconsistent. Exogenous surfactant is a lipid protein mixture synthesized by alveolar type II cells and secreted into the hypophase of the lung and is an essential component that allows for normal lung functions [15,27]. The exogenous surfactants used in clinical trials are from artificial or natural origins. One of the most important characteristics is its specific phospholipid and protein composition. Approximately 85% of the surfactants consist of phospholipids. The main

K. T. Atanassov et al. (Eds.): BOS/SOR/IWIFSGN 2022, LNNS 793, pp. 207–215, 2023.
https://doi.org/10.1007/978-3-031-45069-3_19

functions of different surfactants are to form surface tension reducing lipid film at the air-liquid interface of the alveolar surface [11,15,29]. Thus stabilized the lung preventing alveolar collapse and allows for inflections. There are a lot of factors which influence the administration of exogenous surfactants. The efficacy of exogenous application depends also on the timing of the treatment and as well as on the host's response [19]. The mode of the mechanical ventilation is also a very important event. The positive result of exogenous surfactant depends on its interaction with the components of the host's alveolar surfactant [10,26] as well as the nature and the type of lung injury [16,20,21,26].

Thus, the surfactant therapy is very complicated and it is necessary to evaluate different treatments in animal models of lung injury as a first step for clinical trials.

In this paper, a Generalized Net (GN; see [1,4,5]) model of ARDS will be described. In Sect. 2, short remarks on GNs are given, in Sect. 3 - the process of animal treatment, and in Sect. 4 - the GN model is described.

2 Short Remarks on Generalized Nets

The concept of a *Generalized Net* (GN) is described in details in [4,5]. GNs are extensions of the classical Petri net, as well as of all their extensions, available by the moment. A given GN may not have some of the components, and such GNs give rise to special classes of GNs called "reduced GNs". For the needs of the present research we shall use (and describe) one of the reduced types of GNs. The way of defining the GNs is principally different from the ways of defining the other types of Petri nets. The first basic difference between GNs and the ordinary Petri nets is the "place—transition" relation. Here, transitions are objects of a more complex nature.

Formally, every GN-transition is described by a seven-tuple, but here we shall discuss the following reduced form of it:

$$Z = \langle L', L'', r \rangle,$$

where L' and L'' are finite, non-empty sets of places (the transition's input and output places, respectively); r is the transition's condition determining the tokens which will transfer from the transition's inputs to its outputs; it has the form of an index matrix (see, e.g., [6]):

$$r = \begin{array}{c|ccccc} & l''_1 & \dots & l''_j & \dots & l''_n \\ \hline l'_1 & & & & & \\ \vdots & & & r_{i,j} & & \\ l'_i & & (r_{i,j} & - predicates) & & \\ \vdots & & (1 \le i \le m, & 1 \le j \le n) & & \\ l'_m & & & & & \end{array}$$

(i, j) denotes the element which corresponds to the i-th input and j-th output places; these elements are predicates and when the truth value of the (i, j)-th element is true, the token from the i-th input place can be transferred to the j-th output place; otherwise, it is not possible. The ordered four-tuple (also for the reduction case):

$$E = \langle A, K, X, \Phi \rangle$$

is called a *(reduced) Generalized Net* (GN), if: A is a set of transitions; K is the set of the GN's tokens; X is the set of all initial characteristics the tokens can receive up on entering the net; Φ is a characteristic function which gives new characteristic to every token when it makes a transfer from an input to an output place of a given transition.

During last years, the GNs were used as tool for modelling of different processes in a lot of areas, e.g., in artificial intelligence (see, e.g., [7]), medicine (see, e.g., [8]), economics, industry, etc.

3 Animal Treatment and Investigation of Gas Exchange, Blood Parameters and Biophysical Analysis of Bronchoalveolar Lavage

In our previous investigation, we have demonstrated phospholipid multi-llamelar liposomes prepared from soybean phosphatidylcholine and cholesterol by reversed phases method [14], which possess specific lung and alveolar surfactant accumulation after i.v. (intravenus) application [25].

In this study HCL-injured rabbits were used as an animal model for ARDS. The HCL was instilled through a side-port adaptor of the endotracheal tube during the inspiratory phase of respiration in order to induce an acute lung injury. After improvement of lung injury tested by blood measurements by AVL OMNI (AVL, Austria), phospholipid–cholesterol liposomes were administrated in small boluses during the inspiratory phase of ventilation trough a side-port adaptor. After ventilation for 60 min Arterial Blood Gas (ABG) analysis of control, HCL-injured and liposome-treated animals was performed. The animals were killed and lavaged for recovery of BronchoAlveolar Lavage (BAL) for analysis of protein and lipid content and composition, as well as investigation of adsorbed monolayers formed in a Langmuir trough (Biegler-Electronic, Austria). The administration of HCL leads to dramatic changes of physiological parameters of treated animals, as compared to the control. After liposome administration PaO_2/FiO_2 (arterial oxygen pressure/fraction of expired oxygen) was improved almost to the control value, as well as the blood pH, saturation and other blood parameters returned to the normal value. The surface pressure area curves of monolayers obtained from BAL during compression/decompression was always higher for the HCL-treated animals and lower for liposome-treated group in comparison to the control group. The hysteresis area was lower for the HCL-treated group, comparable to the other two groups (control and liposome-treated animals) in each cycle. The surface tension values increased after HCL-treatment and sharply decreased after the liposomes treatment. The surface behavior of surfactant components measured under dynamic conditions show recovery of these value to the normal one. These results demonstrated the recovery potential of the phospholipid-cholesterol liposomes and their ability to increase surface activity of BAL under *in vivo* conditions. The changes of hysteresis area of compression/decompression loop show a decrease after HCL treatment and the shape of the loop was changed, which was connected to the decline of lung function. After liposome instillation the hysteresis area increased in parallel with the recovery of gas exchange. The investigations of the surface tension and of the dynamic characteristics of monolayers of BAL and the surface activity of BAL *in vitro* is recovered to the normal values after liposomes treatment.

The results of this investigation demonstrated an *in vivo* improvement of gas exchange after liposome administration. The liposome intratracheal administration was performed in small doses in accordance with the literature data that application of surfactant in small doses led to significant smaller lung injury score [28]. On the basis of these results we may speculate that it is quite likely that the used liposomes in this study might be used for improvement of oxygenation after acid aspiration induced ARDS.

4 A Generalized Net Model

The GN model contains 6 transitions, 16 places and 11 types of tokens. The sense of the places is the following.

l_1 - mouth (for inspiration),
l_2 - trachea,
l_3 - mouth (for expiration),
l_4 - blood vessel,
l_5 - trachea for BAL,
l_6 - expiration air,
l_7 - lung,
l_8 - cardiovascular system,
l_9 - blood for physiological parameters,
l_{10} - information obtained from adsorbed monolayers from BAL,
l_{11} - Langmuir trough,
l_{12} - equipment for blood measurements,
l_{13} - blood for physiological parameters,
l_{14} - information about blood measurements after HCL application during the inspiration phase of respiration,
l_{15} - endotracheal tube,
l_{16} - decision maker.

The tokens are the following.

φ - fresh air,
β - blood,
σ - BAL,
λ - lung,

κ - cardiovascular system,
α_1 - Langmuir trough,
α_2 - apparatus for blood measurements,
ι_1 - information about biophysical parameters of BAL,
ι_2 - information about blood parameters,
γ' - 0.2N HCl-administration,
γ'' - phospholipid liposomes,
δ - decision maker.

The forms of the transitions are the following (see Fig. 1).

In each time-moment, a token φ with a characteristic

"air for inspiration"

enters place l_1 and another token φ with a characteristic

"air for expiration"

enters place l_6.

In some moment, token γ' enters place l_{15} with a characteristic

"Hidrochloric Acid (HCl), quantity"

In a next moment (after 45 min), token γ'' enters the same place l_{15}, but with a characteristic

"phospholipid liposomes, quantity".

$$Z_1 = \langle \{l_1, l_6, l_{15}\}, \{l_2, l_3\}, \begin{array}{c|cc} & l_2 & l_3 \\ \hline l_1 & true & false \\ l_6 & true & false \\ l_{15} & false & true \end{array} \rangle.$$

Token φ from place l_1 enters place l_2 and token φ from place l_6 enters place l_3, both without new characteristics.

When there is a token in place l_{15}, it enters place l_2 without new characteristics.

In each time-moment token β from place l_8 and token φ from place l_2 enter place l_7 and unite with token λ.

$$Z_2 = \langle \{l_2, l_7, l_{13}\}, \{l_4, l_5, l_6, l_7\}, \begin{array}{c|cccc} & l_4 & l_5 & l_6 & l_7 \\ \hline l_2 & false & false & false & true \\ l_7 & true & W_{7,5} & true & true \\ l_{13} & false & false & false & true \end{array} \rangle,$$

where

$W_{7,5}$ = "45 min after entering token γ' in place l_7 are performed".

Tokens φ from place l_2 and β from l_{13} enter place l_7 and unites with token λ.

On the next time-step, token λ splits to three tokens - φ that enters place l_6, β that enters place l_4, the original token λ that stays permanently in place

l_7 and in case that predicate $W_{7,5}$ is *true*, token λ generates one more, fourth, token - σ that enters place l_5 with a characteristic

"BAL, quantity".

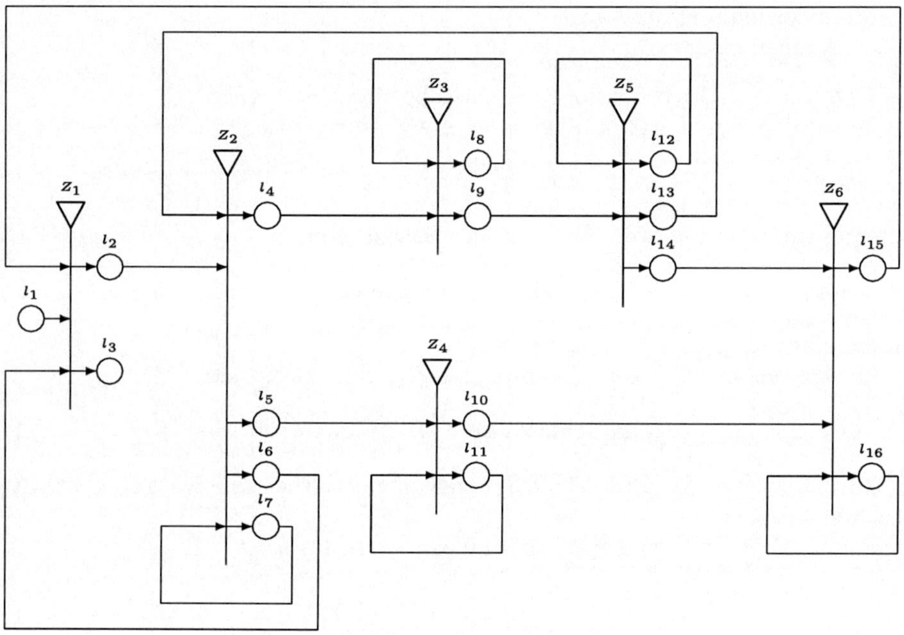

Fig. 1. GN-model

$$Z_3 = \langle \{l_4, l_8\}, \{l_8, l_9\}, \begin{array}{c|cc} & l_8 & l_9 \\ \hline l_4 & false & true \\ l_8 & true & false \end{array} \rangle.$$

Token β from place l_4 enters place l_8 and unites with token κ. In a next more detailed model, token κ will obtain specific characteristics, related to the cardio-vascular system, but here, for simplicity, it will not obtain any characteristic. On the next time-moment, token κ splits to the original token κ, which continues to stay in place l_8, and token β which enters place l_9.

$$Z_4 = \langle \{l_5, l_{11}\}, \{l_{10}, l_{11}\}, \begin{array}{c|cc} & l_{10} & l_{11} \\ \hline l_5 & false & true \\ l_{11} & true & false \end{array} \rangle.$$

Token σ from place l_5 enters place l_{11} and unites with token α_1. On the next time-step, token α_1 splits to two tokens - the original token α_1, which continues to stay in place l_{11}, and token ι_1, which enters place l_{10} with a characteristic

"information of surface properties of BAL".

$$Z_5 = \langle \{l_9, l_{12}\}, \{l_{12}, l_{13}, l_{14}\}, \begin{array}{c|ccc} & l_{12} & l_{13} & l_{14} \\ \hline l_9 & true & true & false \\ l_{12} & true & false & true \end{array} \rangle.$$

On each time-moment, token β splits to two tokens - β' that enters place l_{12} and unites with token α_2 and token β that enters place l_{13}. Both β-tokens, obtain no new characteristics. On the next time-moment, token α_2 splits to two tokens - the original token α_2 that continues to stay in place l_{12}, and token ι_2 that enters place l_{14} with a characteristic

"information about blood analysis".

$$Z_6 = \langle \{l_{10}, l_{14}, l_{16}\}, \{l_{15}, l_{16}\}, \begin{array}{c|cc} & l_{15} & l_{16} \\ \hline l_{10} & false & true \\ l_{14} & false & true \\ l_{16} & W_{16,15} & true \end{array} \rangle,$$

where
$W_{16,15} =$ "the moment for inhalation of 0.2N HCl ,or liposomes arise".

Tokens ι_1 and ι_2 enter place l_{16} and unite with token δ. On some time-moment token δ splits to two tokens - the same token δ that continues to stay in place l_{16}, and token γ' that enters place l_{15} with a characteristic

"0.2N HCl, quantity".

On some next time-moment, token δ splits to two tokens - the same token δ that continues to stay in place l_{16}, and token γ'' that enters place l_{15} with a characteristic

"liposomes, quantity".

5 Conclusion

The present GN-model could be used for several purposes: for investigation of the changes of ABG-analysis after inhalation of liposomes for improvement of gas exchange after administration of HCl; surface properties of BAL of HCl-treated and liposome treated animals, as well as for the proof of improvement the gas exchange after liposome administration.

Moreover, the model could be used for training medical students to take decisions in cases of ARDS pathology after HCl-inhalation. Given enough information about particular patient, the model could be used also for some surfactant and liposome based therapy for improvement of oxygenation after acid aspiration-induced ARDS.

In [8], a GN-model of a human body and GN-models of the separate human body systems are described. All models are related to the healthy human body. In future, on the basis of these models, GN-modes of pathological processes in the human body will be discussed. The present model will be a part of these models.

Acknowledgment. The authors acknowledge the support from the project UNITe BG05M2OP001-1.001-0004 /28. 02.2018 (2018–2023).

References

1. Alexieva, J., Choy, E., Koycheva, E.: Review and bibliography on generalized nets theory and applications. In: Choy, E., Krawczak, M., Shannon, A., Szmidt, E. (eds.) A Survey of Generalized Nets, Raffles KvB Monograph No. 10, pp. 207–301 (2007)
2. Anzueto, A., et al.: Aerosolized surfactant in adults with sepsis-induced acute respiratory distress syndrome. Nat. Engl. J. Med. **334**, 1417–1421 (1996)
3. Anzueto, A., et al.: Effect of aerosolized surfactant in patients with stable chronic bronchitis: a prospective randomized controlled trial. J. Am. Med. Assoc. **278**, 1426–1431 (1997)
4. Atanassov, K.: Generalized Nets. World Scientific, Singapore (1991)
5. Atanassov, K.: On Generalized Nets Theory. "Prof. M. Drinov" Academic Publishing House, Sofia (2007)
6. Atanassov, K.T.: Index Matrices: Towards an Augmented Matrix Calculus. SCI, vol. 573. Springer, Cham (2014). https://doi.org/10.1007/978-3-319-10945-9
7. Atanassov, K.: Generalized nets as a tool for the modelling of data mining processes. In: Sgurev, V., Yager, R.R., Kacprzyk, J., Jotsov, V. (eds.) Innovative Issues in Intelligent Systems. SCI, vol. 623, pp. 161–215. Springer, Cham (2016). https://doi.org/10.1007/978-3-319-27267-2_6
8. Atanassov, K., Chakarov, V., Shannon, A., Sorsich, J.: Generalized Net Models of the Human Body. "Prof. M. Drinov" Academic Publishing House, Sofia (2008)
9. Bernard, G., et al.: The American-European Consensus Conference on ARDS. Definitions, mechanisms, relevant outcomes, and clinical trial coordination. Am. J. Respir. Crit. Care Med. **149**, 818–824 (1994)
10. Brackenbury, A., et al.: Evaluation of exogenous surfactant in HCL-induced lung injury. Am. J. Resp. Crit. Care Med. **163**, 1135–1142 (2001)
11. Frerking, I., Gunter, A., Seeger, W., Pison, U.: Pulmonary surfactant: functions, abnormalities and therapeutic opsins. Intesive Care Med. **27**, 1699–1717 (2001)
12. Greenfield, L., Singleton, R., Mc Carffree, D., Coalson, J.: Pulmonary effects of experimental graded aspiration of hydrochloric acid. Ann. Surg. **170**, 74–86 (1969)
13. Gregory, T., et al.: Surfactant chemical composition and biophysical activity in acute respiratory distress syndrome. Clin. Invest. **63**, 1981–1991 (1976)
14. Gregoriadis, G., Allison, A.: Chester. Wiley, Brisbane (1983)
15. Goerke, J.: Pulmonary surfactant: function and molecular composition. Biochim. Biophys. Acta **1408**, 79–89 (1998)
16. Haitsma, J., Papadakos, P., Lachman, B.: Surfactant therapy for acute lung injury/acute respiratory distress syndrome. Curr. Opin. Crit. Care **10**, 18–22 (2004)

17. Kobauashi, T., Ganzuka, M., Taniguchi, J., Nitta, K., Murakami, S.: Lung lavage and surfactant replacement for hydrochloric acid aspiration in rabbits. Anesthesiol. Scand. **34**, 216–221 (1990)
18. Lewis, J., Ikegami, M., Jobe, A., Absonlom, D.: Physiologic responses and distribution of aerosolized surfactant (Survanta) in a nonuniform pattern of lung injury. Am. Rev. resp. Dis. **147**, 1364–1370 (1993)
19. Lewis, J., Veldhuizen, A.: Factors influencing efficacy of exogenous surfactant in acute lung injury. Biol. Neonate **67**, 48–60 (1995)
20. Lewis, J., Veldhuizen, A.: The role of exogenous surfactant in the treatment of acute lung injury. Annu. Rev. Physiol. **65**, 613–642 (2003)
21. Lewis, J., Veldhuizen, A.: The future of surfactant therapy during ALI/ARDS. Sem. Respir. Crit. Care Med. **27**, 377–388 (2006)
22. Luce, J.: Acute lung injury and the acute respiratory distress syndrome. Crit. Care Med. **26**, 369–376 (1998)
23. Matthay, M., Ware, L., Zimmerman, G.: The acute respiratory distress syndrome. J. Clin. Invest. **122**, 2731 2740 (2012)
24. Montgomery, A., Stager, M., Carrico, C., Hudson, E.: Causes of mortality in patients with the adult respiratory distress syndrome. Am. Rev. Respir. Dis. **132**, 485–489 (1985)
25. Ninio, S., Kovacheva, S., Neicheva, T., Petkova, D.: Mechanism of accumulation of double labeled 3H- 99mTc liposomes in lung and alveolar surfactant. Pharmacy XL15, pp. 18–22 (1997)
26. Puligandla, P., et al.: Alveolar environment influences the metabolic and biophysical properties of exogenous surfactants. J. Appl. Physiol. **88**, 1061–1071 (2001)
27. Van Golde, L.M.G., Batenburg, J., Robertson, B.: The pulmonary surfactant system. News Physiol. Sci. **9**, 13–20 (1994)
28. Wolf, S., et al.: Small dose of exogenous surfactant combined with partial liquid ventilation in experimental acute lung injury: effects on gas exchange, haemodynamics, lung mechanics, and lung pathology. Br. J. Anaesthesia **87**, 593–601 (2001)
29. Zuo, Y.Y., Veldhuizen, R.A.W., Neumann, A.W., Petersen, N.O., Possmayer, F.: Current perspective in pulmonary surfactant - inhibition, enhancement and evaluation. BBA-Biomembrane **1778**, 1947–1997 (2008)

Generalized Net Model of Rehabilitation Algorithm for Patients with Shoulder Impingement Syndrome

Simeon Ribagin[1,2](✉) and Gergana Angelova-Popova[2]

[1] Department of Bioinformatics and Mathematical Modelling, Institute of Biophysics and Biomedical Engineering, Bulgarian Academy of Sciences, Sofia, Bulgaria
simribagin@gmail.com

[2] Department of Health and Pharmaceutical Care, Medical College, University "Prof. D-r Asen Zlatarov", Burgas, Bulgaria
giggi_popa@abv.bg

Abstract. Shoulder pain related to the impingement syndrome is a common disorder and despite the growing evidence of the importance of rehabilitation, in particular active exercise physiotherapy, little data is available to guide the treatment process. The purpose of the present paper is to present an example of Generalized Nets application in orthopedics and traumatology rehabilitation and the model describes a possible algorithm protocol for rehabilitation treatment of the patients with shoulder impingement syndrome. The proposed model can be implemented in the decision making support systems, telerehabilitation platforms, optimization of the physiotherapy protocols for shoulder impingement syndrome and better rehabilitation strategies.

Keywords: Shoulder impingement syndrome · Rehabilitation algorithm · Generalized Nets · GN-model

1 Introduction

Shoulder impingement syndrome is a painful condition of the upper extremity resulting from a structural narrowing of the so called "subacromial space", which is the space between the humeral head inferiorly, the anterior edge and under surface of the anterior third of the acromion, coracoacromial ligament, and the acromioclavicular joint superiorly. This condition was first coined by Charles Neer in 1972 [16, 17]. Neer described this as a progressive syndrome with three stages, beginning with chronic bursitis and proceeding to partial and complete tears of the supraspinatus tendon, which may extend to rupture of other parts of the rotator cuff and may also involve the long biceps tendon. Neer was also one of the first to classify the impingement lesions and described three stages depending on the amount of damage in the subacromial space and the age of the patient [8]. According to [15] shoulder impingement results from an "inflammation and degeneration of the anatomical structures in the region of the subacromial space".

K. T. Atanassov et al. (Eds.): BOS/SOR/IWIFSGN 2022, LNNS 793, pp. 216–226, 2023.
https://doi.org/10.1007/978-3-031-45069-3_20

Matsen and Artnz [14] have further defined impingement as the encroachment of the acromion, coracoacromial ligament, coracoid process, or acromioclavicular joint on the rotator cuff mechanism and bursa that passes beneath them as the glenohumeral joint is moved, particularly in flexion and internal rotation [12]. Is the commonest disorder of the shoulder, accounting for 44%–65% of all complaints of shoulder pain [6]. Shoulder impingement syndrome can be categorized according to either the location of the compromised tissues, characterized as external or internal, and/or the underlying cause of the impingement, referred to as primary or secondary impingement. External, or subacromial impingement, results from a mechanical or physical encroachment of the soft tissue located within the subacromial space [10]. Conversely, internal impingement results when the tendons of the rotator cuff encroach between the humeral head and glenoid rim. Internal impingement is most commonly associated with the supraspinatus and infraspinatus tendons [9]. Although impingement symptoms may arise following a traumatic event, the pain more typically develops slowly over a period of days to weeks. The pain is typically localized to the anterolateral acromion and frequently radiates to the lateral middle part of the humerus. Patients usually complain of pain at night, exacerbated by lying on the involved shoulder, or sleeping with the arm overhead [13]. A typical sign in a patient with shoulder impingement is the so called "painful arc" in which the patient experiences pain between 60 and 120 degrees of abduction which reduces once past 120 degrees of abduction. Because of this the normal daily activities become extremely difficult and the overall quality of life status decreases. A systematic approach to shoulder and upper extremity evaluation must be undertaken to identify the specific cause or subtle underlying causes of the impingement. The most used and informative test in practice are the Hawkins-Kennedy test, Neer test, Painful arc, Empty can test, External rotation resistance test, Cross-body adduction test and the Drop arm sign. According to [7] Painful arc, empty can test and external rotation resistance test are the best combination for the diagnosis of subacromial impingement syndrome. Painful arc and external rotation resistance are the best combination for ruling out subacromial impingement syndrome [7]. The initial management of shoulder impingement has traditionally included physical therapy, nonsteroidal anti-inflammatory drugs (NSAIDs), and corticosteroid injection. Conservative treatment is the first line of treatment, and should be considered for up to about a year until improvement and return to function are noticed [19]. Surgery should only be considered if the patient does not respond to exhaustive non-operative treatment [11]. The development of the treatment approach depends on several factors: age, activity level, general health status of the patient and the type of the impingement. The overall goal of the rehabilitation is to reduce pain and regain function of the upper limb. Most of the rehabilitation protocols are based on the different phases of the pathological process and recovery, respectively with different short and long-term therapeutic goals. The phases are: Phase I – maximal protection phase, Phase II – intermediate phase, Phase III - advanced strengthening phase and Phase IV - return to activity phase. The therapeutic goals for the first phase (Phase I) of the rehabilitation treatment are: relieving pain and inflammation, normalizing the range of motion (ROM), reestablishing the muscular balance, improvement of posture, patient education and avoidance of aggravating activities. The therapeutic goals for the second phase (Phase II) of the rehabilitation treatment are: re-establishing the non-painful

ROM, normalizing the athrokinematics of shoulder complex, normalizing the muscular strength, maintaining the reduced inflammation and pain, increasing activities with damaged arm. The therapeutic goals for the third phase (Phase III) of the rehabilitation treatment are: improvement of the muscular strength and endurance, maintaining the flexibility and ROM, maintaining the postural correction and gradual increasing of the functional activity level. The therapeutic goal for the third phase (Phase IV) of the rehabilitation treatment is an unrestricted symptom free activity of the patient. These goals are achieved through regular functional examinations and properly selected therapeutic interventions structured in rehabilitation protocol. Rehabilitation of the patient with shoulder impingement is a complex process that requires a comprehensive evaluation and multifactorial rehabilitation approach. In the aim of this, we propose a mathematical model based on the apparatus of Generalized nets theory of a possible rehabilitation algorithm protocol for patients with proximal humeral fractures.

2 The Apparatus of Generalized Nets

Generalized Nets (GNs, [1–3] were introduced as an extension of Petri Nets (see [18]) and the other their modifications and extensions. Similarly, to all other Petri nets type, they have places (here denoted by *l*), transitions (here denoted by *Z*), and tokens. In GNs the transitions are characterized with more complex structure, having input and output places, moment of activation, duration of the active status, a special matrix, called index matrix (IM, see [4, 5]) of transition condition predicates (here marked by *r*), IM of the capacities of transition arcs and type of the transition. The full definition of a GN contains a set of transitions, functions determining the priorities of the transitions and places, the capacities of the places, of the truth-values of the transition condition predicates, the moments of the transition activations and the duration of them; a set of tokens, their priorities and the moments in which they must enter the net; time-moment in which the GN will start function, the duration of its work and the elementary time-step with which the time will grow; a set of initial token characteristics, a function that give new characteristics of the tokens when they transfer from an input to an output place of a some transition and the maximal number of characteristics that token can have. As it is mentioned in [4, 5], a part of the GN-components can be omitted in respect of the aims of each one concrete model and a such net is called a reduced GN. In the present paper we construct a reduced GN-model of rehabilitation algorithm for patients with shoulder impingement syndrome. The model is based on the time period after the surgical intervention, functional outcomes of the patient and the rehabilitation goals.

3 Results: The Generalized Net Model

The GN model (Fig. 1) has 24 places and the following set of transitions:

$$A = \{Z_1, Z_2, Z_3, Z_4, Z_5, Z_6\}$$

These transitions describe the following processes:

- Transition Z_1 represents the personal data of the patients (age, gender, symptoms, etc.),
- Transition Z_2 represents the current functional status of the patient and the current time period of the development of the impingement,
- Transition Z_3 represents the first phase (Phase I) of the rehabilitation process,
- Transition Z_4 represents the second phase (Phase II) of the rehabilitation process,
- Transition Z_5 represents the third phase (Phase III) of the rehabilitation process,
- Transition Z_6 represents the fourth phase (Phase IV) of the rehabilitation process

The net contains six types of tokens: α, β, μ, η, γ and token φ. Some of the model transitions contain a so called "special place" where a token stays and collect information about the specific parts of the screening process and serves as a database storage, which it represents as follows:

- In place l_4, token α^1 stays permanently and collects the overall information obtained from the screening in the personal record (personal data),
- In place l_9, token β stays permanently and collects information about the current status of the patient obtained from the functional evaluation and the time period of the development of the impingement (tissue healing),
- In place l_{13}, token μ stays permanently and collects information about the rehabilitation plan during the Phase I,
- In place l_{17}, token η stays permanently and collects information about the rehabilitation plan during the Phase II,
- In place l_{21}, token γ stays permanently and collects information about the rehabilitation plan during the Phase III,
- In place l_{24}, token φ stays permanently and collects information about the rehabilitation plan during the Phase IV

At the time of duration of the GN-functioning, some of these tokens can split, generating new tokens, that will transfer in the net obtaining respective characteristics, and also in some moments they will unite with some of the tokens β, μ, η, γ and φ.

Token α enters the net with additional characteristics "*patient treated surgically for proximal humeral fracture*" in place l_1.

The transition Z_1 of the GN-model has the following form:

$$Z_1 = <\{l_1, l_4, l_8\}, \{l_2, l_3, l_4\}, r_1>,$$

where:

$r_1 =$

	l_2	l_3	l_4
l_1	*false*	*false*	*true*
l_4	$W_{4,2}$	$W_{1,3}$	*true*
l_8	*false*	*false*	*true*

and,

$W_{4,2}$ = "*the patient has no severe fever, no un resolving numbness/tingling, no uncontrolled pain, no mental health disorders, no high risk of cardiovascular accident*",

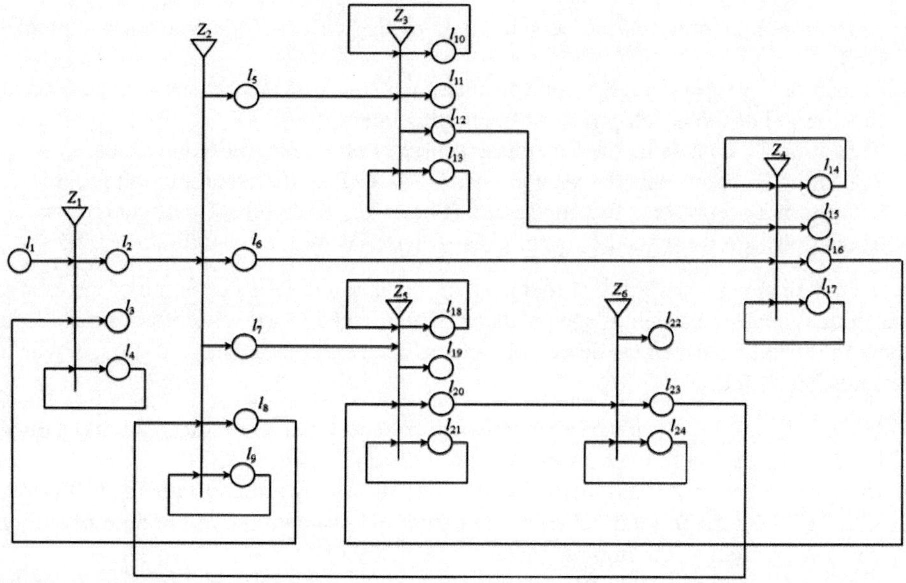

Fig. 1. Generalized Net model of rehabilitation algorithm for patients with shoulder impingement syndrome.

$W_{4,3} = $ "$\neg W_{4,2}$"

The tokens from the three input places of transition Z_1 enter place l_4 and unite with token α^1 with the above mentioned characteristic. On the next time-moment, token α^1 splits to three tokens – the same token α^1 and tokens α_1 *and* α_2. When the predicate $W_{4,2}$ is true, token α_1 enters place l_2 and there it obtains a characteristic:

> "*perform an initial functional examination and set the rehabilitation goals for the current patient*".

When the predicate $W_{4,2}$ is true, token α_2 enters place l_2 and there it obtains a characteristic:

> "*consider the presents of "red flags": ask for medical consultation and diagnostic imaging*"

The transition Z_2 has the following form:

$$Z_2 = \langle \{l_2, l_9, l_{23}\}, \ \{l_5, l_6, l_7, l_8, l_9\}, r_2 \rangle,$$

where:

$r_2 =$

	l_5	l_6	l_7	l_8	l_9
l_2	*false*	*false*	*false*	*false*	*true*
l_9	$W_{9,5}$	$W_{9,6}$	$W_{9,7}$	$W_{9,8}$	*true*
l_{23}	*false*	*false*	*false*	*false*	*true*

and,

$W_{9,5}$ = ***"The patient is in the acute phase (Phase I)"***;
$W_{9,6}$ = ***"The patient is in the intermediate phase (Phase II)"***;
$W_{9,7}$ = ***"The patient is in the advanced strengthening phase (Phase III)"***
$W_{9,8}$ = ***"the patient has undergone a course of rehabilitation for Phases I to IV"***

The tokens from the three input places of transition Z_2 enter place l_9 and unite with token β with the above mentioned characteristic. On the next time-moment, token β splits to five tokens – the same token β and tokens α_1, α_2, α_3 and α_4. When the predicate $W_{9,5}$ is true, token α_1 enters place l_5 and there it obtains a characteristic:

"*select the therapeutic interventions for the first phase of the rehabilitation process based on the current rehabilitation goals and the functional outcomes*".

When the predicate $W_{9,6}$ is true, token α_2 enters place l_6 and there it obtains a characteristic:

"*select the therapeutic interventions for the second phase of the rehabilitation process based on the current rehabilitation goals and the functional outcomes*".

When the predicate $W_{9,7}$ is true, token α_3 enters place l_7 and there it obtains a characteristic:

"*select the therapeutic interventions for the third phase of the rehabilitation process based on the current rehabilitation goals and the functional outcomes*"

When the predicate $W_{9,8}$ is true, token α_4 enters place l_8 and there it obtains a characteristic:

"*give patient a set of home based therapeutic exercises and sports activities*"

The transition Z_3 has the following form:

$$Z_3 = \langle \{l_5, l_{10}, l_{13}\},\ \{l_{10}, l_{11}, l_{12}, l_{13}\}, r_3 \rangle,$$

where:

$r_3 =$

	l_{10}	l_{11}	l_{12}	l_{13}
l_5	*false*	*false*	*false*	*true*
l_{10}	*true*	*true*	*true*	*true*
l_{13}	$W_{13,10}$	$W_{13,11}$	$W_{13,12}$	*true*

and,

$W_{13,10}$ = "*the patient has tolerable pain, abnormal scapulohumeral movement pattern or painful arc*"
$W_{13,11}$ = "*the patient has severe pain, difficult breathing, dizziness or lightheadedness or there is abnormal hemodynamics*";
$W_{13,12}$ = "*the patient has decreased pain and/or symptoms, normal ROM, no painful arc, improved muscular balance*";

The tokens from the three input places of transition Z_3 enter place l_{13} and unite with token μ with the above mentioned characteristic. On the next time-moment, token μ splits to four tokens – the same token μ and tokens α_1, α_2, and α_3. When the predicate $W_{13,10}$ is true, token α_1 enters place l_{10} and there it obtains a characteristic:

"*perform: cryotherapy, reflexology and segmental massage, inferior and posterior glides to the GH joint in scapular plane, rhythmic stabilization exercises for ER/IR, rhythmic stabilization drills Flex/Ext, external rotation strengthening and scapular strengthening exercises, strengthen scapular muscles (depressors, retractors & protractors), stretch pectoralis minor (corner stretch), wall circles*"

When the predicate $W_{13,11}$ is true, token α_2 enters place l_{11} and there it obtains a characteristic:

"*medical attention is needed*".

When the predicate $W_{13,12}$ is true, token α_3 enters place l_{12} and there it obtains a characteristic:

"*progress to the second phase of the rehabilitation process*".

The transition Z_4 has the following form:

$$Z_4 = \langle \{l_6, l_{12}, l_{14}, l_{17}\},\ \{l_{14}, l_{15}, l_{16}, l_{17}\}, r_4 \rangle,$$

where:

r_4 =

	l_{14}	l_{15}	l_{16}	l_{17}
l_6	*false*	*false*	*false*	*true*
l_{12}	*false*	*false*	*false*	*true*
l_{14}	*true*	*true*	*true*	*true*
l_{17}	$W_{17,14}$	$W_{17,15}$	$W_{17,16}$	*true*

and,

$W_{17,14}$ = "*the patient has undergone a course of rehabilitation for Phase I*";
$W_{17,15}$ = "*the patient has severe pain, difficult breathing, dizziness or lightheadedness or there is abnormal hemodynamics*";

$W_{17,16}$ = "*the patient has full non-painful ROM, no pain or tenderness, strength test fulfills criteria, satisfactory clinical examination*"

The tokens from the four input places of transition Z_4 enter place l_{17} and unite with token η with the above mentioned characteristic. On the next time-moment, token η splits to four tokens – the same token η and tokens α_1, α_2, and α_3. When the predicate $W_{17,14}$ is true, token α_1 enters place l_{14} and there it obtains a characteristic:

"*perform: exercises from the I phase, inferior, anterior and posterior glides, combined glides, self-capsular stretching, scapular neuromuscular exercises, complete shoulder strengthening exercise program,*".

When the predicate $W_{17,15}$ is true, token α_2 enters place l_{15} and there it obtains a characteristic:

"*medical attention is needed*".

When the predicate $W_{17,16}$ is true, token α_3 enters place l_{16} and there it obtains a characteristic:

"*progress to the third phase of the rehabilitation process*"

$$Z_5 = \langle \{l_7, l_{16}, l_{18}, l_{21}\},\ \{l_{18}, l_{19}, l_{20}, l_{21}\}, r_5 \rangle,$$

where:

$$r_5 = \begin{array}{c|cccc} & l_{18} & l_{19} & l_{20} & l_{21} \\ \hline l_7 & false & false & false & true \\ l_{16} & false & false & false & true \\ l_{18} & true & true & true & true \\ l_{21} & W_{21,18} & W_{21,19} & W_{21,20} & true \end{array}$$

and,

$W_{21,18}$ = "*the patient has undergone a course of rehabilitation for Phase II*";
$W_{21,19}$ = "*the patient has severe pain, difficult breathing, dizziness or lightheadedness or there is abnormal hemodynamics*";
$W_{21,20}$ = "*the patient has full non-painful ROM, no pain or tenderness, strength test fulfills criteria, satisfactory clinical examination*";

The tokens from the four input places of transition Z_5 enter place l_{21} and unite with token γ with the above mentioned characteristic. On the next time-moment, token γ splits to four tokens – the same token γ and tokens α_1, α_2, and α_3. When the predicate $W_{21,18}$ is true, token α_1 enters place l_{18} and there it obtains a characteristic:

"perform: exercises from the II phase, tubing ER/IR, exercises with weights, side lying ER, prone horizontal abduction, prone extension, wall slides, biceps/triceps resisting exercises, scapular neuromuscular control drills".

When the predicate $W_{21,19}$ is true, token α_2 enters place l_{19} and there it obtains a characteristic:

"medical attention is needed".

When the predicate $W_{21,20}$ is true, token α_3 enters place l_{20} and there it obtains a characteristic:

"progress to the fourth phase of the rehabilitation process"

The transition Z_6 has the following form:

$$Z_6 = \langle \{l_{20}, l_{24}\}, \ \{l_{22}, l_{23}, l_{24}\}, r_6 \rangle,$$

where:

$$r_6 = \begin{array}{c|ccc} & l_{22} & l_{23} & l_{24} \\ \hline l_{20} & false & false & true \\ l_{24} & W_{24,22} & W_{24,23} & true \end{array}$$

and,

$W_{24,22}$ = *"the patient has severe pain, difficult breathing, dizziness or lightheadedness or there is abnormal hemodynamics"*;
$W_{24,23}$ = *"the patient has adequate dynamic stability, appropriate rehabilitation progression to this point"*;

The tokens from the four input places of transition Z_6 enter place l_{24} and unite with token φ with the above mentioned characteristic. On the next time-moment, token φ splits to three tokens – the same token φ and tokens α_1 and α_2. When the predicate $W_{24,23}$ is true, token α_1 enters place l_{23} and there it obtains a characteristic:

"perform: exercises from the III phase, initiate interval sport program, interval return to normal activities".

When the predicate $W_{24,22}$ is true, token α_2 enters place l_{19} and there it obtains a characteristic:

"medical attention is needed".

4 Conclusions

The proposed model gives the possibility of development of more complex and detailed model of rehabilitation protocol for patients with shoulder impingement syndrome, allowing implementation in decision making support systems, telerehabilitation platforms, optimization of the physiotherapy protocols for shoulder pain rehabilitation and better rehabilitation strategies. The GN-model may provide a framework that can be used by physiotherapists and other medical specialists, to guide the rehabilitation treatment process for such patients, enabling more accurate and efficient management of that condition and would assist in optimizing patient outcomes. This model can be complicated and detailed, which will significantly improve the reliability of the proposed algorithm.

Acknowledgements. The authors are grateful for the support provided by the University "Prof. Dr. Asen Zlatarov"-Burgas, Scientific research sector "Scientific and artistic activity" approved by № NIH - 475 /2022 "Generalized nets models in orthopedic and traumatology rehabilitation".

References

1. Atanassov, K.: Theory of generalized nets (an algebraic aspect). Adv. Model. Simul. **1**(2), 27–33 (1984)
2. Atanassov, K.: Generalized Nets. World Scientific, Singapore (1991)
3. Atanassov, K.: On Generalized Nets Theory. Prof. M. Drinov Academic Publishing House, Sofia (2007)
4. Atanassov, K.: Generalized index matrices. Comptes rendus de l'Academie Bulgare des Sciences **40**(11), 15–18 (1987)
5. Atanassov, K.: Index Matrices: Towards an Augmented Matrix Calculus. Springer, Cham (2014)
6. Bhattacharyya, R., Edwards, K., Wallace, A.W.: Does arthroscopic sub-acromial decompression really work for sub-acromial impingement syndrome: a cohort study. BMC Musculoskelet. Disord. **15**, 324 (2014)
7. Chen, C.W., Pan, Z.E., Zhang, C., Liu, C.L., Chen, L.: Clinical research on the efficiency of physical examinations used for diagnosis of subacromial impingement syndrome. Zhongguo Gu Shang **29**(5), 434–438 (2016). (in Chinese)
8. Consigliere, P., Haddo, O., Levy, O., Sforza, G.: Subacromial impingement syndrome: management challenges. Orthop. Res. Rev. **23**(10), 83–91 (2018). https://doi.org/10.2147/ORR.S157864
9. Cools, A.M., Cambier, D., Witvrouw, E.E.: Screening the athlete's shoulder for impingement symptoms: a clinical reasoning algorithm for early detection of shoulder pathology. Br. J. Sports Med. **42**(8), 628–635 (2008)
10. Creech, J.A., Silver, S.: Shoulder impingement syndrome. In: Treasure Island (FL). StatPearls Publishing (2022)
11. Diercks, R., Bron, C., Dorrestijn, O., Meskers, C., Naber, R., de Ruiter, T., et al.: Guideline for diagnosis and treatment of subacromial pain syndrome. Acta Orthop. **85**(3), 314–322 (2014)
12. Escamilla, R.F., Hooks, T.R., Wilk, K.E.: Optimal management of shoulder impingement syndrome. Open Access J. Sports Med. **28**(5), 13–24 (2014). https://doi.org/10.2147/OAJSM.S36646
13. Koester, M.C., George, M.S., Kuhn, J.E.: Shoulder impingement syndrome. Am. J. Med. **118**(5), 452–455 (2005). https://doi.org/10.1016/j.amjmed.2005.01.040

14. Matsen, F.A., III., Arntz, C.T.: Subacromial impingement. In: Rockwood, C.A. Jr., Matsen, F.A. III. (eds). The Shoulder. II, pp. 623–646. Saunders, Philadelphia (1990)
15. Michener, L.A., Walsworth, M.K., Doukas, W.C., Murphy, K.P.: Reliability and diagnostic accuracy of 5 physical examination tests and combination of tests for subacromial impingement. Arch. Phys. Med. Rehabil. **90**(11), 1898–1903 (2009)
16. Neer, C.S.: 2nd Anterior acromioplasty for the chronic impingement syndrome in the shoulder: a preliminary report. J. Bone Joint Surg. Am. **54**(1), 41–50 (1972)
17. Palvanen, M., Kannus, P., Niemi, S.: Parkkari, Update in the epidemiology of proximal humeral fractures. J. Clin. Orthop. Relat. Res. **442**, 87–92 (2006)
18. Petri, C.-A.: Kommunication mit Automaten, Ph.D. dissertation, Univ. of Bonn, 1962; Schriften des Inst. fur Instrument. Math., No. 2, Bonn (1962)
19. Rhon, D.I., Boyles, R.E., Cleland, J.A., Brown, D.L.: A manual physical therapy approach versus subacromial corticosteroid injection for treatment of shoulder impingement syndrome: a protocol for a randomized clinical trial. BMJ Open (2011)

Application of the InterCriteria Analysis Aproach to a Burnout Syndrome Data

Evdokia Sotirova[1(✉)], Valentin Stoyanov[1,2], Sotir Sotirov[1], Zlatina Mirincheva[1,3], Hristo Bozov[1,4], and Todor Kostadinov[1]

[1] Prof. Asen Zlatarov University, 1 Prof. Yakimov Str., 8010 Burgas, Bulgaria
{esotirova,ssotirov}@btu.bg, {drvstoyanov,drmirincheva}@abv.bg
[2] Department of Ophthalmology and Otorhinolaryngology, Trakia University, Stara Zagora, Bulgaria
[3] Nephrology, UMBAL, Burgas, Bulgaria
[4] Oncology Complex Center - Burgas, 86 Demokratsiya Blvd, Burgas 8000, Bulgaria

Abstract. In this study a statistical data related to burnout syndrome among the medical employees was investigated. The data were collected through a survey among the staff in 5 medical centers in Bulgaria (2 University General Hospitals for Active Treatment, 1 General hospital for active treatment (municipal), and 1 Specialized rehabilitation hospital). The dependencies between the various parameters describing the studied objects are studied by InterCriteria approach (ICA).

Keywords: Intercriteria analysis method · Index matrix · Intuitionistic fuzzy sets · Burnout syndrome

1 Introduction

Burnout syndrome is the increased feeling of intellectual, emotional and physiological exhaustion. It is seen as a professional phenomenon that affects all aspects of the personality - physiological, emotional, behavioral, including the manifestation of suicidal moods, stroke, heart attack, colitis, ulcer, gastritis, obesity, migraine, asthma, sterility.

Research in the last few years covers an increasingly wide range of professions– doctors, nurses, police officers, teachers, researchers, etc. The National Association of General Practitioners reports that 70% of doctors in Bulgaria suffer from this syndrome.

In this paper, we will use the ICA [1, 2] method for detection of dependencies between parameters to establish the level of Burnaut syndrome among doctors and medical specialists in Bulgaria.

Using the ICA approach, the multicriteria decision making can be applied. The aim is to obtain from the matrix containing the data of the measurements of m in number evaluated objects - by n in number evaluation criteria, based on pairwise comparisons of objects and criteria, a new matrix with dimensions $n \times n$, providing in the form of intuitionistic fuzzy couples the correlations between any two criteria. The proposed approach calculates the degrees of correlation between all possible pairs of criteria in

K. T. Atanassov et al. (Eds.): BOS/SOR/IWIFSGN 2022, LNNS 793, pp. 227–236, 2023.
https://doi.org/10.1007/978-3-031-45069-3_21

the form of intuitionistic fuzzy pairs of values in the [0, 1]-interval [3], which means that there can be a rediscovery of relations already known from literature (and established by other methods), as well as a discovery of new, hitherto unknown, connections leading to the generation of new knowledge. The correlations between the criteria are called "positive consonance", "negative consonance" or "dissonance". Here we use the scale used in previous studies that is shown in [2].

The InterCriteria Analysis approach is applied over medical employees from 5 medical centers - two University Hospitals for Active Treatment (UHAT), one Municipal Hospital for Active Treatment (MHAT), and one Specialized Rehabilitation Hospital (SRH).

The ICA approach has been applied for analyzing data and decision making medical investigations [4, 5, 8–15], Genetic Algorithm [6, 7], and etc.

2 Materials and Methods

The data on the level of Burnout syndrome among employees in medical facilities for hospital care were collected by applying a survey method in five medical centers - three University Hospitals for Active Treatment (UHAT), one Municipal Hospital for Active Treatment (MHAT), and one Specialized Rehabilitation Hospital (SRH) in three different cities in Bulgaria. The questionnaire contains 64 questions. The first 10 questions are for obtaining a general information about the respondent: gender, age, profession, marital status, work experience, type of medical facility where the respondent works, duty hours. Each of the next 54 questions in the questionnaire is evaluated on a 5-point scale (1- completely disagree, 2–disagree, 3–unsure, 4–agree, 5–completely agree). These questions are related with the degree of an exhaustion and fatigue, depersonalization and personal accomplishments of the respondents. The next 54 questions in the questionnaire are connected with the exhaustion, inefficacy, depersonalization and personal accomplishments of the respondents.

The respondents are distributed as follows:

- UHAT 1 (in Burgas city)–11 responders (3 men, 8 women); 1 is under the age of 30, 3 are in the age range of 30–40, 5 are in the 41–50, 1 is in the 51–60, and 1 is over 61; 7 are married, 4 are single, divorced or widowed; 3 doctors with specialty, 2 doctors without specialty, 6 nurses, 0 rehabilitators; work experience: 3 with 1–5 years, 1 with 6–10 years, 0 with 11–15 years, 2 with 16–20 years, 1 in 21–25 years, 1 in 26–30 years, 3 over 30 years; 11 with surgical profile, 0 with therapeutic profile; 9 with day and night shifts; 7 with duty on Saturday/Sunday and 5 with 24 h availability;
- UHAT 2 (in Plovdiv city)–39 responders (20 men, 19 women); 9 is under the age of 30, 13 are in the age range of 30–40, 7 are in the 41–50, 10 is in the 51–60, and 0 over 61; 21 are married, 18 are single, divorced or widowed; 15 doctors with specialty, 16 doctors without specialty, 8 nurses, 0 rehabilitators; work experience: 12 with 1–5 years, 7 with 6–10 years, 6 with 11–15 years, 0 with 16–20 years, 6 in 21–25 years, 4 in 26–30 years, 4 over 30 years; 31 with surgical profile, 8 with therapeutic profile; 35 with day and night shifts; 31 with duty on Saturday/Sunday and 33 with 24 h availability;

- UHAT 3 (in Plovdiv city)–18 responders (6 men, 12 women); 5 are under the age of 30, 5 are in the age range of 30–40, 6 are in the 41–50, 1 is in the 51–60, and 1 is over 61; 8 are married, 10 are single, divorced or widowed; 2 doctors with specialty, 7 doctors without specialty, 9 nurses, 0 rehabilitators; work experience: 5 with 1–5 years, 2 with 6–10 years, 4 with 11–15 years, 0 with 16–20 years, 3 in 21–25 years, 2 in 26–30 years, 2 over 30 years; 18 with surgical profile, 0 with therapeutic profile; 14 with day and night shifts; 15 with duty on Saturday/Sunday and 5 with 24 h availability;
- MHAT–30 responders (7 men, 23 women); 2 are under the age of 30, 4 are in the age range of 30–40, 8 are in the 41–50, 9 are in the 51–60, and 7 are over 61; 16 are married, 14 are single, divorced or widowed; 7 doctors with specialty, 8 doctors without specialty, 15 nurses, 0 rehabilitators; work experience: 1 with 1–5 years, 1 with 6–10 years, 4 with 11–15 years, 4 with 16–20 years, 2 in 21–25 years, 5 in 26–30 years, 13 over 30 years; 17 with surgical profile, 13 with therapeutic profile; 26 with day and night shifts; 16 with duty on Saturday/Sunday and 20 with 24 h availability;
- SRH–41 responders (2 men, 39 women); 0 are under the age of 30, 2 are in the age range of 30–40, 16 are in the 41–50, 17 are in the 51–60, and 6 are over 61; 14 are married, 27 are single, divorced or widowed; 0 doctors with specialty, 2 doctors without specialty, 14 nurses, 25 rehabilitators; work experience: 0 with 1–5 years, 0 with 6–10 years, 4 with 11–15 years, 0 with 16–20 years, 6 in 21–25 years, 12 in 26–30 years, 19 over 30 years; 11 with surgical profile; 39 with day and night shifts; 0 with duty on Saturday/Sunday and 5 with 24 h availability (Figs. 1, 2, 3, 4, 5, 6 and 7).

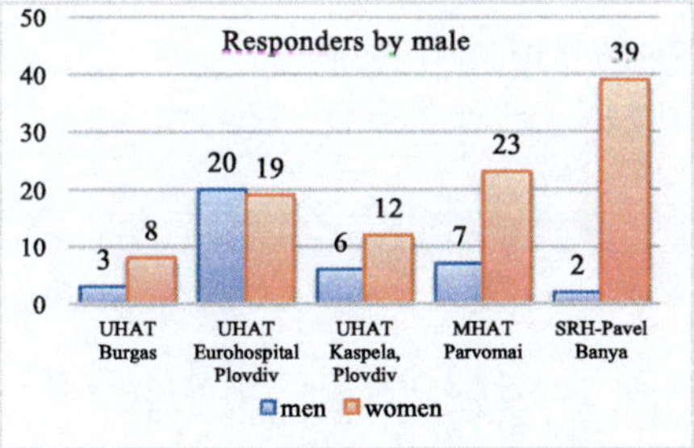

Fig. 1. Distribution of the responders by male

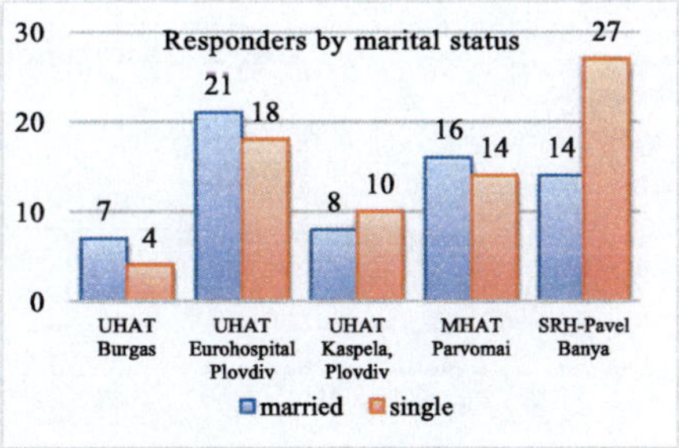

Fig. 2. Distribution of the responders by marital status

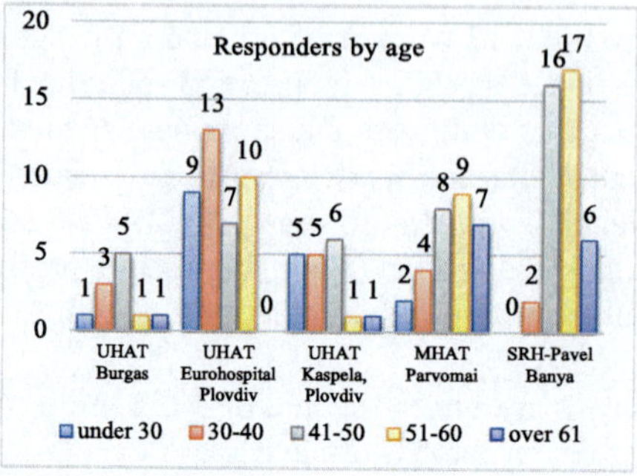

Fig. 3. Distribution of the responders by age

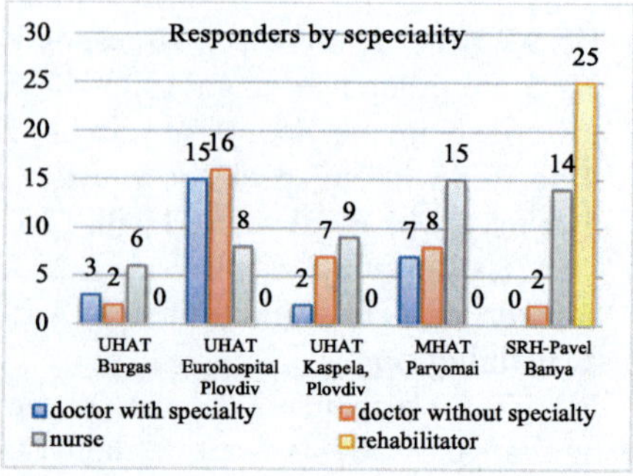

Fig. 4. Distribution of the responders by scpeciality

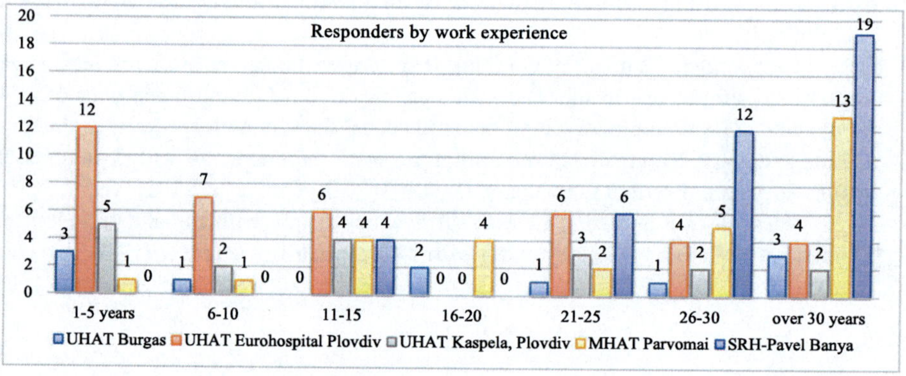

Fig. 5. Distribution of the work experience

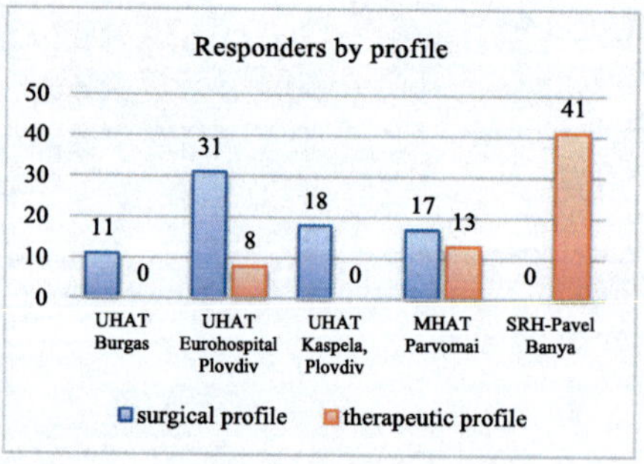

Fig. 6. Distribution of the responders by profile

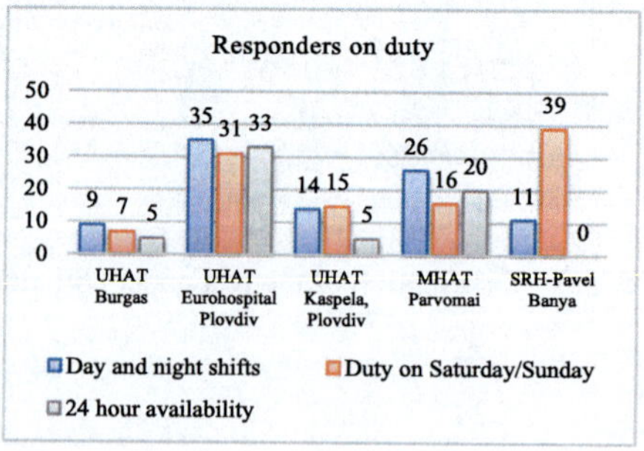

Fig. 7. Distribution of the responders on duty

3 An application of the ICA

In our research, we are interested in discovering patterns and trends between health parameters related to burnout syndrome of the investigated respondents. For this purpose, we used the developed ICA software (freely available online at: http://intercriteria.net/software) consistently for the five medical centers. The obtained results are presented below in this section. Five index matrices that contain 64 rows (for questions in the questionnaire) and 11 columns (for responders) were used. After each testing with the software, two tables with membership part and non-membership part of the 4096 intuitionistic fuzzy pairs were obtained. This IFPs present the relations between every pair of questions in the questionnaire (criteria).

For interpretation the obtained results we have used the scale for determining of consonance or dissonance between pairs of criteria presented in [4] (Fig. 8): Strong Negative Consonance (SNC) [0; 0,05), Negative Consonance (NC) [0,05; 0,15), Weak Negative Consonance (WNC) [0,15; 0,25), Weak Dissonance (WD) [0,25; 0,33), Dissonance (D) [0,33; 0,43), Strong Dissonance (SD) [0,43; 0,57), Dissonance (D) [0,57; 0,67), Weak Dissonance (WD) [0,67; 0,75), Weak Positive Consonance (WPD) [0,75; 0,85), Positive Consonance (PC) [0,85; 0,95), Strong Positive Consonance (SPC) [0,95; 1].

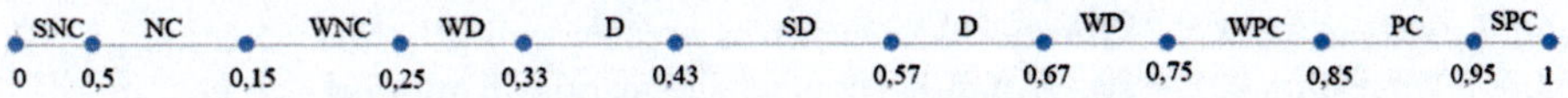

Fig. 8. Scale for determining of consonance or dissonance between pairs of criteria

3.1 Applying ICA approach for University Hospital for Active Treatment Burgas

The visualization of the obtained IFPs for the UHAT-1 in the IF triangle is presented on Fig. 9.

In SPC there are seven pairs:

- Five pairs of criteria with evaluation ⟨1; 0⟩; : "Type of specialty"- "Type Medical facility for hospital care"; "Question 43. Do you notice a decrease in body weight?"- "Question 51. Do you resort to using more "intoxicating" substances to affect your mood? (alcohol; tranquilizers)?"; "Question 50. Do you feel more helpless than before?"- "Question 52. Do you feel softer?"; "Question 50. Do you feel more helpless than before?"- "Question 55. Do you notice even a partial loss of your sense of humor?"; "Question 52. Do you feel softer?"- "Question 55. Do you notice even a partial loss of your sense of humor?";
- Two pairs of criteria with evaluation ⟨0,964; 0⟩;: "Question 7. I feel at the end of my strength."- "Question 11. I feel that patients blame me for their problems."; "Question 9. I feel frustrated with my work."- "Question 38. Do you lose interest in your work?".

In WPC there are fourteen pairs.

The other pairs are in WD, D, SD, WNC, NC or SNC which means that they are not dependent on each other.

Fig. 9. UHAT Burgas

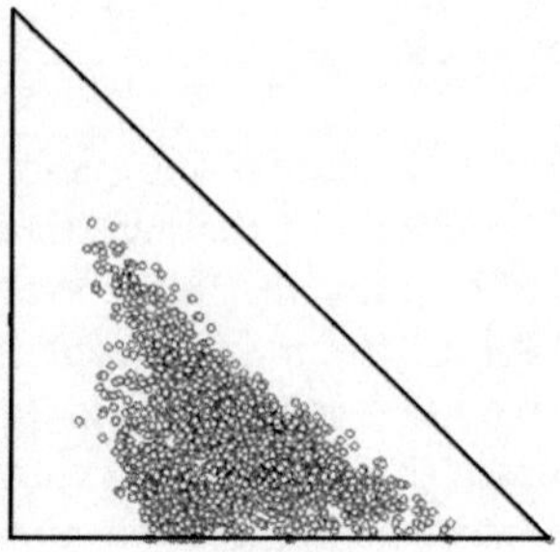

Fig. 10. UHAT Eurohospital

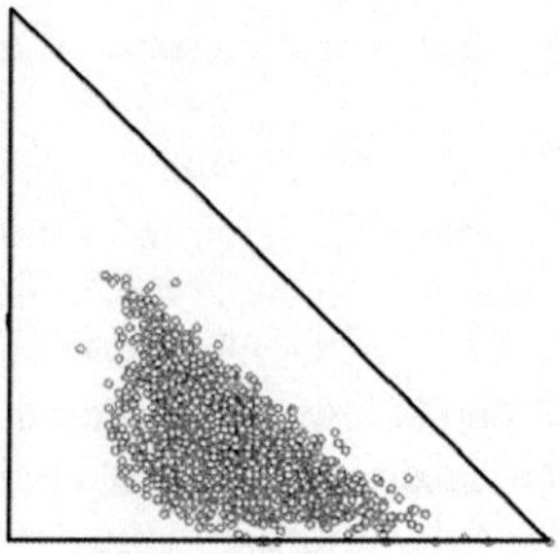

Fig. 11. UHAT Kaspela

3.2 Applying ICA approach for University Hospital for Active Treatment Eurohospital

The visualization of the obtained IFPs for the UHAT-2 in the IF triangle is presented on Fig. 10.

In WPC there are eleven pairs: "Do you give shifts (day, night)"- "Do you give shifts (Saturday Sunday)" with evaluation ⟨0,811; 0⟩; "Question 39. Is stress increasing in your activity at work?"- "Question 41. Do you have a headache?" with evaluation ⟨0,799; 0,006⟩; "Question 4. The stress of direct contact with people in my work "comes" more."- "Question 46. Do you often change your mood?" with evaluation ⟨0,802; 0,018⟩; "Question 46. Do you often change your mood?"- "Question 47. Are you easily irritated?" with evaluation ⟨0,814; 0,018⟩; "Age"- "Work experience" with evaluation ⟨0,773; 0,031⟩; "Question 31. Difficulties related to specializations and scientific work stress me out."- "Question 41. Do you have a headache?" with evaluation ⟨0,76; 0,024⟩; "Question 6. I feel like my job is destroying me."- "Question 7. I feel at the end of my strength." with evaluation ⟨0,781; 0,051⟩; "Question 30. The negative impact of work on my family relationships stresses me out"- "Question 31. Difficulties related to specializations and scientific work stress me out." with evaluation ⟨0,738; 0,016⟩; "Question 25. The fear of sanctions (NHOC, hospital management, RZI) stresses me out."- "Question 54. Do you think you work more and achieve less?" with evaluation ⟨0,75; 0,03⟩; "Question 2. I feel like I'm working too much."- "Question 31. Insufficient (in my opinion) time for daily and weekly rest stresses me out." with evaluation ⟨0,757; 0,049⟩; "Question 4. The stress of direct contact with people in my work "comes" more."- "Question 39. Is stress increasing in your activity at work?" with evaluation ⟨0,751; 0,054⟩; "Question 9. I feel frustrated with my work."- "Question 43. Do you notice a decrease in body weight?" with evaluation ⟨0,752; 0,076⟩;.

The other pairs are in WD, D, SD, WNC, NC or SNC.

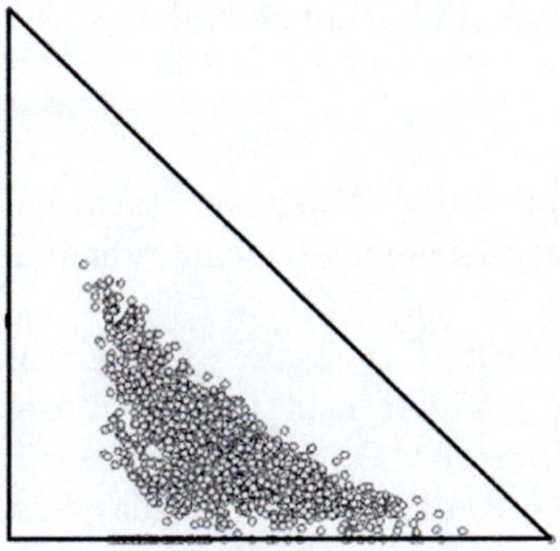

Fig. 12. MHAT Parvomai

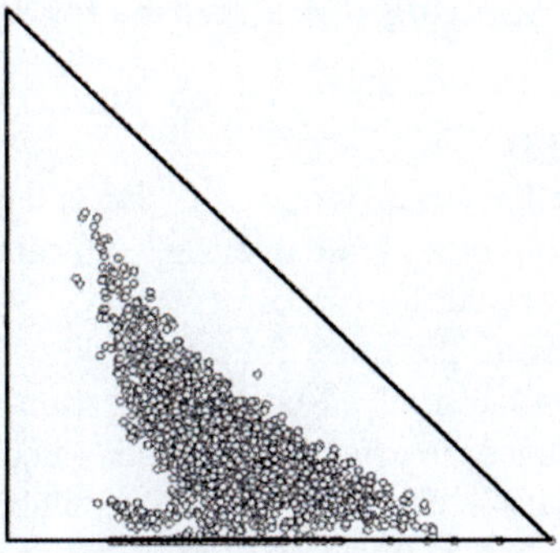

Fig. 13. SRH-Pavel Banya

3.3 Applying ICA approach for University Hospital for Active Treatment Kaspela

The visualization of the obtained IFPs for the UHAT-3 in the IF triangle is presented on Fig. 11.

In PC there is one pair: "Do you give shifts (day, night)"- "Do you give shifts (Saturday Sunday)" with evaluation ⟨0.889; 0⟩;.

In WPC there are five pairs: "Age"- "Work experience" with evaluation ⟨0,843; 0,007⟩; "Question 6. I feel like my job is destroying me."- "Question 7. I feel at the end of my strength." with evaluation ⟨0,791; 0⟩; "Question 37. Do you lose some of your initiative at work?"- "Question 49. Are you more suspicious than before?" with evaluation ⟨0,758; 0,026⟩; "Question 36. Does it reduce your work efficiency?"- "Question 43. Do you notice a decrease in body weight?" with evaluation ⟨0,771; 0,052⟩; "Question 17. I feel that through my work I positively influence the health and quality of life of my patients."- "Question 18. I feel that I have achieved very significant results in my work." with evaluation ⟨0,752; 0,059⟩;.

The other pairs are in WD, D, SD, WNC, NC or SNC.

3.4 Applying ICA approach for Municipal Hospital for Active Treatment

The visualization of the obtained IFPs for the MHAT in the IF triangle is presented on Fig. 12.

In WPC there are seven pairs: "Age"- "Work experience" with evaluation ⟨0,839; 0,016⟩; "Question 22."- "Question 24." with evaluation ⟨0,8; 0⟩; "Question 32. Do you lose some of your initiative at work?"- "Question 34." with evaluation ⟨0,791; 0,018⟩; "Question 32. Do you lose some of your initiative at work?"- "Question 33." with evaluation ⟨0,754; 0,014⟩; "Question 29."- "Question 30." with evaluation ⟨0,795; 0,048⟩; "Type Medical facility for hospital care"-"Do you give shifts (day, night)" with evaluation ⟨0,761; 0,0⟩;.

The other pairs are in WD, D, SD, WNC, NC or SNC.

3.5 Applying ICA approach for Specialized Rehabilitation Hospital

The visualization of the obtained IFPs for the SRH in the IF triangle is presented on Fig. 13.

In SPC there is one pair: “Do you give 24 h’ availability”-”Question 51. Do you resort to using more “intoxicating” substances to affect your mood? (alcohol; tranquilizers)” with evaluation ⟨1; 0⟩;.

In PC there are four pairs with evaluation ⟨0,905; 0⟩;: “Male”- “Do you give 24 h’ availability”, “Male”- “Question 51. Do you resort to using more “intoxicating” substances to affect your mood? (alcohol; tranquilizers)”, “Do you give shifts (Saturday Sunday)”- “Do you give 24 h’ availability” and “Do you give shifts (Saturday, Sunday)”- “Question 51. Do you resort to using more “intoxicating” substances to affect your mood? (alcohol; tranquilizers)”.

In WPC there are nine pairs: “Male”-“Do you give 24 h’ availability” with evaluation ⟨0,82; 0⟩; “Speciality”-”Do you provide shifts (day, night)” with evaluation ⟨0,771; 0⟩; “Question 3. I feel like I’m already emotionally drained from my job”- “Question 6. I feel like my job is destroying me.” with evaluation ⟨0,78; 0,008⟩; “Question 10. I do my daily work with patients with reluctance.”- “Question 36. Does it reduce your work efficiency?” with evaluation ⟨0,78; 0,02⟩; “Question 6. I feel like my job is destroying me.”- “Question 7. I feel at the end of my strength.” with evaluation ⟨0,777; 0,024⟩; “Question 46. Do you often change your mood?”- “Question 47. Are you easily irritated?” with evaluation ⟨0,766; 0,024⟩; “Question 49. Are you more suspicious than before?”- “Question 50. Do you feel more helpless than before?” with evaluation ⟨0,75; 0,027⟩; “Question 20. I feel full of energy.”- “Question 21. I feel fresh after constantly working with patients.” with evaluation ⟨0,767; 0,057⟩;.

The other pairs are in WD, D, SD, WNC, NC or SNC.

4 Conclusion

The InterCriteria Analysis method was applied to the study of the burnout syndrome among doctors in 5 hospitals University Hospitals for Active Treatment (UHAT), one Municipal Hospital for Active Treatment (MHAT), and one Specialized Rehabilitation Hospital (SRH) in three different cities in Bulgaria. A burnout survey questionnaire was developed.

The results are commented from different points of view: relationship between gender and age of patients and relationship between gender and marital status, degree of an exhaustion and fatigue, depersonalization and personal accomplishments of the respondents.

In this way, disease can be predicted and dependencies between individual parameters related to constraints such as time and resources can be sought.

Acknowledgment. This research was funded in part by the European Regional Development Fund through the Operational 268 Programme “Science and Education for Smart Growth" under contract UNITe . BG05M2OP001–1.001–0004 269 (2018–2023).”

References

1. Atanassov, K., Mavrov, D., Atanassova, V.: Intercriteria decision making: a new approach for multicriteria decision making, based on index matrices and intuitionistic fuzzy sets. Issues Intuitionistic Fuzzy Sets Generalized Nets **11**, 1–8 (2014)
2. Atanassov, K., Atanassova, V., Gluhchev, G.: InterCriteria analysis: ideas and problems. Notes Intuitionistic Fuzzy Sets **21**(1), 81–88 (2015)
3. Atanassov, K., Szmidt, E., Kacprzyk, J.: On intuitionistic fuzzy pairs. Notes Intuitionistic Fuzzy Sets **19**(3), 1–13 (2013)
4. Hristov, S., Baltov, A., Sotirova, E., Bozov, H.: Intuitionistic Fuzzy Evaluations for Analysis of the Proximal Humerus Fractures. In: Sotirov, S.S., Pencheva, T., Kacprzyk, J., Atanassov, K.T., Sotirova, E., Staneva, G. (eds) Contemporary Methods in Bioinformatics and Biomedicine and Their Applications. BioInfoMed 2020. Lecture Notes in Networks and Systems, vol. 374, pp. 279–284. Springer, Cham. (2022). https://doi.org/10.1007/978-3-030-96638-6_30
5. Krumova, S., et al.: Intercriteria analysis of calorimetric data of blood serum proteome. Biochimica et Biophysica Acta – General Subjects, vol. 1861, no. 2 (2017). https://doi.org/10.1016/j.bbagen.2016.10.012. ISSN: 0304–4165
6. Pencheva, T., Angelova, M.: InterCriteria analysis of simple genetic algorithms performance. In: Georgiev, K., Todorov, M., Georgiev, I. (eds.) Advanced Computing in Industrial Mathematics. Studies in Computational Intelligence, vol. 681, pp. 147–159. Springer, Cham (2017).https://doi.org/10.1007/978-3-319-49544-6_13
7. Roeva, O., Vassilev, P., Fidanova, S., Paprzycki, M.: InterCriteria analysis of genetic algorithms performance. In: Fidanova, S. (ed.) Recent Advances in Computational Optimization. SCI, vol. 655, pp. 235–260. Springer, Cham (2016). https://doi.org/10.1007/978-3-319-40132-4_14
8. Sotirova, E., Bozova, G., Bozov, H., Sotirov, S., Vasilev, V.: Application of the intercriteria analysis method to a data of malignant melanoma disease for the Burgas region for 2014–2018. In: Atanassov, K.T., et al. (eds.) IWIFSGN 2019 2019. AISC, vol. 1308, pp. 166–174. Springer, Cham (2021). https://doi.org/10.1007/978-3-030-77716-6_15
9. Sotirov, S., Bozova, G., Vasilev, V., Krawczak, M.: Clustering of intercriteria analysis data using a malignant neoplasms of the digestive organs data. In: Atanassov, K.T., et al. (eds.) IWIFSGN 2019 2019. AISC, vol. 1308, pp. 193–201. Springer, Cham (2021). https://doi.org/10.1007/978-3-030-77716-6_18
10. Sotirov, S., et al.: Application of the intuitionistic fuzzy intercriteria analysis method with triples to a neural network preprocessing procedure. Comput. Intell. Neurosci. 9. Hindawi (2017). https://doi.org/10.1155/2017/2157852.Article ID 2157852,
11. Sotirov, S., Atanassova, V., Sotirova, E., Bureva, V., Mavrov, D.: Application of the intuitionistic fuzzy intercriteria analysis method to a neural network preprocessing procedure. In: 9th Conference of the European Society for Fuzzy Logic and Technology (EUSFLAT), 30.06–03.07.2015, Gijon, Spain, pp. 1559–1564 (2015). https://doi.org/10.2991/ifsa-eusflat-15.2015.222
12. Sotirova, E., Vasilev, V., Bozova, G., Bozov, H., Sotirov, S.: Application of the intercriteria analysis method to a dataset of malignant neoplasms of the digestive organs for the Burgas region for 2014–2018. In: 2019 Big Data, Knowledge and Control Systems Engineering (BdKCSE), pp. 1–6. IEEE (2019)
13. Todinova, S., et al.: Blood plasma thermograms dataset analysis by means of intercriteria and correlation analyses for the case of colorectal cancer. Int. J. Bioautomation **20**(1), 115–124 (2016)

14. Zaharieva, B., Doukovska, L., Ribagin, S., Michalíková, A., Radeva, I.: Intercriteria analysis of behterev's kinesitherapy program. Notes Intuitionistic Fuzzy Sets **23**(3), 69–80 (2017)
15. Zaharieva, B., Doukovska, L., Ribagin, S., Radeva, I.: InterCriteria approach to Behterev's disease analysis. Notes Intuit. Fuzzy Sets **23**(2), 119–127 (2017)

Decision Making, Optimization and Problem Solving

On the Use of 'Ideal Structures' in Opinion Profile Identification

Jan W. Owsiński(✉)

Systems Research Institute, Polish Academy of Sciences, Newelska 6, 01-447 Warszawa, Poland
owsinski@ibspan.waw.pl

Abstract. The concept of 'ideal structures', introduced by the author in some preceding studies, is shortly presented along with some considerations, referring to the potential use and significance in multi-modal aggregation of observations, generated through some mechanism, which offers by itself little insight into the potential structure of the resulting data set. An example very much to the point is opinion analysis, based, e.g., on some survey, involving several response dimensions. The aim is to gain knowledge of the structure of the set of opinions with emphasis on the potential variety of coherent opinions. Connections with cluster analysis are indicated and the distinction from this domain is also shown. Some preliminary algorithmic considerations are provided as well.

Keywords: opinion aggregation · distance · similarity · clusters · opinion profiles

1 Introduction: The Problem Considered

We are addressing in this paper an apparently quite common situation of a set of multi-dimensional objects, denoted x_i, $i = 1,\ldots,n$, where n is the number of such objects, and we wish to gain some kind of knowledge, concerning the structure of the entire set of x_i, denoted X. We are particularly interested in the potential existence of well-pronounced subsets of objects, such that inside these subsets the objects are possibly mutually similar. This sounds almost exactly as the essential problem of cluster analysis, namely: to divide X into subsets in such a way that objects, belonging to the same subset (cluster) are possibly similar, while objects belonging to different subsets (clusters) are possibly dissimilar.

The essential difference is, though, that we do not aim here at the complete division of the set X, but primarily at the identification of existence of the potential "clusters", which, therefore, do not have to exhaust X. It is our hope, of course, that these identified "clusters" shall account for the decisive majority of objects (ultimately, after all, exhausting the set X), but, actually, we cannot assume anything about this.

Our willingness to study this problem is motivated by the scenario, in which x_i represent some kind of "opinions", and we are interested not so much in determining an aggregate for the entire X as in exactly identifying the clearly consistent groups of similar "opinions" ("opinion profiles").

K. T. Atanassov et al. (Eds.): BOS/SOR/IWIFSGN 2022, LNNS 793, pp. 239–249, 2023.
https://doi.org/10.1007/978-3-031-45069-3_22

Actually, within the framework of this setting, we may pose the proposition that, ultimately, a single aggregate may be, and indeed often is, completely off the point in terms of representing the given set of objects. While, of course, the case of, say, national presidential elections is out of question (the "model", including the rules and purpose, being different), a single aggregate may very well not represent anything sensible exactly in the political arena – see the deep divisions between the political parties and their partisans. Hence, the considerations, concerning the social choice theory or the voting theory, like those presented in references [1, 13] or [19] are not applicable here (nor is the search for the "median order" of Kemeny's, [9]).

Our goal, namely, is to better catch the actual variety of tangible opinions, where "tangible" means being represented by at least some perceptible or significant proportion of the set. This vague statement leads to quite some practical consequences:

- the number of the partial ("clusterwise") aggregates cannot be too high, for both the more pragmatic reasons (what knowledge do we gain from a multitude of such aggregates?) and the more substantive ones (if the respective clusters are small, what is their ontological status?)
- we are ready to neglect a part of the opinions from X which do not seem to form clusters, being, in particular, very far from any of them (like when the opinions are internally inconsistent, or are very "strange"); the proportion neglected should not, of course, exceed some threshold, but this threshold cannot be set a priori, in general.

Hence, what we need at this point is the appropriate definition of similarity or distance, preferably applying not only to single objects, but also to their sets, and some sort of criterion for accepting the groups of opinions, if we do not refer directly to the methods of cluster analysis.

At the end of this introduction we wish to provide an illustration, which is shown Fig. 1. The illustration of Fig. 1 implies very clearly the existence of distinct groups of opinions (between 3 and 5 of such groups), as well as some "outliers", which would not be accounted for when trying to gain insight into the "structure of opinions". This illustration provides also other important hints, related to the problem here considered, namely:

- there is a possibility of existence of some kind of "overall aggregate", possibly of a different nature than that routinely looked for – here in the form of price-quality relation, which would have to be identified through different kind of approaches;
- the exercise, leading to this kind of results, has to be very well prepared, with a lot of knowledge on the side of responding participants, including such items as: (i) appropriate scaling of values (normalization, reference points, etc.) for the variables accounted for; (ii) securing of comparability among opinions; (iii) possibly including checks for internal consistency; (iv) adequate choice of samples (responding persons, objects of evaluation, etc.); all this adds up to formation of a quite specific context of the problem here considered.

In these circumstances, not only the search for the global aggregate is out of question, but also more specific methodologies, oriented at the reconstruction of group-wise rankings, preference relations etc., like those analyzed and methodologically equipped in the earlier works of the present author and associates, see [4, 14, 15], since they actually

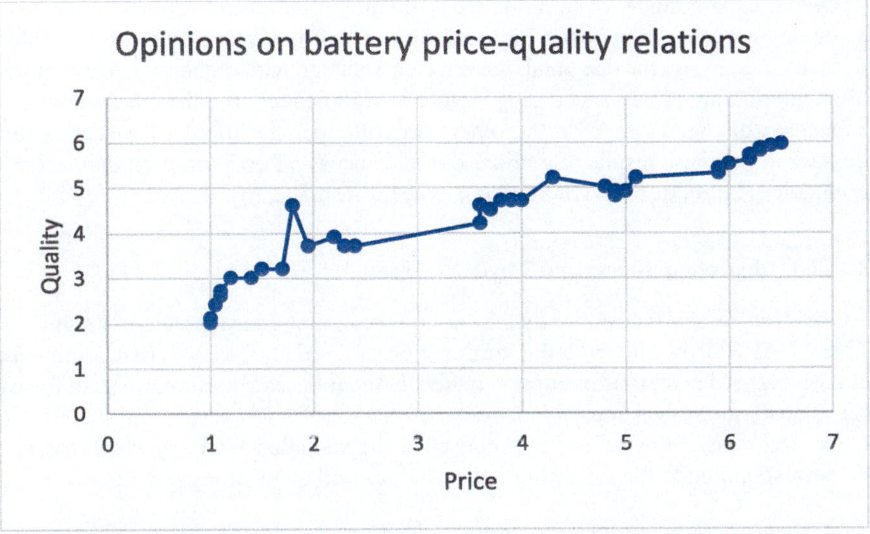

Fig. 1. An illustration for the case considered: a small set of opinions (dots) on the relation of prices and qualities for electric batteries.

aim at a well-defined structure inside the individual groups, and assume the complete partition of the set of objects X.

2 The Outline of Approach

2.1 The General Setting

Thus, we deal with a set of objects (opinions), denoted x_i, $i = 1,\ldots,n$, this set being denoted X and the set of indices being denoted I. Each object description x_i is a vector of values of definite variables. We shall generally assume that we deal with a multivariate situation, although the problem may be quite non-trivial also for unidimensional cases (see the case of a murder trial, mentioned in a previous paper of this author, [17]). We shall denote the subsets of the objects from X with A, B, C,… or A_q, $q = 1, 2, \ldots$ With P we shall denote a partition of X (or, equivalently, of I) into the subsets A_q, $q = 1, 2,\ldots,p$, meaning, at least, that together the subsets from A_1 to A_p exhaust the set X (or I).

We assume also, of course, that we dispose of a distance measure $d(.,.)$, with values $d(x_i,x_j)$, or, simpler, d_{ij}, for $i, j \in I$. It is of little importance what this measure exactly is, as long as it satisfies the basic properties of distance. No other notion shall be used to deal with similarity and dissimilarity at the level of individual objects (opinions).

Regarding the groups of opinions, though, we shall also be considering the "diameters" of the groups A_q, $d^q = \max d_{ij}$, $i, j \in A_q$ (the "diameter" of the entire set being denoted d^X), as well as the distances between groups, denoted $d^{qq'}$ for groups A_q and $A_{q'}$, and defined as $\min d_{ij}$, $i \in A_q, j \in A_{q'}$ (actually, $d^{qq'}$ could also be defined in some other manner, like, e.g. the distance between the means or medians of the respective groups).

Given the main model of situation we deal with ("opinion collection and analysis") we assume that n is not very big, say – at the order of thousands rather than millions. This is justified by, on the one hand, the effort, associated with organization and running of the undertaking, aimed at securing results of significance (see the conclusions from the example of Fig. 1), and, on the other hand, the technicalities of the data analysis necessary to get these results (e.g. the use of distances and operations on them). For an introductory text on this kind of situation, see, for instance, [5].

2.2 The 'Ideal Structures' and Their Meaning

In this section we shall introduce the key notions of our approach, previously signaled in [17] and [16]. Side by side with the introduction of these notions, we shall comment on them, with regard to their properties, and then, in the following section, we shall forward some algorithmic suggestions.

The very first of the structures proposed is the so-called 'globally ideal structure', GIS, corresponding to the following quite obvious and quite strong condition:

"*any distance between the opinions, which belong to the same group (cluster), is smaller than any of the distances between the opinions from different groups*",

this intuitive requirement being expressed as:

$$\max_q d^q < \min_{q \neq q'} d^{qq'}. \tag{1}$$

It appears obvious that if subsets (clusters) A_q, corresponding to the condition (1), form a partition of X (or I), then we can consider this a partition into truly cohesive and similar groups of opinions. (It would also be accepted if (1) were satisfied for a weak inequality. In further course of our considerations we shall not be distinguishing the two.) Yet, the strength and significance of (1) is just an impression, to which quite some reservations ought to be ascribed.

Actually, this kind of condition has been formulated long ago as a basis for clustering (see, e.g., [7] and [8]) and can be considered to constitute an attempt at 'defining a cluster' or 'defining clusters'. As we know, such attempts were abandoned to the advantage of the approaches, in which clusters are not explicitly defined, but determined through some heuristic procedure. Regarding the particular proposal of (1) this failure is explained by two simple facts:

1. if we assume that $d(x,x) = 0$ for any x, i.e. $d_{ii} = 0$ for any $i \in I$, which is rather obvious, then (1) is satisfied for the partition, formed by n singular, one-object "clusters" ($p = n$); let us denote this partition P^0;

2. Then, given the partition P^0 as above, any partition, obtained from it through merging of the pairs of mutually the closest objects, i.e. all pairs ($i*$,$j*$) such that.

$$\min_j d_{i*j} = d_{i*j*} = \min_i d_{ij*} \tag{2}$$

does also satisfy (1).

Thus, the structure, implied by (1) is by no means unique, even if it still preserves the intuitive appeal as to its properties, important for the purpose here aimed at.

On the other hand, though, it is, though, quite strict and demanding, and one can easily imagine that for a lot of empirical circumstances it would yield only few groups (other than the trivial ones, implied by points 1 and 2 above, if at all), possibly accounting for only a minor proportion of the data sets considered.

Hence, we may wish to somewhat alleviate the condition (1), and for this purpose we can take another intuitively obvious, but this time more local condition, which is at the object level expressed as:

$$\forall i : j' \in A(i) \wedge j'' \notin A(i) \Rightarrow d_{ij'} < d_{ij} \tag{3}$$

where $A(i)$ denotes the group, to which opinion i was assigned, condition (3) verbally corresponding to: "*for any object, all its distances to objects from the same cluster are smaller than any distance to objects from different (other) clusters*". Condition (3) actually implies another formulation.

$$\forall q : i, j \in A_q, j' \notin A_q : d_{ij} < d_{ij} \tag{4}$$

that is, the one, in which we perceive the level of groups or clusters: "*for every cluster, the distances inside the cluster are all smaller than all of the distances between the objects from that cluster and objects outside of it*".

Still another formulation of essentially the same condition, even more explicitly at the cluster level, takes on the form.

$$\forall q : \max_{i,j \in A_q} d_{ij} < \min_{i \in A_q, j' \notin A_q} d_{ij}. \tag{5}$$

Now, we can yet simplify the formulation of this condition, obtaining.

$$\forall q : d^q < \min_{q'} d^{qq'}. \tag{6}$$

That is: "*for every cluster, its diameter is smaller than any of the distances between this cluster and any other cluster*". Although the latter formulation is not strictly equivalent to the one we started with, it entails, in fact, the same consequences, considering that the remaining clusters (other than A_q) may be of any configuration, including the partition of the remaining part of the set X into single-object clusters.

The condition, expressed through formulae (3) through (6) will be referred to as LIS (local ideal structures), even if they differ not only as to the form, but also as to their strength. This class of local conditions seems to be altogether similar to (1), but is actually not (a series of simple pertinent negative graphical examples is provided in both [16] and [17]).

Having the above, we can also formulate a condition "dual" to LIS at the cluster level, namely:

$$\forall q \neq q' : \max\left\{d^q, d^{q'}\right\} < d^{qq'}. \tag{7}$$

This condition ("*for any two clusters their distance is bigger than the diameters of any of the two*") can be derived directly from (3).

Once we have the condition (7), we can also formulate a less demanding condition, also at the cluster level, which is as follows:

$$\forall q, q' : \min\left\{d^q, d^{q'}\right\} < d^{qq'} \tag{8}$$

which can be referred to as LBSs (local biased structure), i.e. "*for any two clusters, their distance is bigger than the smaller of their diameters*".

It is quite evident that despite the consecutive weakening of this sequence of conditions, there is still a certain sense to the latter version. In fact, "small" clusters, in terms of the diameter, and not necessarily the number of objects in them, may happen to lie "close" to the "big" ones. However, it should be kept in mind that the "very local" character of this particular condition makes it susceptible to various "exotic" configurations, in which, even if it is fulfilled, they would by no means be intuitively accepted as representing actual clusters or coherent groups (see, once again, the examples provided in [16] and [17]).

We now turn to the last of the conditions considered. It appears to go very far in the direction of weakening of requirements on the structure, but, in reality, not necessarily so. The conditions is.

$$\forall q : \delta^q_{max} < d^{qq'} \forall q' \neq q \tag{9}$$

where $\delta^q{}_{\max} = \max_{i \in Aq} \min_{i' \in Aq\text{-}i} d_{ii}$, that is – the biggest nearest neighbor distance in cluster q, which was used here to replace the diameter from the previously formulated LISs.

2.3 Some Algorithmic Considerations

As noted already, both the problem considered and the potential ways of treating it are very much reminding the essential problem of cluster analysis and its algorithms. We can, namely, start from the observation from the preceding section that the partition of the set of opinions analyzed into the single object clusters is equivalent to the partition into ideal structures, even if this partition, and the structures, are trivial. Next, for almost all of the ideal structure conditions formulated, the key of the definitional nature was to find the closest neighbors in the set of observations, in order to check whether in this manner we do not form the subsequent ideal structure. If so, we are very close to the precepts of a wide class of clustering algorithms, notably those from the very popular hierarchical aggregation family (see the classical references [11] and [12] for the fundamental work in the domain, or [18] for a review, or yet [6] for the very first algorithm from this group), all of which function according to the following scheme:

1. treat all opinions as separate clusters, forming the initial "trivial" partition P^0; step index $t = 0$;
2. for the given partition P^t find the closest clusters, in the sense of some inter-cluster distance measure, $d(A_q,A_{q'})$, say $q1$ and $q2$;
3. merge these closest clusters, A_{q1} and A_{q2}, to form a new cluster, a hence a new partition, P^{t+1};

4. if necessary, recalculate in the proper manner the distances between clusters in the new partition;
5. $t := t + 1$, if $t + 1 = n$ then stop – all clusters have been already merged into one cluster, otherwise go to step 2.

The algorithms in this class differ mainly as to step 4 in the above procedure, producing numerous variants. For our purposes, we modify this procedure in an appropriate manner. Namely, an important conjecture from our previous considerations was that *we should look for the partition that fulfils the ideal structure condition (e.g. (1)) for the minimum value of p*. This can be done, at least as an approximation, with the use of a concrete hierarchical merger algorithm, namely the *complete linkage* (see [10] and [11]) which uses, actually, similar definitions to the ones assumed here, especially regarding GIS. The algorithm would have to include *the step of verifying whether we actually (still) deal with ideal structures, after a merger*. This is insofar important as it was demonstrated in [16] and [17] that a sequence of mergers may produce ideal structures even after having failed to do so at some preceding step.

The most computationally burdensome aspect of this procedure is distance calculation and re-calculation (if necessary). Although we can apply definite simplifications that somehow alleviate this problem, the procedure, as it is known, is prohibitive for the really large datasets. It remains really feasible for at most thousands of objects. Yet, for the purposes of opinion analysis more than that is seldom necessary.

The respective algorithm would, therefore, take on the form of:

1. treat all opinions as separate clusters, forming the initial "trivial" partition P^0; step index $t = 0$;
2. for the given partition P^t find the closest clusters, in the sense of some inter-cluster distance measure, $d(A_q,A_{q'})$, say $q1$ and $q2$;
3. merge these closest clusters, A_{q1} and A_{q2}, to form a new cluster, a hence a new partition, P^{t+1};
4. if necessary, recalculate in the proper manner the distances between clusters in the new partition;
5. check the clusters formed for the ideal structure condition, and if it satisfied, register it;
6. $t := t + 1$, if $t + 1 = n$ then go to 7 – all clusters have been already merged into one cluster, otherwise go to step 2.
7. reconstruct the last partition, satisfying the ideal structure condition for the particular clusters (or the entire set, depending upon the nature of condition).

In case we are obliged to consider a bigger number of opinions, we might use another clustering paradigm, i.e. the density-based algorithms, e.g. the very popular DBSCAN (see [3], or for a remote original work, [21]). The density-based techniques group locally nearby objects using some pragmatic precepts, which can be brought to a definite similarity with the ideal structure conditions. Ultimately, though, it is not advised to form bigger groups with algorithms like DBSCAN, for then the essential flavor of the ideal structures would get lost.

To conclude with the connections to the clustering algorithms, if we specifically focus on the local structures, then we can find of use the clustering algorithms from the very popular and effective k-means family (see, for instance, [2] and [12], or the very first formulation of the concept behind this group of algorithms in [20]). The general outline of the respective algorithm is as follows:

1. having an initial partition P^0 (thereafter a current partition P^t) determine the representatives of each cluster, e.g. as means or medians of the group (also in the case of the singleton groups), denoted x^q,
2. assign each opinion x_i from the set of opinions to the closest representative x^q, thereby forming a new partition, $t{:} = t + 1$;
3. if the stopping criterion is not satisfied, return to step 1, otherwise stop.

Theoretically, the stopping criterion is constituted by the sufficient similarity of partitions before and after an iteration, but it is common to adopt just a number of iterations as the stopping criterion, since the algorithm converges very quickly (sometimes in just a couple of iterations) to a local solution, largely influenced by the starting point of the algorithm.

In order to make an appropriate use of this kind of algorithm, we could proceed according to the following scheme: after having obtained, say, from DBSCAN, small local groups of opinions, we could use this result as a partition, constituting the starting point for the k-means, and then realize the procedure as above, with appropriate checks as to the ideal structures, i.e.

1. having an initial partition P^0 (thereafter a current partition P^t) determine the representatives of each cluster, e.g. as means or medians of the group (also in the case of the singleton groups), denoted x^q;
2. assign each opinion x_i from the set of opinions to the closest representative x^q, thereby forming a new partition, $t{:} = t + 1$;
3. check the clusters for the ideal structure condition(s) and retain (locally) those satisfying these conditions;
4. if the stopping criterion is not satisfied, return to step 1, otherwise stop;
5. reconstruct the biggest clusters (cardinality-wise) satisfying the ideal structure conditions.

Let us add here that the representatives of clusters can also be constituted by the opinions from the set, those closest to the respective cluster mean or median. This is of special importance, when we deal with discrete spaces of variables, which is often the case in the studies we consider here.

2.4 Remarks on the Practical Aspects

If we aim at gaining knowledge on the basis of a set of opinions by identifying the groups of similar opinion profiles (partial aggregates), we encounter a broad variety of potential circumstances as regards the multidimensional distribution of the opinions. It may, namely, be so that the ideal structures, even the weakest ones, and even if acceptable account for, e.g., only 15% of the entire set, while the rest would have to be treated as "trivial singleton ideal structures". On the other hand, even if the ideal structures encompass, say, 80% of the set, they may be small or very small (e.g., composed of 2–3 opinions each of them), and hence providing very little of constructive information.

We do assume, though, that a rationally designed exercise, involving rational respondents, should produce an important proportion of similar, and possibly also internally consistent, opinions within a not too big number of groups.

At the same time, it is quite natural that in any case a part of opinions would have to be treated, for "statistical" and/or "substantive" reasons, as "noise" (especially when internally inconsistent, the possibility of qualification of inconsistency being one of the features of a well-designed exercise). It appears to be reasonable to assume the level of "acceptable noise" at 10–20% of the entire data collection.

In exactly this context, we may be interested, for an obtained partition, composed of groups A_q, this partition not fulfilling the ideal structure condition (1), or perhaps also some other ones, in knowing the proportion (minimum number) of opinions that have to be deleted from X in order to secure the fulfilment of the selected conditions. This can be done by using algorithms either oriented at detection of anomalous observations, or those allowing for an easy and effective detection of the elements most distant from the center of a group (like k-means). On the other hand, if the proportion of (statistically and substantively significant) ideal structures is sufficiently high, exceeding, e.g., 75% of the total, we might treat ("automatically") the rest as either "noise" or "anomaly", depending upon the distances from the identified ideal structures.

3 Summary and Conclusions

We have presented here an outline for an approach, aimed at acquiring deeper knowledge from quite standard situation of "opinion collection", in cases when the respective opinions can be represented through numbers.

When summarizing, we would like to deal with a question, which definitely has to come to the mind of almost any Reader: given all these references to clustering and the formulation of the problem, one might ask why not simply apply any of the here mentioned clustering algorithms, doing away with the "ideal structures", their search and verification?

The response is associated with the very purpose of the exercise here considered, namely gaining of knowledge on the basis of multiple opinions – the purpose of cluster analysis is to provide the structure of the entire set of data, while we are interested, first, in the structure of its part, even if constituting majority, but not necessarily the entirety, and, second, we wish to have the intuitively clear ("ideal") structures that might be easily interpreted. We wish to identify the groups of similar opinions that are pronounced and well discernible from the other ones. Existence of "noise" and "isolated" opinions (which may in general be of interest, of course) is here of lesser importance.

And that is why we propose to use the clustering algorithms as tools, but complemented with appropriate checks, and not necessarily yielding the complete clustering results. The resulting computational burden involved is roughly the same as for the clustering algorithms, and, given the magnitude of expected datasets, should not be prohibitive.

It is fully feasible to compare the proposed schemes of the modified algorithms to gradual construction of high density regions in the space of opinions, stopping when the density peaks show the signs of formation of valleys between them, or altogether plains, these being of less or even no interest to us. Such an image makes the proposed approach somewhat similar to those characterized in [21] and [22], with, however, very pronounced differences not only in terms of purpose, but also as to the very course of the respective procedures.

In the light of the prevailing practice in organization of the exercises we aim at ("opinion collection"), the proposal we forward may lead to a much more effective use of the results obtained, meaning deeper knowledge concerning the different and well pronounced, coherent opinion groups or profiles.

References

1. Arrow, K.J.: Social Choice and Individual Values, 2nd ed. Wiley, New York (1963). (First edition: 1951)
2. Ball, H., Hall, D.: ISODATA, a novel method of data analysis and pattern classification. Technical report NTIS AD 699616. Stanford Research Institute, Stanford, CA (1965)
3. Ester, M., Kriegel, H.-P., Sander, J., Xu, X.-W.: A density-based algorithm for discovering clusters in large spatial databases with noise. In: Simondis, E., Han, J., Fayyad, U.M. (eds.) Proceedings of the Second International Conference on Knowledge Discovery and Data Mining (KDD-96), pp. 226–231. AAAI Press (1996)
4. Fedrizzi, M., Kacprzyk, J., Owsiński, J.W., Zadrożny, Sł.: A DSS for consensus reaching using fuzzy linguistic majority and clustering of preferences. In: Nahorski, Z., Owsiński, J.W. (eds.) Support Systems for Decision and Negotiation Processes, Special Issue of Annals of Operations Research, vol. 51, pp. 127–135 (1994)
5. Filip, F.G.: Collaborative decision-making: concepts and supporting information and communication technology tools and systems. Int. J. Comput. Commun. Control **17**(2), 4732 (2022)
6. Florek, K., Łukaszewicz, J., Perkal, J., Steinhaus, H., Zubrzycki, S.: Taksonomia Wrocławska (The Wrocław Taxonomy; in Polish). Przegląd Antropologiczny **17** (1956)
7. Kacprzyk, J., Stańczak, W.: On an extension of the method of minimally interconnected subnetworks. Control Cybern. **5**(4) (1976)
8. Kacprzyk, J., Stańczak, W.: On a further extension of the method of minimally interconnected subnetworks. Control. Cybern. **7**(2), 17–31 (1978)
9. Kemeny, J., Snell, L.: Mathematical Models in the Social Sciences. Ginn, Boston (1961)
10. Lance, G.N., Williams, W.T.: A generalized sorting strategy for computer classifications. Nature **212**, 218 (1966)
11. Lance, G.N., Williams, W.T.: A general theory of classification sorting strategies. 1. hierarchical systems. Comput. J. **9**, 373–380 (1967)
12. MacQueen, J.: Some methods for classification and analysis of multivariate observations. In: LeCam, L.M., Neyman, J. (eds.) Proceedings of the 5th Berkeley symposium on Mathematical Statistics and Probability, vol. 1, pp. 281–297. University of California Press, Berkeley (1967)
13. Nurmi, H.: Social choice, stable outcomes and deliberative democracy. Control Cybern. **51**(2) (2022). https://doi.org/10.2478/candc-2022-0011
14. Owsiński, J.W.: A simple software system for eliciting structured sets of notions from a group of experts (methods and experiences). In: Schader, M., Gaul, W. (eds.) Knowledge, Data and Computer-Assisted Decisions, NATO ASI F61, vol. 61, pp. 369–378. Springer, Berlin (1990). https://doi.org/10.1007/978-3-642-84218-4_29
15. Owsiński, J.W.: Aggregation and clustering of preferences: opinion vs. action. In: Bachem, A., Derigs, U., Jűnger, M. (eds.) Operations Research 1993, pp. 377–384. Physica-Springer, Heidelberg (1994). https://doi.org/10.1007/978-3-642-46955-8_93

16. Owsiński, J.W.: Ideal structures in crowdsoucing opinion aggregation: some basic observations. Procedia Comput. Sci. **214**, 42–49 (2022)
17. Owsiński, J.W.: Group opinion structure: the ideal structures, their relevance and effective use. In: Baier, D., Wernecke, K.-D. (eds.) Innovations in Classification, Data Science, and Information Systems, pp. 471–481. Springer, Berlin (2004). https://doi.org/10.1007/3-540-26981-9_54
18. Owsiński, J.W.: Data Analysis in Bi-partial Perspective: Clustering and Beyond. Studies in Computational Intelligence, vol. 818. Springer, Switzerland (2020). https://doi.org/10.1007/978-3-030-13389-4
19. Sosnowska, H.: Transformation into anti-manipulation method in voting. Changes in properties. Control Cybern. **51**(4) (2022)
20. Steinhaus, H.: Sur la division des corps matériels en parties. Bull. l'Academie Polonaise Sci. **IV**(C1.III), 801–804 (1956)
21. Tremolières, R.: The percolation method for an efficient grouping of data. Pattern Recogn. **11**, 255–262 (1979)
22. Yager, R.R., Filev, D.P.: Approximate clustering via the mountain method. IEEE Trans. Syst. Man Cybern. **24**, 1279–1284 (1994)

Estimation of the Strict Preference Relation on the Basis of Pairwise Comparisons for Moderate and Large Size Sets

Leszek Klukowski(✉)

Systems Research Institute Polish Academy of Sciences, Newelska 6, 01-447 Warsaw, Poland
Leszek.Klukowski@ibspan.waw.pl

Abstract. The paper presents two approaches for determining of estimates of the preference relation on the basis of multiple binary pairwise comparisons with random errors. Obtaining of the estimates requires an optimal solution of a discrete programming problem which minimizes sum of differences between relation form and comparisons. The problem is NP hard and can be solved with the use of exact algorithms for moderate size of sets, i.e. below 100 elements. In the case of larger sets, i.e. at least 200–300 comparisons for each element, it is necessary to apply heuristic algorithms. The paper presents the results, which allow determining of optimal or suboptimal solution with acceptable computational cost. They include: statistical procedure and test producing "new" comparisons with low probabilities of errors; the comparisons provides efficiency of heuristic algorithms. Thus, the proposed approach guarantees applicability of the estimators for any size of set.

Keywords: estimation of the strict preference relation for large sets · binary pairwise comparisons with random errors · nearest adjoining order method

1 Introduction

The estimators of strict preference relation based on multiple binary pairwise comparisons with random errors, proposed by Slater (1961) (see also David 1988, Bradley (1984), Klukowski (2021)), require optimal solutions of a discrete programming problem. The problem minimizes differences between relation form, determined in appropriate way, and comparisons. The estimates are consistent under non-restricted assumptions about comparisons errors; the speed of convergence is of exponential type - for increasing number of independent comparisons of each pair. The optimization problems can be solved with the use of appropriate algorithms: the complete enumeration – for sets including not more than 20 elements, discrete mathematical programming – up to 100 elements (assuming single comparison of each pair – see David 1988 Chapt. 2, Bezembinder (1981), Philips (1967, 1969), Remage and Thomson (1966)), heuristic approach - for sets exceeding 100, especially in the case of multiple comparisons of each pair. Heuristic algorithms reduce computational costs, but can provide questionable solutions in the case of probabilities of comparisons errors not close to zero. However, large number of comparisons of each element, i.e. at least 200–300 (taking into account multiple

K. T. Atanassov et al. (Eds.): BOS/SOR/IWIFSGN 2022, LNNS 793, pp. 250–260, 2023.
https://doi.org/10.1007/978-3-031-45069-3_23

comparisons of each pair), can be advantageous. It allows producing of "new" comparisons with significantly reduced probabilities of errors. Proposed approach is based on statistical test and procedure, presented earlier (procedure) in Klukowski 2021, Chapt. 10. Two versions of the test are proposed – the first one for the case when multiple comparisons (of each pair) are replaced by their median (distribution of median is not the same as distribution of individual comparison). The second version is based on whole set of comparisons; this approach requires higher computation costs, but has also additional statistical properties. The results of applications of the tests are comparisons of each pair with reduced probabilities of errors. These comparisons can be not independent, in stochastic way, but comparisons of different pairs are independent. Therefore the estimates based on "new" comparisons are consistent (see Klukowski 2021). Comparisons with reduced probabilities of errors can be used as the effective base for heuristic algorithms; also exact algorithms work efficiently with the comparisons including few errors. The computational cost of "combined" approach, i.e.: application of the test and heuristic algorithm, is rather low. Similar approach has been proposed by the author (see Klukowski 2017) for equivalence relation. These features make the approach, based on nearest adjoining order idea (Slater 1961), highly efficient and applicable for any size of a set.

The paper consists of six sections. The second section presents the estimation problem, assumptions about pairwise comparisons and the form of estimator. In the third section is proposed: statistical test reducing probabilities of errors of comparisons (in the form of medians from multiple comparisons of each pair) and the procedure based on the test, which determines parameters of "new" comparisons (direction of preference and probability of errors). In next section are determined examples of application of the test, which shows the range of reduction of probability of errors of comparisons for some number of elements. The fifth section presents the test based on whole set of comparisons, i.e. not medians from comparisons, and properties of this approach. Last section summarizes the results.

2 Estimation Problem, Assumptions About Comparisons, form of Estimator

We are given a finite set of elements $\mathbf{X} = \{x_1, \ldots, x_m\}(3 \leq m < \infty)$. It is assumed that there exists in the set $\mathbf{X}$ the strict preference relation with unknown form. The relation generates a family of subsets $\chi_1^*, \ldots, \chi_m^*$; each subset includes succeeding element of the relation.

The relation $\mathbf{X}$ can be expressed by the values $T(x_i, x_j)(< i,j >\in R_m; R_m = \{i,j|1 \leq i,j \leq m; j > i\}$, defined as follows:

$$T(x_i, x_j) = \left\{ \begin{array}{l} -1 \; if \; x_i \in \chi_r^*, x_j \in \chi_s^*, r \; lt; s; \\ \; 1 \; if \; x_j \in \chi_r^*, x_i \in \chi_s^*, r \; \; gt; s. \end{array} \right\}. \tag{1}$$

The relation has to be estimated on the basis of pairwise comparisons $g_k(x_i, x_j)$ $(k = 1, \ldots, N; (N \geq 1), < i,j >\in R_m)$ with random errors, defined as follows:

$$g_k(x_i, x_j) = \left\{ \begin{array}{l} -1 \; if \; g_k(x_i, x_j) indicates \; that \; x_i \in \chi_r^*, x_j \in \chi_s^*, r \; lt; s; \\ \; 1 \; if \; g_k(x_i, x_j) indicates \; that \; x_j \in \chi_r^*, x_i \in \chi_s^*, r \; \; lt; s, \end{array} \right\}, \tag{2}$$

where:

- the probability of correct result of each comparison satisfy the condition:

$$P\left(g_k\left(x_i, x_j\right) = T\left(x_i, x_j\right)\right) \geq 1 - \delta\left(\delta \in \left(0, \frac{1}{2}\right)\right), \tag{3}$$

- comparisons $g_k\left(x_i, x_j\right)(k = 1, \ldots, N; < i, j > \in R_m)$ are independent random variables.

In the case of $N > 1$ (N – odd) it is also discussed the case of medians $g^{(me)}\left(x_i, x_j\right)$ from comparisons of each pair $g_k\left(x_i, x_j\right)(k = 1, \ldots, N)$; the median is equivalent to majority in the set $\{-1, 1\}$. It is obvious that the median satisfies the condition (see David (1970), Chapt. 2):

$$P\left(g^{(me)}\left(x_i, x_j\right) = T\left(x_i, x_j\right)\right) > 1 - \delta. \tag{4}$$

In the case of even N a number of comparisons: -1 and 1 can be equal; in this case the approach based on medians can be replaced by using whole set of comparisons (see Sect. 5 below).

The estimator $\breve{\chi}_1, \ldots, \breve{\chi}_m$ (or $\breve{T}\left(x_i, x_j\right)$) of the preference relation $\chi_1^*, \ldots, \chi_m^*$, based on medians from comparisons of each pair, is obtained on the basis of the following minimization problem

$$\min_{F_{\mathbf{X}}} \left\{ \sum\nolimits_{<i,j> \in R_m} \left| g^{(me)}\left(x_i, x_j\right) - t\left(x_i, x_j\right) \right| \right\}, \tag{5}$$

where:

$F_{\mathbf{X}}$ - the feasible set, i.e. the family of all strict preference relations in the set $\mathbf{X}$,

$t(x_i, x_j)$ - the values describing the elements of the family $F_{\mathbf{X}}$, defined in the same way as $T(x_i, x_j)$.

The estimator $\breve{\chi}_1, \ldots, \breve{\chi}_n$ is consistent (see Klukowski 1994, 2021) for $N \to \infty$ and

$$P\left(\breve{\chi}_1, \ldots, \breve{\chi}_m = \chi_1^*, \ldots, \chi_m^*\right) \geq 1 - 2\exp\{-2N\left(\frac{1}{2} - \delta_{me}\right)^2\}, \tag{6}$$

where:

δ_{me} – the probability $P(g^{(me)}\left(x_i, x_j\right) \neq T\left(x_i, x_j\right))$ resulting from the values of: N, δ in the case of uneven N.

The inequality (6) (see Klukowski 2021) is based on the Hoeffding inequality for zero-one random variables and Chebyshev inequality for expected value.

Some other properties of the estimator are presented in the book Klukowski (2021). Especially, the estimator is also consistent for $m \to \infty$, under some weak assumptions. Moreover, the assumption about independency of comparisons $g^{(me)}\left(x_i, x_j\right)(j \neq i)$ can be replaced by the assumptions that comparisons of different pairs, i.e. $g^{(me)}\left(x_i, x_j\right)$ and $g^{(me)}(x_r, x_s)$ are independent for $r \neq i, j$; $s \neq i, j$ and that (multiple) comparisons of the same pair are independent.

3 Test Reducing Probabilities of Errors of Comparisons Based on Medians from Comparisons

The probability δ_{me} of errors of comparisons can be significantly reduced in the case of large value of m and adequate value of difference of ranks of elements $x_i \in \chi_r^*$ and $x_j \in \chi_s^*$, i.e. $|r - s|$. The proposed approach is based on statistical test and the procedure using the test in sequential way. The first two hypotheses of the procedure have the form:

- the null hypothesis $H_{0,-1}$: the difference of ranks between elements x_i and x_r is equal -1, under alternative $H_{1,l}$ $(l \leq -2)$, stating that it is lower than -1;
- the null hypothesis $H_{0,-1}$: the difference of ranks between elements x_i and x_r is equal -1, under alternative $H_{1,l}$ $(l \geq 1)$, stating that it is equal or greater than 1.

The test for such the hypotheses, presented below, has exact binomial distribution and asymptotic Gaussian distribution; the asymptotic distribution is adequate to large number of comparisons case. The procedure based on verifying of the hypotheses consists of three steps and is finite (see Klukowski 2021, Chapt. 10). It has the form:

1^0. The first step has a form of two excluding hypotheses for a pair $((x_i, x_j); <i,j> \in R_m)$: $H_{0,-1}$ vs $H_{1,-2}$, i.e. difference of ranks equal -1 vs equal or lower than -2 and $H_{0,-1}$ vs $H_{1,1}$, i.e. difference equal -1 vs equal or greater than 1. If the hypothesis $H_{0,-1}$ is not rejected in both cases, then difference of ranks -1 is accepted for a pair (x_i, x_j). Rejecting of $H_{0,-1}$ in the first case (the alternative $H_{1,-2}$) indicates the next step 2^0, in the second case (alternative $H_{1,1}$) – the next step 3^0.

2^0. It is performed the test $H_{0,-2}$ vs $H_{1,-3}$; if $H_{0,-2}$ is not rejected, then it is assumed difference of ranks equal -2. In opposite case, the test is repeated in the same form, i.e. $H_{0,-3}$ vs $H_{1,-4}$; the last possible hypothesis has the form $H_{0,-(m-2)}$ vs $H_{1,-(m-1)}$.

3^0. It is performed the test $H_{0,1}$ vs $H_{1,2}$; if $H_{0,1}$ is not rejected, then it is assumed difference of ranks equal 1. In opposite case, the test is repeated in the same form, i.e. $H_{0,2}$ vs $H_{1,3}$; the last possible hypothesis has the form $H_{0,m-2}$ vs $H_{1,m-1}$.

The results of the above (finite) procedure determine the preference in each pair (x_i, x_r) (also difference of ranks) and allows evaluating of probability of error of each comparison. The probability can be determined in the following way. The procedure determines finally the evaluation of difference of ranks between any pair of elements (x_i, x_r), namely ρ_{ir} $(\rho_{ir} \in \{\pm 1, \ldots, \pm(m-1)\})$. For the difference it can be verified the hypothesis $H_{0,-1}$ versus $H_{1,\rho_{ir}}$ $(\rho_{ir} < 0)$ or $H_{0,1}$ versus $H_{1,\rho_{ir}}$ $(\rho_{ir} > 0)$. The test of such the (simple) hypotheses is uniformly powerful and allows determining of the probability of error $\delta_{ir}^{(test)}$, assuming that probability of first and second type error are equal. Such the probability is assumed as evaluation of the probability of error of comparison of the pair (x_i, x_r). The result of the test and the probability $\delta_{ir}^{(test)}$ replace: the comparison $g^{(me)}(x_i, x_r)$ and probability of its error δ_{me} - in the case

$$\delta_{ir}^{(test)} < \delta_{me}. \tag{7}$$

In opposite case the comparison $g^{(me)}(x_i, x_r)$ and probability δ_{me} is not changed. It can be shown that the inequality (7) is satisfied for difference of ranks greater than some value; for large difference the probability $\delta_{ir}^{(test)}$ is close to zero (see an example below).

However, the probabilities of errors of individual comparisons, which satisfy (7) are not equal. It should be noted that comparisons resulting from the test can be not independent, but comparisons of different pairs (x_i, x_j) $(x_r, x_s)(r \neq i, j;\ s \neq i, j)$ are independent. In such the case the estimates resulting from (5) are still consistent (see Klukowski 1994, 2021).

It can be also considered some alternative system of hypothesis, i.e. verification of successive hypotheses of the form: $H_{0,-1}$ vs $H_{0,l}(l \leq -2)$ or $H_{0,-1}$ vs $H_{0,l}(l \geq 2)$. However, such the approach can lead to incompatible results, e.g. rejecting $H_{0,-1}$ vs $H_{0,-2}$, indicates next hypotheses $H_{0,-1}$ vs $H_{0,-3}$, i.e. verification of rejected hypothesis $H_{0,-1}$. Moreover, acceptation of rejected hypothesis $H_{0,-1}$ is also not excluded. The probabilities $\delta_{ir}^{(test)}$ are not estimates, especially they can be higher than actual.

The test statistic for any pair (x_i, x_r) is based on properties of comparisons $g^{(me)}(x_i, x_j)$ and $g^{(me)}(x_r, x_j)(r \neq i; j = 1, \ldots, m)$, indicated by actual values $T(x_i, x_j), T(x_r, x_j)$. The properties state equal or different values of comparisons – equal with the probability $(1-\delta_{me})^2+\delta_{me}^2$, for $T(x_i, x_j) = T(x_r, x_j)$, not equal with the probability $2\delta_{me}(1-\delta_{me})$. The probabilities for the case $T(x_i, x_j) \neq T(x_r, x_j)$ have "reversal" form – equal values with probability $2\delta_{me}(1-\delta_{me})$ and different – with probability $(1-\delta_{me})^2+\delta_{me}^2$. The test statistic is proposed in the form:

$$\gamma_{ir}^{(me)} = \frac{1}{m}\Big(\sum_{\substack{j \neq i, r \\ r \neq i}} (g^{(me)}(x_i, x_j) - g^{(me)}(x_r, x_j)) - g^{(me)}(x_r, x_i) + g^{(me)}(x_i, x_r)\Big). \tag{8}$$

The statistic has a following properties for individual hypothesis. Let's start from the case $H_{0,-1}$ vs $H_{1,-2}$; differences

$$g^{(me)}(x_i, x_j) - g^{(me)}(x_r, x_j) \tag{9}$$

have the following parameters, depending on the values $T(x_i, x_j), T(x_r, x_j)$. In the case of true $H_{0,-1}$ and $j \neq i, r$ the comparisons $g^{(me)}(x_i, x_j)$ and $g^{(me)}(x_r, x_j)$ have the same distributions. Therefore, any difference (9) assumes values from the set {-2,0,2} and has expected value and variance:

$$E\Big(g^{(me)}(x_i, x_j) - g^{(me)}(x_r, x_j)|H_{0,-1}\Big) = 0\ (j \neq i, r) \tag{10}$$

$$Var(g^{(me)}(x_i, x_j) - g^{(me)}(x_r, x_j)) = 8\delta_{me}(1-\delta_{me})(j \neq i, r). \tag{11}$$

In the case $j = i$ and $j = r$ it is assumed $g^{(me)}(x_i, x_i) = g^{(me)}(x_r, x_r) \equiv 0$ and the difference (9) have expected value and variance equal – respectively:

$$E\Big(g^{(me)}(x_i, x_j) - g^{(me)}(x_r, x_j)|H_{0,-1}\Big)) = -1(1-2\delta_{me})(j = i, j = r), \tag{12}$$

$$Var(g^{(me)}(x_i, x_j) - g^{(me)}(x_r, x_j)) = 4\delta(1-\delta_{me})(j = i, j = r). \tag{13}$$

Thus, the expected value and variance of the random variable $\gamma_{ir}^{(me)}$ has the following parameters under $H_{0,-1}$:

$$E\left(\gamma_{ir}^{(me)} | H_{0,-1}\right)) = \frac{2}{m}(2\delta_{me} - 1) < 0, \tag{14}$$

$$Var(\gamma_{ir}^{(me)}) = \frac{8(m-1)}{m^2}\delta_{me}(1 - \delta_{me}) < \frac{8}{m}\delta_{me}(1 - \delta_{me}). \tag{15}$$

It can be shown that under $H_{1,-2}$ these parameters have a form:

$$E\left(\gamma_{ir}^{(me)} | H_{0,-2}\right)) = \frac{2}{m}(2\delta_{me} - 1) + \frac{2}{m}(2\delta_{me} - 1) = \frac{4}{m}(2\delta_{me} - 1), \tag{16}$$

$$Var(\gamma_{ir}^{(me)}) = \frac{8(m-1)}{m^2}\delta_{me}(1 - \delta_{me}). \tag{17}$$

The expected value corresponding to the hypothesis $H_{1,-2}$ is lower than corresponding to $H_{0,-1}$, i.e. $2(2\delta_{me} - 1)/m < 4(2\delta_{me} - 1)/m$, while variance is the same in both cases. It can be shown that if the alternative has a form $H_{1,-3}$ then the expected value has the form:

$$\frac{6}{m}(2\delta_{me} - 1), \tag{18}$$

i.e. is lower than for $H_{1,-2}$. Therefore, rejecting $H_{0,-1}$ vs $H_{1,-2}$ indicates also that some hypothesis $H_{1,-l} (l \geq 3)$ can be true. As a result of rejecting the hypothesis $H_{0,-1}$, the next step of the procedure $1^0 - 3^0$ has a form $H_{0,-2}$ vs $H_{1,-3}$.

For large m, usually above 200, the statistic γ_{ir} has asymptotic Gaussian distribution with the above parameters (expected value and variance). The result of application of the sequence of tests in the procedure 1^0–3^0 generates evaluation of difference of ranks between elements x_i and x_r, which determines direction of preference in this pair. It also allows evaluation of probability of error for a pair under consideration. The considerations about the second system of hypothesis, i.e. $H_{0,-1}$ vs $H_{1,1}$ are similar.

In the case $H_{0,-1}$ vs $H_{1,1}$ and rejected null hypothesis, it can be shown that the expected value and variance of the test statistic $\gamma_{ir}^{(me)}$ are equal:

$$E\left(\gamma_{ir}^{(me)} | H_{1,1}\right) = \frac{2}{m}(1 - 2\delta_{me}), \tag{19}$$

$$Var(\gamma_{ir}^{(me)}) = \frac{8(m-1)}{m^2}\delta_{me}(1 - \delta_{me}). \tag{20}$$

It is clear that

$$E\left(\gamma_{ir}^{(me)} | H_{1,1}\right) > E\left(\gamma_{ir}^{(me)} | H_{0,-1}\right), \tag{21}$$

while the variances in both cases are the same. Moreover, it can be shown that if the alternative has a form $H_{1,2}$, then the expected value has a form (variance the same):

$$E\left(\gamma_{ir}^{(me)} | H_{1,2}\right) = \frac{4}{m}(1 - 2\delta_{me}), \tag{22}$$

satisfying the condition:

$$E\left(\gamma_{ir}^{(me)}|H_{1,2}\right) > E\left(\gamma_{ir}^{(me)}|H_{1,1}\right) > E\left(\gamma_{ir}^{(me)}|H_{0,-1}\right). \tag{23}$$

Thus, rejecting of the $H_{0,-1}$ vs $H_{1,1}$ indicates also that some alternative $H_{1,l}(l > 1)$ can be true. As a result of rejecting the hypothesis $H_{0,-1}$, the next step of the procedure $1^0 - 3^0$ has a form $H_{0,1}$ vs $H_{1,2}$. Acceptance of a hypothesis $H_{0,l}(l \geq 2)$, in the procedure, indicates that in the case

$$\delta_{ir}^{(test)} < \delta_{me} \tag{24}$$

the median $g^{(me)}(x_i, x_r)$ has to be replaced by the value 1 (one) and the probability of error of comparison is assumed equal $\delta_{ir}^{(test)}$. It is clear that for large difference of ranks, i.e. greater than some value l_e, the probability $\delta_{ir}^{(test)}$ is close to zero, i.e. significantly lower than δ_{me} (see next point for details).

4 Effects of Application of Proposed Test

The results of application of the procedure presented above have a following form:

- it can be determined "new" comparisons with probabilities of errors lower than δ_{me} (i.e. $\delta_{ir}^{(test)} < \delta_{me}$) – on the basis of parameters of appropriate hypothesis: $H_{0,-1}$ vs $H_{1,-l-1}$ or $H_{0,-1}$ vs $H_{1,l}(l \geq 1)$ (using limiting Gaussian distribution); such the results and their probabilities of errors replace comparisons $g^{(me)}(x_i, x_j)(< i, j > \in R_m)$; it is clear that reduced probabilities of errors indicate lower errors of the estimator (5) of the relation;
- it can be determined comparisons which have significantly lower probabilities of errors, i.e.: $\delta_{ir}^{(test)} \ll \delta_{me}$; such the comparisons correspond to high differences of ranks (see example below)

The probabilities of errors of comparisons (based on limiting Gaussian distribution) result from expected values (14), (16), (18) and variance (17) of the test statistic $\gamma_{ir}^{(me)}$ corresponding to null and alternative hypothesis (assuming equal probabilities of the errors of the first and second type). They show the cases $\delta_{ir}^{(test)} < \delta_{me}$ and $\delta_{ir}^{(test)} \ll \delta_{me}$ for large number of elements m (higher than 200).

The approximate approach for detecting indices $< i, r >$ indicating $\delta_{ir}^{(test)} \ll \delta_{me}$ is based on difference of expected values of the test statistic $\gamma_{ir}^{(me)}$ for null and alternative hypothesis and standard deviation of the statistic. More precisely, the probability of error in the test is low, namely equal to 0.00135, for difference of these values equal to three standard deviations. These values can be determined easily on the basis of the formulas: (14), (16), (17) and (18) (for the hypotheses $H_{0,-1}$ vs $H_{1,-l-1}$). Some examples of the probabilities are presented below for the case: $m = 500$ and $\delta_{me} \in \{0.2, 0.1, 0.05, 0.025, 0.01, 0, 005\}$. For these parameters are determined the differences of ranks indicating difference of expected values higher than three standard deviations (see table below), i.e. satisfying the condition:

$$\left|E\left(\gamma_{ir}|H_{0,-1}\right) - E\left(\gamma_{ir}|H_{1,-l-1}\right)\right| \geq \frac{6\sqrt{2(m-1)}}{m}\left(\delta_{me}(1-\delta_{me})\right)^{0,5}. \tag{25}$$

The comparisons having difference of ranks equal or higher than shown in the table can be assumed as errorless in practice; for the examples considered in the table ($m = 500$) the majority of comparisons have this property. For example, for $\delta_{me} = 0.1$ the comparisons with difference of ranks higher than 33 are in practice errorless - the expected number of errors corresponding to these differences of ranks of any element $x_i (1 \leq i \leq 500)$ is lower than one, while for the value $\delta_{me} = 0.1$ is equal to 46,7 (Table 1).

Table 1. Differences of ranks satisfying the condition (25).

δ_{me}	0.005	0.01	0.025	0.05	0.1	0.2
Difference of ranks	8	11	17	23	33	49

Computations of the author.

For the differences higher than 33 the probabilities of errors are lower than 0.00135.

The difference of ranks between elements x_i and x_r can be evaluated on the basis of the hypothesis $H_{0,-1}$ vs $H_{1,-l-1}$ or $H_{0,-1}$ vs $H_{1,l} (l \geq 1)$, which completes the procedure $1^0 - 3^0$. The set of comparisons $\{g^{(proc)}(x_i, x_j)(< i, j > \in R_m)\}$ (resulting from the procedure) which includes significant number of comparisons having low probabilities of errors, is efficient basis for heuristic optimization of the problem:

$$\min_{F_X}\{\sum\nolimits_{<i,j>\in R_m}\left|g^{(proc)}(x_i, x_j) - t(x_i, x_j)\right|\}. \tag{26}$$

Such the solution of the problem can be next used as a starting point for discrete mathematical programming problems – the papers presented efficient algorithms are presented e.g. in Philips (1967, 1969) (for other papers see David 1988, Sect. 2.2). Another approach is to solve the problem on the basis of comparisons with decreased probabilities of errors only.

It was mentioned above that the probabilities of errors of comparisons $g^{(proc)}(x_i, x_j)(< i, j > \in R_m)$, resulting from the procedure $1^0 - 3^0$, have different values. More precisely, some part of these comparisons have probability δ_{me}, while remaining $\delta_{ir}^{(test)}$ - lower than δ_{me}. Such the probabilities guarantee consistency of the estimator, obtained on the basis of the problem (5), also in the case of independency (only) of comparisons of different pairs (x_i, x_j) and $(x_r, x_s)(r \neq i, j; s \neq i, j)$. It is clear that reduction of probabilities of errors indicates higher precision of an estimate.

5 Procedure and Test Reducing Probabilities of Errors of Comparisons Based on Whole Set of Comparisons

The approach based on whole set of comparisons exploits the some idea as median approach, but is performed for $k = 1, \ldots, N$. Now the test statistics $\gamma_{ir}^{(me)}$ has to be replaced by $\gamma_{kir}(k = 1, \ldots, N)$:

$$\gamma_{kir} = \frac{1}{m}(\sum_{\substack{j \neq i, r \\ r \neq i}} g_k(x_i, x_j) - g_k(x_r, x_j)) - g_k(x_r, x_i) + g_k(x_i, x_r)). \quad (27)$$

The statistic γ_{kir} has analogous properties as $\gamma_{ir}^{(me)}$, i.e. the form of expected value and variance for null and alternative hypotheses, but the probability δ_{me} has to be replaced by δ. The application of the procedure 1^0–3^0 produces the set of "new" comparisons $g_k^{(proc)}(x_i, x_j)(k = 1, \ldots, N)$, i.e. the procedure has to be performed N times. The probabilities of errors of these comparisons are equal δ or $\delta_{irk}^{(test)}$; they satisfy the condition $\delta < \delta_{irk}^{(test)}$. The comparisons $g_k^{(proc)}(x_i, x_j)$ are used as a base of the estimator $\widehat{\chi}_1, \ldots, \widehat{\chi}_n$ (or $\widehat{T}(x_i, x_j)$) resulting from the discrete optimization problem:

$$\min_{F_{\mathbf{X}}}\{\sum_{<i,j>\in R_m} \sum_{k=1}^{N} \left| g_k^{(proc)}(x_i, x_j) - t(x_i, x_j) \right| \}. \quad (28)$$

The problem (28) can be also solved with the use of heuristic algorithm. It is clear that computation cost of the approach is higher than in the median approach. Moreover, probabilities of errors δ or $\delta_{irk}^{(test)}$ are higher than - respectively δ_{me} or $\delta_{ir}^{(test)}$. Advantageous feature of the approach is possibility of application of the Hoeffding (1963) inequality, in the form:

$$P(\sum_{k=1}^{N} Y_k - \sum_{k=1}^{N} E(Y_k) \geq Nt) \leq \exp\{-2Nt^2/(b-a)^2\},$$

where:

$$P(a \leq Y_k \leq b) = 1, a < b, t > 0,$$

For evaluation of the precision of the estimate. The inequality is applied to the random variables

$$Y_{ijk} = \left| g_k^{(proc)}(x_i, x_j) - T(x_i, x_j) \right| - \left| g_k^{(proc)}(x_i, x_j) - \tilde{T}(x_i, x_j) \right| (\langle i, j \rangle \in R_m), \quad (29)$$

where:

$$\tilde{T}(x_i, x_j) \neq T(x_i, x_j), Y_{ijk} \in \{-2, 0, 2\},$$

Which provides the evaluation of probability of errorless estimate. The Hoeffding inequality indicates the following relationship:

$$P\left(\left(\sum_{<i,j>\in R_m}\sum_{k=1}^{N}\left|g_k^{(proc)}(x_i,x_j)-T(x_i,x_j)\right|<\left(\sum_{<i,j>\in R_m}\sum_{k=1}^{N}\left|g_k^{(proc)}(x_i,x_j)-\tilde{T}(x_i,x_j)\right)\right|\right)\right.$$
$$\geq 1-\exp\left\{-2N\frac{\left(\sum_{<i,j>\in R_m}E(Q_{i,j})^2\right)}{(2\vartheta(m-1))^2}\right\}, \quad (30)$$

where:

$$Q_{ij}=\left|g_k^{(proc)}(x_i,x_j)-T(x_i,x_j)\right|-\left|g_k^{(proc)}(x_i,x_j)-\tilde{T}(x_i,x_j)\right|,$$

ϑ – number of cases $T(x_i,x_j)\neq\tilde{T}(x_i,x_j)$ in the set R_m.

Moreover, it can be shown that:

$$E(Q_{i,j})=2\left(2\delta_{ijk}^{(proc)}-1\right),$$

where:

$\delta_{ijk}^{(proc)}$ – probabilities resulting from application of the procedure 1^0–3^0, for comparisons $g_k(x_i, x_j)$.

The relationship (30) indicates that the probability of errorless estimate, i.e. $P(T(x_i,x_j)=\widehat{T}(x_i,x_j))(<i,j>\in R_m)$, converges to one for $N\to\infty$ (consistency of the estimator). It is obvious that reduction of probabilities of errors of comparisons δ increases probability of errorless estimate.

6 Summary and Conclusions

The paper presents statistical tools (test and procedure), which allow application of the NAO method for estimation of strict preference relation on the basis of multiple pairwise comparisons in binary form - in the case of large sets of elements (for low and medium sets are available exact mathematical programming algorithms). These tools guarantee also obtaining good statistical properties of an estimate – i.e. the form of preference relation. Application of the test and procedure is not complex and does not require high computation costs. The estimate obtained with the use of heuristic algorithm can be also used as starting point for exact algorithm; it is necessary in the case of value of the criterion functions (5, 28) significantly higher than zero.

References

Bezembinder, T.G.: Circularity and consistency in paired comparisons. Brit. J. Stat. Psychol. **34**, 16–37 (1981)

Bradley, R.A.: Paired comparisons: some basic procedures and examples. In: Krishnaiah, P.R., Sen, P.K. (eds.) Handbook of Statistics, vol. 4, pp. 299–326. North-Holland, Amsterdam (1984)

David, H.A.: The Method of Paired Comparisons. 2nd edn. Ch. Griffin, London (1988)

David, H.A.: Order Statistics. Wiley, New York (1970)

Hoeffding, W.: Probability inequalities for sums of bounded random variables. JASA **58**, 13–30 (1963)

Klukowski, L.: Some probabilistic properties of the nearest adjoining order method and its extensions. Ann. Oper. Res. **51**, 241–261 (1994)

Klukowski, L.: Estimation and verification of relations and trees on the basis of multiple binary and multivalent pairwise comparisons. Systems research institute polish academy of sciences, Series: Systems Research, vol. 79. Warsaw (2011)

Klukowski, L.: Determining an estimate of an equivalence relation for moderate and large sized sets. Oper. Res. Decis. **27**, 45–58 (2017)

Philips J.P.N.: A procedure for determining Slater's i and all nearest adjoining orders. Brit. J. Stat. Psychol. **20**, 217–225 (1967)

Philips J.P.N.: A further procedure for determining Slater's i and all nearest adjoining orders. Brit. J. Stat. Psychol. **22,** 97–101 (1969)

Remage, R., Jr., Thomson W.A., Jr.: Maximum-likelihood paired comparisons rankings. Biometrika 53, 143–149 (1966)

Slater, P.: Inconsistencies in a schedule of paired comparisons. Biometrika **48**, 303–312 (1961)

Improved CSO Algorithm in Practical Applications

Hubert Zarzycki(✉)

Tadeusz Kosciuszko Military Academy of Land Forces, Wroclaw, Poland
hubert.zarzycki@awl.edu.pl

Abstract. The paper presents the CSO algorithm in the area of solutions inspired by nature. The algorithm was applied to the problem of creating a diet. After implementing the algorithm, research was carried out and interesting solutions were obtained. In the analysis of the results, the advantages and disadvantages of the CSO approach were indicated. The article helps to understand how the CSO algorithm can be used and adapted to practical applications such as diet selection.

Keywords: CSO · swarm intelligence · recommendation of diets · nature-inspired algorithms

1 Introduction

The progressive development of the application of artificial intelligence methods in practice prompts modern researchers to look for new solutions. The inspiration here are the phenomena often occurring in the world of nature, models of behavior and the observation of adaptive abilities that enabled organisms to survive even in extreme conditions. For many generations, behaviors have been transformed, becoming more and more perfect. Today, the world of nature is full of natural solutions and empirical data that enable the discovery and construction of new solutions and algorithms, the application of which can bring measurable benefits [11].

The article analyzes one of the nature-inspired algorithms. The potential of such algorithms is very large, because the number of natural solutions that are a source of inspiration around us is practically unlimited. Nature-based algorithms make it possible to find optimal ways to solve complex problems. Nature-inspired algorithms among other the genetic algorithm (GA). There is a subgroup of these solutions based on swarm intelligence (SI). These include the bee colony algorithm (ABC), the ant algorithm (ACO), the particle swarm optimization (PSO) [9], the firefly algorithm (FA), the whale algorithm (WOA), the bat algorithm (BA) [1, 6]. One of the relatively new algorithms is Cat Swarm Optimization, which is the subject of this study.

The author of the article has been working on the practical use of artificial intelligence methods for a long time. This paper is an extension of work on the various use of artificial intelligence in the field of nature-inspired algorithms [23, 25], swarm intelligence [8, 26, 27], ordered fuzzy numbers [7, 20–22], optimization methods[4, 5, 17] and performance analysis of information systems [18, 24].

K. T. Atanassov et al. (Eds.): BOS/SOR/IWIFSGN 2022, LNNS 793, pp. 261–268, 2023.
https://doi.org/10.1007/978-3-031-45069-3_24

2 CSO Algorithm

Cat Swarm Optimization is a meta-heuristic based algorithm inspired by the behavior of cats. Many agents (cats) are used in the cat algorithm. Each of them performs part of the task. This makes it possible to solve many optimization problems [14].

There are more than 30 species of cats, including lmpart, tiger, lion, cheetah, etc. A special behavior of these animals has been observed. As it turns out, domestic cats also behave a bit differently from their cousins. When cats are awake, they are usually inactive, moving lazily. However, their vigilance is very high, even when they are resting. Lying down, they look around or even stare into the eyes. They watch every moving object. They can quickly go into a state of high activity. In order to model these behaviors, two modes of cat operation were proposed, i.e. the tracking mode and the searching mode. Chu et al. [2], after building and testing the algorithm, found its high efficiency in solving optimization problems.

The input data include among other:
M – number of dimensions,
$\left[p_{i,j}^{min}, p_{i,j}^{max}\right]$ – variability range for j-th dimension and i-th agent (cat)
MR – mixture ratio,
N - total number of agents (cats),
SMP - seeking memory pool,
CDC - count of dimensions to change,
SRD - selected dimension seeking range,
SP C - the self-position consideration,
max – maximum number of iterations,
c_1 - constant used to modify the velocity of the cats,
N- number of cats
G_{best} – returned as the result the best position obtained by one of the agents (cats) [16].

Below is a presentation of the CSO algorithm:

1. N cats are randomly distributed in M-dimensional space. The placements must not exceed the limits of the initial value. Velocity is generated for each dimension. The cats are divided into two groups according to the MR factor. MR determines how many cats are to move in seeking mode. The other cats move in tracing mode. A movement (seeking/tracing) flag is set for each cat.
2. Evaluating the fitness value of cats by running the fitness function for the coordinates of individual cats. The fitness function represents the characteristics of the problem being solved.

Efficiency is calculated for each cat. The coordinates (G_{best}) and the fitness value of the cat with the best score are stored.

3. Read the status of the traffic flag - track or search mode. Move the cats according to the flag.
4. Reset agent (cat) movement flag. The MR factor divides cats into a group in tracking mode and a group in search mode.

5. The program goes to the end state if the end condition is met. In case of completion, the coordinates of the best solution are displayed. Otherwise, it returns to step number 2 of the algorithm (Fig. 1).

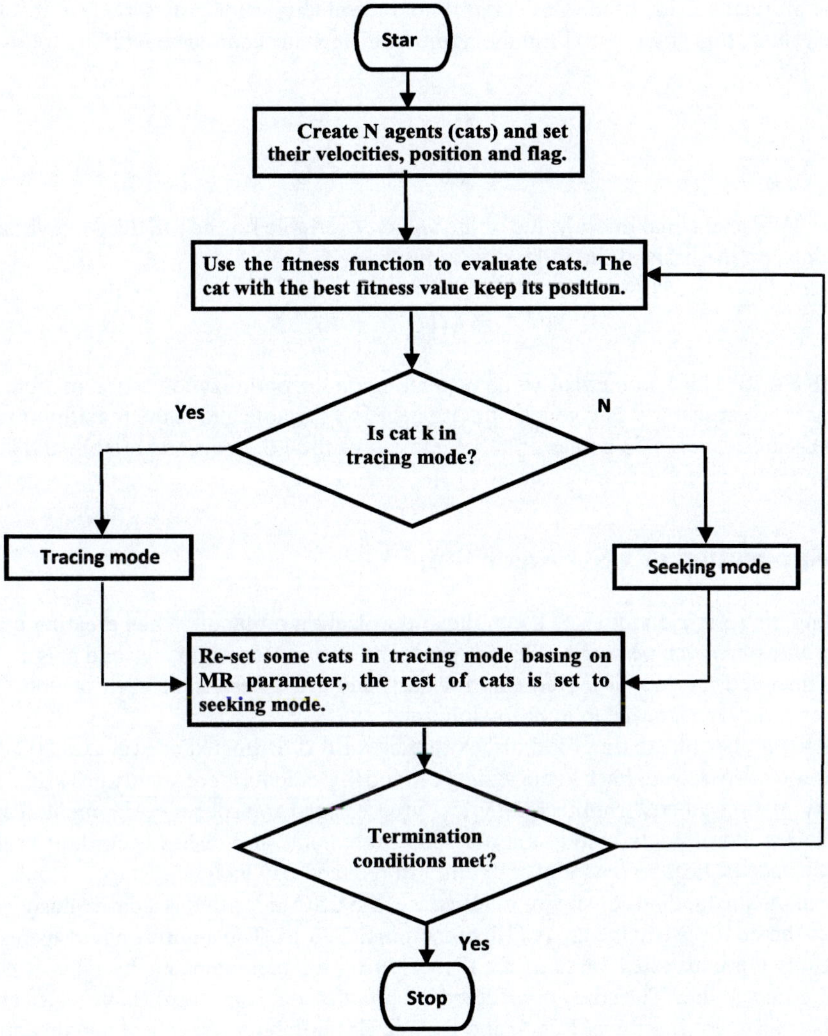

Fig. 1. CSO algorithm.

There are several solutions based on cat behavior;
Varieties of CSO algorithm:

- Harmonious Cat Swarm Optimization (HCSO) [10],
- Binary Cat Swarm Optimization (BCSO) [3],

- Parallel Cat Swarm Optimization (PCSO) [19],
- Crazy Cat Swarm Optimization (CCSO) [15].

3 Improvement of the CSO Algorithm

In the original version of the CSO algorithm, the velocity value grows too fast, reaching huge values. It is possible to limit the number of steps in accordance with the following formulas:

$$V_{i,j} = \begin{cases} V_{i,j} if \left|V_{i,j}\right| < V_j^{max} \\ V_j^{max} if \left|V_{i,j}\right| \geq V_j^{max} \end{cases}$$

where V_j^{max} is the maximum value of the velocity variable for the j-th decision variable. Obtained on the basis of the following formula:

$$V_j^{max} = \delta x\left(P_j^{max} - P_j^{min}\right)$$

where $\delta \in [0, 1]$ is a numerical value depending on the optimization problem, obtained by trial and error. P_j^{max} is given for the jth decision variable and is the maximum value in the search domain. Whereas P_j^{max} is defined for the j-th decision variable and is the minimum value in the search domain [16].

4 Application of CSO Algorithm

Finding diets for individuals is a complex optimization problem. When creating a diet, the profile of a given person is taken into account, as well as weight, age and height. The recommended nutritional ingredients for each day are different for each person. This creates a very large space to look for solutions.

The number of existing food dishes in the world is estimated at over 200,000. The number of dishes available and the variety of food ingredients is constantly growing. This results, among others, from technological progress and transport development. Today, people are increasingly able to choose food ingredients and dishes according to their choices, needs, health constraints or ability to reduce body weight [12].

Among the methods of swarm intelligence, the CSO algorithm is a particularly well-suited choice for use in the area of diet recommendations. The algorithm returns results regarding the nutritional value of the dishes. Diets are homogeneous by type, it is the Mediterranean diet. The study considered a daily diet that consists of four meals. These meals are Breakfast, Lunch, Lunch and Dinner. Each of them consists of a main course, second course and dessert. The figure below visualizes meals for one day [13] (Fig. 2).

Using CSO to recommend diets is the right approach because it generates acceptable near-optimal solutions in a short time. Searching the entire space of all solutions would be an extremely long process. CSO's heuristic algorithm returns results quickly, and these are different sets of solutions each time it is run. Thanks to this, the diversity of the meals offered is ensured. CSO has two modes of searching for solutions. These are agents (cats) in search mode and in track mode. This works well in the case study of finding

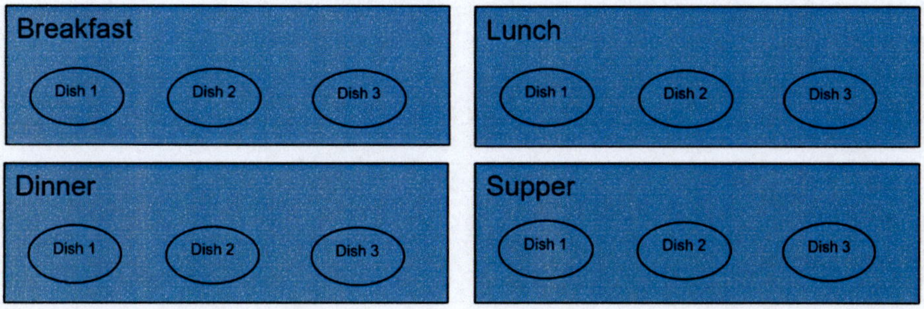

Fig. 2. Daily diet that consists of four meals. Each of them of a main course, second course and dessert [13].

solutions to a homogeneous (Mediterranean) diet. Differentiation of the uniformity of created diets can be done by appropriate adjustment of parameters such as the number of CDC dimensions, SPC, i.e. the flag determining whether the cat's position is taken into account, and the base MR parameter determining how many cats are in the search mode and how many in the tracking mode.

5 Application of CSO Algorithm

In the work, cases of dishes based on the Mediterranean diet are considered. The following data were used for calculations: the search space contains $12 \times 4 = 48$ dimensions. The number of generations is 50. The number of generations can be determined experimentally by observing whether the results from the previous and the next generation change sufficiently (e.g. by 0.1 0.001). Another variable N representing the number of agents (cats) is initially set to 10. The MR factor is equal to 0.3. This means that each generation of the algorithm is based on 3 (0.3×10) cats in the tracking mode and 7 cats in the search mode [13].

The SRD value is set to 0.2. An SMP of 5 means that 5 copies of the cat are generated in search mode. The coefficient c1 contains the number 0.5.

For the i-th cat in each of the j-th dimension z, the range of variation $\left[P_{i,j}^{min}, P_{i,j}^{max}\right]$ corresponds to the minimum and maximum ranges of nutritional properties.

6 Obtained Results

The study below considered a female, age 55, weight 71 kg, height 173 cm. The resulting recommended values for a single day matched to the profile of the model woman were as follows: 1924 energy (calories), 134 proteins (g), 198 carbohydrates (g) and 63 lipids (g) (Table 1).

The study below considered a male, age 75, weight 84 kg, height 178 cm. The obtained recommended values for a single day matched to the reference male profile were as follows: 2145 energy (calories), 157 proteins (g), 224 carbohydrates (g) and 73 lipids (g) (Table 2).

Table 1. Nutrient data per day for an example person in terms of maximum, minimum and optimum values

Neutrient	Minimum Value/Dish	Maximum Value/Dish	Optimal Value/Day
Proteins (g)	*0.3*	*39*	*134*
Lipids (g)	*0.1*	*20.8*	*63*
carbohydrates (g)	*6.5*	*35.4*	*198*
Energy (cal)	*27*	*358*	*1924*

Table 2. Nutrient data per day for an example person in terms of maximum, minimum and optimum values

Neutrient	Minimum Value/Dish	Maximum Value/Dish	Optimal Value/Day
Proteins (g)	*1.7*	*42.1*	*157*
Lipids (g)	*0.4*	*21.2*	*73*
carbohydrates (g)	*0.6*	*35.9*	*224*
Energy (cal)	*47*	*849*	*2145*

7 Conclusions

In this paper the CSO algorithm was presented as well as velocity clamping - a modifications of the CSO algorithm for diets recommendation were demonstrated and discussed. The CSO algorithm was applied to a recommendation of diets problem.

The CSO algorithm makes it possible to obtain near-optimal results in a short time. By adjusting parameters such as the number of generations, the number of agents (cats), the number of dimensions or the MR mixture ratio, you can profile the results and control the diversity of diets. In future works, it will be possible to take into account additional parameters, such as the exclusion of selected ingredients from the diet, limiting the scope regarding the price and quality of products, defining groups related to taste and smell. This can be done e.g. by initiating agents (cats) to specific diet ranges and searching for agents in their neighborhood. This will allow you to create diets with similar parameters.

References

1. Panigrahi, B.K., Shi, Y., Lim, M.H., Red., Handbook of Swarm Intelligence, t. 8, Adaptation, Learning, and Optimization. Springer, Heidelberg (2011)
2. Chu, S.-C., Tsai, P.-W., Pan, J.-S.: Cat swarm optimization. In: Yang, Q., Webb, G. (eds.) PRICAI 2006. LNCS (LNAI), vol. 4099, pp. 854–858. Springer, Heidelberg (2006). https://doi.org/10.1007/978-3-540-36668-3_94
3. Crawford, B., Soto, R., Berrios, N., Johnson, F., Paredes, F.: Binary cat swarm optimization for the set covering problem. In: 10th Iberian Conference on Information Systems and Technologies (CISTI), pp. 1–4 (2015)

4. Czerniak, J.M., Zarzycki, H., Apiecionek, Ł., Palczewski, W., Kardasz, P.: A cellular automata-based simulation tool for real fire accident prevention. Math. Probl. Eng. **2018**, 12 (2018). Article ID 3058241
5. Dobrosielski, W.T., Czerniak, J.M., Zarzycki, H., Szczepański, J.: Fuzzy numbers applied to a heat furnace control, theory and applications of ordered fuzzy numbers - a tribute to Professor Witold Kosiński. Stud. Fuzziness Soft Comput. **356**, 69–288 (2017)
6. Engelbrecht, A.P.: Fundamentals of Computational Swarm Intelligence. Wiley, Hoboken (2005)
7. Ewald, D., Czerniak, J.M., Zarzycki, H.: OFNBee method used for solving a set of benchmarks. In: Kacprzyk, J., Szmidt, E., Zadrożny, S., Atanassov, K., Krawczak, M. (eds.) Advances in Fuzzy Logic and Technology 2017. EUSFLAT IWIFSGN 2017 2017. Advances in Intelligent Systems and Computing, vol. 642, pp. 24–35. Springer, Cham (2018). https://doi.org/10.1007/978-3-319-66824-6_3
8. Ewald, D., Zarzycki, H., Apiecionek, Ł, Czerniak, J.M.: Ordered fuzzy numbers applied in bee swarm optimization systems. J. Univ. Comput. Sci. **26**(11), 1475–1494 (2020)
9. Imran, M., Hashim, R., Khalid, N.E.A.: An overview of particle swarm optimization variants. Procedia Eng. **53**, 491–496 (2013)
10. Lin, K.C., Zhang, K.Y., Hung J.C.: Feature selection of support vector machine based on harmonious cat swarm optimization. In: 7th International Conference on Ubi-Media Computing and Workshops, pp. 205–208 (2014)
11. Merkle D.: Swarm Intelligence: Introduction and Application, Springer, Heidelberg (2008). https://doi.org/10.1007/978-3-540-74089-6
12. Mohammed, H.A., Hagras, H.: Towards developing type 2 fuzzy logic diet recommendation system for diabetes. In: 10th Computer Science and Electronic Engineering (CEEC), pp. 56–59 (2018)
13. Moldovan, D., et al.: Diet generator for elders using cat swarm optimization and wolf search. In: International Conference on Advancements of Medicine and Health Care through Technology, pp. 238–243 (2016)
14. Agarwal, P., Mehta, S.: Nature-inspired algorithms: state-of-art, problems and prospects. Int. J. Comput. Appl. **100**, 14–21 (2014)
15. Sarangi, A., Sarangi, S.K., Mukherjee, M., Panigrahi S.P.: System identification by crazy-cat swarm optimization. In: International Conference on Microwave, Optical and Communication Engineering (ICMOCE), pp. 439–442 (2015)
16. Slowik A.: Swarm INTELLIGENCE ALGORITHMS. Modifications and Applications. CRC Press/Taylor and Francis, Boca Raton (2020)
17. Skubisz, O., Zarzycki, H., Wincewicz-Bosy, M., Ewald, D.: Applying Improved Bee-Inspired Algorithm for the Vehicle Routing Problem, International Business Information Management Association. IBIMA Publishing, Seville (2021)
18. Śmigielski, G., Dygdała, R., Zarzycki, H., Lewandowski, D.: Real-time system of delivering water-capsule for firefighting. In: Kobayashi, S., Piegat, A., Pejaś, J., El Fray, I., Kacprzyk, Janusz (eds.) ACS 2016. AISC, vol. 534, pp. 102–111. Springer, Cham (2017). https://doi.org/10.1007/978-3-319-48429-7_10
19. Tsai, P.W., Pan, J.S., Chen, S.M., Liao, B.Y., Hao S.-P.: Parallel cat swarm optimization. In: International Conference on Machine Learning and Cybernetics, vol. 2, pp. 3328–3333 (2008)
20. Zarzycki, H., Apiecionek, Ł., Czerniak, J.M., Ewald, D.: The proposal of fuzzy observation and detection of massive data DDOS attack threat. Atanassov, K., *et al.* (eds.) Uncertainty and Imprecision in Decision Making and Decision Support: New Challenges, Solutions and Perspectives. IWIFSGN 2018. Advances in Intelligent Systems and Computing, vol. 1081, pp. 363–378. Springer, Cham (2021). https://doi.org/10.1007/978-3-030-47024-1_34

21. Zarzycki, H., Czerniak, J.M., Dobrosielski, W.T.: Detecting nasdaq composite index trends with OFNs. In: Prokopowicz, P., Czerniak, J., Mikołajewski, D., Apiecionek, Ł, Ślęzak, D. (eds.) Theory and Applications of Ordered Fuzzy Numbers. SFSC, vol. 356, pp. 195–205. Springer, Cham (2017). https://doi.org/10.1007/978-3-319-59614-3_11
22. Zarzycki, H., Dobrosielski, W.T., Vince, T., Apiecionek, Ł.: Center of circles intersection, a new defuzzification method on fuzzy numbers. Bulletin of the Polish Academy of Sciences. Technical Sciences (2020)
23. Zarzycki, H., Ewald, D., Skubisz, O., Kardasz, P.: A comparative study of two nature-inspired algorithms for routing optimization. In: Atanassov, K.T., *et al.* Uncertainty and Imprecision in Decision Making and Decision Support: New Advances, Challenges, and Perspectives. IWIFSGN BOS/SOR 2020 2020. Lecture Notes in Networks and Systems, vol. 338, pp. 215–228. Springer, Cham (2021). https://doi.org/10.1007/978-3-030-95929-6_17
24. Zarzycki, H., Czerniak, J.M., Lakomski, D., Kardasz, P.: Performance comparison of CRM systems dedicated to reporting failures to IT department. In: Madeyski, L., Śmiałek, M., Hnatkowska, B., Huzar, Z. (eds.) Software Engineering: Challenges and Solutions. Advances in Intelligent Systems and Computing, vol. 504, pp 133–146. Springer, Cham (2016). https://doi.org/10.1007/978-3-319-43606-7_10
25. Zarzycki, H., Skubisz, O.: A new artificial bee colony algorithm approach for the vehicle routing problem. In: Kahraman, C., Cebi, S., Cevik Onar, S., Oztaysi, B., Tolga, A.C., Sari, I.U. (eds.) INFUS 2021. LNNS, vol. 307, pp. 562–569. Springer, Cham (2022). https://doi.org/10.1007/978-3-030-85626-7_66
26. Zarzycki, H., Skubisz, O.: An application of ant algorithm for routing optimization problem. In: International Business Information Management Association, IBIMA publishing (2021)
27. Zarzycki H.: Comparative study of the firefly algorithm and the whale algorithm. In: Kahraman, C., Tolga, A.C., Cevik Onar, S., Cebi, S., Oztaysi, B., Sari, I.U. (eds.) Intelligent and Fuzzy Systems. INFUS 2022. Lecture Notes in Networks and Systems, vol. 504, pp. 999–1006. Springer, Cham (2022). https://doi.org/10.1007/978-3-031-09173-5_114

On the Quasi-Efficient Frontier of the Set of Optimal Portfolios Under Hybrid Uncertainty with Short Sales Allowed

Stepan A. Rogonov, Ilia S. Soldatenko(✉), and Alexander V. Yazenin

Tver State University, 33 Zhelyabova str., 170100 Tver, Russia
{Rogonov.SA,soldis,Yazenin.AV}@tversu.ru

Abstract. The paper describes methods for constructing a quasi-efficient frontier of minimum risk portfolio under conditions of hybrid uncertainty with allowed short sales. Investor's acceptable level of expected return is defined in crisp and fuzzy forms. Obtained results are illustrated on a model example. The dependence of the quasi-efficient frontier on the value of α-level is investigated.

Keywords: portfolio selection model · hybrid uncertainty · possibility · necessity · quasi-efficient solutions · minimum risk portfolio frontier · fuzzy random variable

1 Introduction

Classical mean-variance portfolio selection model was proposed by Markowitz [1,2] and since then played an important role in the formation of modern portfolio theory. Its basic idea is to characterize a financial asset as a random variable with a probability distribution over its return and to quantify the expected return of portfolio as the investment gain and use its variance as the investment risk.

However, in real life it is usually impossible for investors to get the precise probability distributions of the assets' returns. In real world, there are many non-probabilistic factors that affect the markets. With the introduction of fuzzy sets and possibility theory by Zadeh [3,4], Nahmias [5], Dubois and Prade [6] many scholars began to employ this theory to manage portfolios in a fuzzy environment, because fuzzy approaches are, in general, more appropriate than probabilistic approaches for taking human's subjective opinions into consideration. For example, Tanaka et al. [7] and Inuiguchi and Tanino [8] modeled security returns with fuzzy variables with possibility distributions and proposed the possibilistic portfolio selection models, respectively, Parra et al. [9] proposed a fuzzy goal programming approach for portfolio selection. Ammar and Khalifa [10] proposed a quadratic programming approach for fuzzy portfolio selection problem. Zhang and Nie [11] proposed the admissible efficient portfolio model.

The work was financially supported by RFBR (project No 20-01-00669).

K. T. Atanassov et al. (Eds.): BOS/SOR/IWIFSGN 2022, LNNS 793, pp. 269–280, 2023.
https://doi.org/10.1007/978-3-031-45069-3_25

Bilbao-Terol et al. [12] employed a fuzzy compromise programming technique to deal with the task. Giove et al. [13] constructed a regret function to solve the interval portfolio problem. Vercher [14] employed semi-infinite programming technique to solve the portfolio selection problem with fuzzy returns. And many more authors introduced fuzziness into portfolio theory.

The abovementioned portfolio selection models are mainly based on either probability theory or fuzzy/possibility theory, therefore only one kind of uncertainty is considered. In reality randomness and fuzziness are often combined together and that leads to a hybrid uncertainty. Fuzzy random variables that were introduced by Kwakernaak [15] and then enhanced by Nahmias [16] and other authors are appropriate ways to describe the hybrid uncertainty. Some studies on the fuzzy random programming can be found, for example, in Yazenin [17], Liu [18], Luhandjula [20], Li et al. [21]. Yazenin in [22,23] first introduced formulation of portfolio selection problem under hybrid uncertainty of possibilistic-probabilistic type. Huang [24] employed the random fuzzy theory by Liu [18,19] to study portfolio selection in a random fuzzy environment in which the security returns are assumed to be stochastic variables with fuzzy information.

In the current paper we continue our previous works [23,25–27], where we also immersed Markowitz portfolio model in the context of hybrid uncertainty of the possibilistic-probabilistic type. However, in these studies only individual quasi-efficient solutions to the portfolio optimization problem were built. The question of constructing the whole set of quasi-efficient solutions in an analytical form remained open.

One of the classical methods for constructing an efficient minimum risk portfolio frontier is considered by Barbaumov [28]. In the present work this method is extended to a number of minimal risk portfolio models under conditions of hybrid uncertainty with allowed short sales and with fuzzy level of expected return acceptable to an investor. The obtained results are demonstrated on a numerical model example for a three-dimensional portfolio.

2 Minimal Risk Portfolio with Allowed Short Sales

The model constructed according to the theory of Markowitz [1] does not allow the possibility of opening short positions on securities. In turn, there is no such restriction in the Black approach, that is, the values of the shares of financial assets of the portfolio being formed can be both positive and negative (see [28]). In this paper, we will consider investment portfolio model with allowed short sales.

Let's assume that there are n different assets on the market that an investor is interested in. The investor forms a portfolio by setting the values of the weight vector $w = (w_1, w_2, \ldots, w_n)$, where w_i is the share of capital invested in securities of the i-th type. At the same time, obviously, the normalization condition $\sum_{i=1}^{n} w_i = 1$ must be met.

Let $R_i(\omega)$ is the profitability of the i-th asset, which in classical portfolio analysis is modeled using a random variable. Then the return of the entire portfolio is the weighted sum of its assets: $R(w, \omega) = \sum_{i=1}^{n} R_i(\omega) w_i$. Since the yield is a random variable, the investment portfolio can be described by two numerical characteristics: its mathematical expectation and variance, which are named, respectively, the expected return $\mathbf{E}[R(w, \omega)]$ and risk $\mathbf{D}[R(w, \omega)]$ of the portfolio. Then, in accordance with the classical Markowitz-Black theory, the minimal risk portfolio model with allowed short sales in the most general form can be written as follows:

$$\mathbf{D}[R(w, \omega)] \to \min_{w}, \tag{1}$$

$$\begin{cases} \mathbf{E}[R(w, \omega)] \geq r, \\ \sum_{i=1}^{n} w_i = 1, \end{cases} \tag{2}$$

where r is a level of profitability acceptable to an investor. Sometimes in the model of acceptable portfolios (2) inequality is replaced by equality, which, firstly, somewhat simplifies the solution of the optimization problem itself, and secondly, does not change its essence, because, as is known from classical portfolio optimization, if r is on the efficient frontier of the portfolio, then the minimum value of risk is achieved with equality in (2). We will replace the inequality with equality in the constraint later, because for now it will only complicate immersion of the model into possibility-necessity context.

3 Minimal Risk Portfolio in Conditions of Hybrid Uncertainty

All the concepts and definitions from the theory of possibility that are used in this paper can be found, for example, in the works [23,25–27,30,31].

We will model profitability of i-th financial asset with a fuzzy random variable $R_i(\omega, \gamma)$. For a better intuitive understanding of this model, let's imagine a situation where a financial expert is asked to evaluate profitability of a certain financial asset. Both the profitability and its assessment by an expert are uncertain values. We assume that the uncertainty of market conditions is probabilistic. On the other hand, the uncertainty of an expert's assessment is more naturally described by some distribution of possibilities. For example, one way to model these uncertainties would be the following. An expert gives his estimate of spread for profitability of the asset in the form of, for example, a triangular or trapezoidal fuzzy value. At the same time, boundaries of the spread, its width and relative position can be modified by random components of a fuzzy random variable of the asset. Shift-scale representation of a fuzzy random variable described in Sect. 5 is a convenient tool for constructing such a model.

The exact representation of a fuzzy random variable is not essential at the moment. A specific example will be discussed in Sect. 4.

Let's consider a model of minimizing the risk of a portfolio with restrictions on possibility (necessity) on its expected return. According to the classical model (1)–(2) it is necessary to build a portfolio risk function, and its return should be included in the restrictions system. First, we will identify all the components of the model.

In the conditions of hybrid uncertainty of possibilistic-probabilistic type return of an investment portfolio will be a fuzzy random function:

$$R(w, \omega, \gamma) = \sum_{i=1}^{n} R_i(\omega, \gamma) w_i, \tag{3}$$

which is a linear function of the equity shares $w = (w_1, w_2, \ldots, w_n)$ in the investment portfolio.

We will use the so-called indirect method to solve the optimization problem. The essence of the method is to construct an equivalent deterministic analogue of a possibilistic-probabilistic problem, in this case there will be no more randomness and fuzziness, and therefore it can further be solved by "ordinary" methods. This can be done by stepwise removal of uncertainty. The uncertainty of probabilistic type will be removed based on principle of expected return. To do this, we identify possibility distribution of mathematical expectation of the function $R(w, \omega, \gamma)$. Basing on the properties of mathematical expectation, we obtain the following formula for the expected return of the portfolio:

$$\mathbf{E}[R(w, \omega, \gamma)] = \mathbf{E}\left[\sum_{i=1}^{n} R_i(\omega, \gamma) w_i\right] = \sum_{i=1}^{n} \hat{R}_i(\gamma) w_i, \tag{4}$$

where $\hat{R}_i(\gamma)$ is expected value of a fuzzy random variable $R_i(\omega, \gamma)$.

Next, we need a formula to calculate the covariance of two fuzzy random variables $R_i(\omega, \gamma)$ and $R_j(\omega, \gamma)$. According to Feng's formula [29], it will be determined through the covariance of their α-level sets:

$$\mathrm{cov}(R_i, R_j) = \frac{1}{2} \int_0^1 \Big(\mathrm{cov}\big(R_i^-(\omega, \alpha), R_j^-(\omega, \alpha)\big) + \mathrm{cov}\big(R_i^+(\omega, \alpha), R_j^+(\omega, \alpha)\big)\Big) d\alpha,$$

where $R_k^-(\omega, \alpha)$, $R_k^+(\omega, \alpha)$ are boundaries of an α-level set of fuzzy value $R_k(\omega, \gamma)$. Note that covariance, and, accordingly, variance of a fuzzy random variable, according to this definition, will be crisp quantities.

Given that the variance of a (fuzzy) random variable X is $\mathbf{D}[X] = \mathrm{cov}(X, X)$, with the above-stated assumptions and properties of dispersion determined by Feng [29], we write the formula for $\mathbf{D}[R(w, \omega, \gamma)]$:

$$\mathbf{D}[R(w, \omega, \gamma)] = \sum_{i=1}^{n} w_i^2 \sigma_i^2 + 2 \sum_{i<j}^{n} w_i w_j \sigma_{ij}, \tag{5}$$

where $\sigma_{ij} = \mathrm{cov}(R_i(\omega, \gamma), R_j(\omega, \gamma))$, $\sigma_i^2 = \mathbf{D}\left[R_i(\omega, \gamma)\right]$,

$$\mathbf{D}[R_i(\omega, \gamma)] = \frac{1}{2} \int_0^1 (\mathbf{D}\left[R_i^-(\omega, \alpha)\right] + \mathbf{D}\left[R_i^+(\omega, \alpha)\right]) d\alpha.$$

Since the expected return (4) of a portfolio in conditions of hybrid uncertainty is a fuzzy value, in order to remove the uncertainty of possibilistic type, we will introduce a restriction on possibility (necessity) on the level of expected return acceptable to an investor (which in general may be fuzzy) into the system of restrictions that defines the set of acceptable portfolios. Then the model of the minimum risk portfolio by Black will take the following form with a crisp level r (left constraint) or fuzzy level $r(\gamma)$ (right constraint) of expected return:

$$\mathbf{D}[R(w,\omega,\gamma)] \to \min_{w}, \tag{6}$$

$$\begin{cases} \tau\{\mathbf{E}[R(w,\omega,\gamma)] \geq r\} \geq \alpha \ \ \textbf{OR}\ \tau\{\mathbf{E}[R(w,\omega,\gamma)] \geq r(\gamma)\} \geq \alpha, \\ \sum_{i=1}^{n} w_i = 1, \end{cases} \tag{7}$$

where $\tau \in \{\pi, \nu\}$, π is a measure of possibility, ν is a measure of necessity, α is a level of possibility/necessity.

After that, we will remove the uncertainty of the possibilistic type using an indirect method, namely, we will build an equivalent deterministic analogue of models of acceptable portfolios (7). Based on the results of [30], we obtain the following equivalent deterministic analogues in the various contexts for crisp and fuzzy levels of expected return:

$$\mathbf{D}[R(w,\omega,\gamma)] \to \min_{w}, \tag{8}$$

$$\begin{cases} \sum_{i=1}^{n} \bar{R}_i w_i \geq \bar{r}, \\ \sum_{i=1}^{n} w_i = 1, \end{cases} \tag{9}$$

where

- $\bar{r} = r$ – in case of a crisp level of expected return,
- $\bar{r} = r^{-}(\alpha)$ – in the case of a fuzzy level of expected return in a possibilistic context,
- $\bar{r} = r^{+}(1-\alpha)$ – in the case of an fuzzy level of expected return in the necessity context,
- $\bar{R}_i = \hat{R}_i^{+}(\alpha)$ in the context of possibility measure,
- $\bar{R}_i = \hat{R}_i^{-}(1-\alpha)$ in the context of necessity measure,
- $r^{\pm}(\cdot)$ and $\hat{R}_i^{\pm}(\cdot)$ – are α-levels of corresponding fuzzy values.

4 Quasi-efficient Frontier of Minimum Risk Portfolio

As it is known from the classical portfolio theory, the result of solving the problem of a minimum risk portfolio is its efficient frontier, which is the dependence of the minimum risk of the portfolio on its expected return. In our case, we call this curve a *quasi-efficient* frontier, because due to the uncertainty of possibilistic type that we introduced to the model it now depends on the level of α, which is set by the expert.

In [28] a method for constructing an efficient frontier of the classical minimal risk portfolio (1)–(2) based on the Lagrange multiplier method is considered. We generalize it to the case of a minimal risk portfolio in conditions of hybrid uncertainty of possibilistic-probabilistic type (8)–(9). To do this, first we replace the inequality in (9) with equality. As mentioned above, if $\bar{r}$ is on the efficient portfolio frontier, then the minimum risk value is achieved when the equality in (9) is met. As a result, the Lagrange function will look like this:

$$\mathcal{L}(w,\omega,\gamma) = \mathbf{D}[R(w,\omega,\gamma)] + \lambda_1\left(\sum_{i=1}^{n}\overline{R}_i w_i - \bar{r}\right) + \lambda_2\left(\sum_{i=1}^{n} w_i - 1\right).$$

After substituting formula for the variance (5) we get:

$$\mathcal{L}(w,\omega,\gamma) = \sum_{i=1}^{n} w_i^2\sigma_i^2 + 2\sum_{i<j}^{n} w_i w_j \sigma_{ij} + \lambda_1\left(\sum_{i=1}^{n}\overline{R}_i w_i - \bar{r}\right) + \lambda_2\left(\sum_{i=1}^{n} w_i - 1\right). \quad (10)$$

Next we differentiate (10) by variables w_i, $i = 1,\dots,n$, λ_1, λ_2, the results of differentiation we equate to zero and obtain the following system of linear equations:

$$\begin{cases} 2\sigma_1^2 w_1 + 2\sigma_{12}w_2 + \dots + 2\sigma_{1n}w_n + \lambda_1\overline{R}_1 + \lambda_2 = 0, \\ 2\sigma_{21}w_1 + 2\sigma_2^2 w_2 + \dots + 2\sigma_{2n}w_n + \lambda_1\overline{R}_2 + \lambda_2 = 0, \\ \dots \\ 2\sigma_{n1}w_1 + 2\sigma_{n2}w_2 + \dots + 2\sigma_n^2 w_n + \lambda_1\overline{R}_n + \lambda_2 = 0, \\ w_1 + w_2 + \dots + w_n = 1, \\ \overline{R}_1 w_1 + \dots + \overline{R}_n w_n = \bar{r}. \end{cases} \quad (11)$$

Having solved the (11) system using the Gauss-Jordan method, we get the result

$$w^*(\bar{r}) = \{a_1 + b_1\bar{r},\ a_2 + b_2\bar{r},\ \dots,\ a_n + b_n\bar{r}\}, \quad (12)$$

where numbers a_i and b_i $(i = 1,2,\dots,n)$ are the coefficients that depend only on mathematical expectations, variances and covariances between the returns of various assets and do not depend on the level of expected return $\bar{r}$.

Substituting (12) into (5), we find a quasi-efficient frontier of the set of investment portfolios in the context of hybrid uncertainty with allowed short sales:

$$D^*(\bar{r}) = \sqrt{A\bar{r}^2 + B\bar{r} + C}, \text{ where } \bar{r} \geq -\frac{B}{2A}, \quad (13)$$

$$A = \sum_{i=1}^{n}\sigma_i^2 b_i^2 + 2\sum_{i<j}\sigma_{ij}b_i b_j, \quad B = 2\sum_{i=1}^{n}\sigma_i^2 a_i b_i + 2\sum_{i<j}\sigma_{ij}(a_i b_j + b_i a_j),$$

$$C = \sum_{i=1}^{n}\sigma_i^2 a_i^2 + 2\sum_{i<j}\sigma_{ij}a_i a_j.$$

A visualization of the resulting frontier can be seen in Sect. 5.

5 Specification of a Model for One Class of Fuzzy Random Variables

We specify a minimal risk portfolio model under conditions of hybrid uncertainty and a method for constructing its quasi-efficient frontier for one particular class of fuzzy random variables $R_i(\omega, \gamma)$, which we will model using a shift-scale representation:

$$R_i(\omega, \gamma) = a_i(\omega) + \delta_i(\omega)X_i(\gamma).$$

Here the fuzzy values $X_i(\gamma) \in Tr(c_i, d_i)$ are mutually min-related, and the shift and scale coefficients $a_i(\omega)$, $\delta_i(\omega)$ are independent random variables uniformly distributed over the segments $[l_i^a, r_i^a]$ and $[l_i^\delta, r_i^\delta]$, respectively. For convenience, let's put $l_i^\delta \geq 0$, in order not to introduce the absolute value operation into the fuzzy coefficient of the value $\delta_i(\omega)X_i(\gamma)$. Recall that $Tr(a, d)$ is a class of triangular fuzzy values with a modal value of a, a fuzziness coefficient of d.

To simplify the writing, we denote:

$$\hat{a}_i = \mathbf{E}[a_i(\omega)] = \frac{l_i^a + r_i^a}{2}, \qquad \hat{\delta}_i = \mathbf{E}[\delta_i(\omega)] = \frac{l_i^\delta + r_i^\delta}{2},$$

$$\tilde{a}_i = \mathbf{D}[a_i(\omega)] = \frac{(r_i^a - l_i^a)^2}{12}, \qquad \tilde{\delta}_i = \mathbf{D}[\delta_i(\omega)] = \frac{(r_i^\delta - l_i^\delta)^2}{12}.$$

We also have a fuzzy level of expected return $r(\gamma) \in Tr(c_r, d_r)$.

In accordance with calculus of possibilities of fuzzy values (see [30,31]), we have the following parametrized representation for $R_i(\omega, \gamma)$:

$$R_i(\omega, \gamma) \in Tr\big(a_i(\omega) + \delta_i(\omega)c_i, \delta_i(\omega)d_i\big) \tag{14}$$

and based on the linearity of mathematical expectation from (14), we get a parameterized representation of the expected return of ith asset:

$$\hat{R}_i(\gamma) = \mathbf{E}[R_i(\omega, \gamma)] \in Tr(\hat{a}_i + \hat{\delta}_i c_i, \hat{\delta}_i d_i). \tag{15}$$

In order to construct a model of acceptable portfolios (9), we define boundaries of α-level set of fuzzy random variables (14):

$$R_i^{\mp}(\omega, \alpha) = a_i(\omega) + \delta_i(\omega)c_i \mp \frac{\delta_i(\omega)d_i}{2}(1 - \alpha).$$

Replacing in the above formulas random variables $a_i(\omega)$ and $\delta_i(\omega)$ with their expected values $\hat{a}_i$ and $\hat{\delta}_i$, we get boundaries of α-level sets of fuzzy values (15). We will also write out boundaries of α-level set of fuzzy level of expected return:

$$r^-(\alpha) = c_r - \frac{d_r}{2}(1 - \alpha), \qquad r^+(1 - \alpha) = c_r + \frac{d_r}{2}\alpha.$$

Now we specify the portfolio risk function (8). With the assumptions we made earlier and based on (5) it takes the following form:

$$\mathbf{D}[R(w, \omega, \gamma)] = \sum_{i=1}^{n} w_i^2 \left(\tilde{a}_i + \tilde{\delta}_i \left(c_i^2 + \frac{d_i^2}{12} \right) \right).$$

After determining all the components, the equivalent deterministic analogue of the minimum risk portfolio model (8)-(9) can be written in the following form in possibility/necessity contextes with a fuzzy level of expected return:

$$\sum_{i=1}^{n} w_i^2 \left(\tilde{a}_i + \tilde{\delta}_i \left(c_i^2 + \frac{d_i^2}{12}\right)\right) \to \min_{w},$$

$$\begin{cases} \sum_{i=1}^{n} \left(\hat{a}_i + \hat{\delta}_i c_i + \frac{\hat{\delta}_i d_i}{2}(1-\alpha)\right) w_i \geq c_r - \frac{d_r}{2}(1-\alpha) \text{ (possibility context)}, \\ \text{or } \sum_{i=1}^{n} \left(\hat{a}_i + \hat{\delta}_i c_i - \frac{\hat{\delta}_i d_i}{2}\alpha\right) w_i \geq c_r + \frac{d_r}{2}\alpha \text{ (necessity context)}, \\ \sum_{i=1}^{n} w_i = 1. \end{cases} \quad (16)$$

By replacing the right-hand sides of the inequalities in (16) with r, we get models with a crisp level of expected return.

Using Python packages for symbolic mathematics, we have derive formulas for the optimal solution and quasi-efficient frontier in a closed analytical form. The optimal vector of weights is equal to:

$$w_i = \eta_i \left(\mathcal{A} + \frac{\mathcal{B} - s_i}{\mathcal{C}} \mathcal{D}\right),$$

and dependence of risk on the level of expected return in a fuzzy form:

$$\mathbf{D}[R(w, \omega, \gamma)] = \sum_{i=1}^{n} w_i^2 v_i = \frac{1}{\mathcal{C}^2} \left(\mathcal{D}^2 \sum_{i=1}^{n} \eta_i s_i^2 - 2\mathcal{D}\mathcal{X} \sum_{i=1}^{n} \eta_i s_i + \mathcal{X}^2 \sum_{i=1}^{n} \eta_i\right),$$

where

- $\eta_i = \frac{1}{v_i}$, $v_i = \frac{1}{48}\left(4(r_i^a - l_i^a)^2 + \frac{1}{3}(12c_i^2 + d_i^2)(r_i^\delta - l_i^\delta)^2\right)$,
- $\mathcal{A} = \frac{1}{\sum_{i=1}^{n} \eta_i}$, $\mathcal{B} = \mathcal{A} \sum_{i=1}^{n} \eta_i s_i$, $\mathcal{C} = \sum_{i=1}^{n} \eta_i \cdot \frac{\sum_{i=1}^{n} \eta_i s_i^2}{\sum_{i=1}^{n} \eta_i s_i} - \sum_{i=1}^{n} \eta_i s_i$,
- $\mathcal{D} = 1 - \frac{(c_r - \beta d_r) \sum_i \eta_i}{\sum_{i=1}^{n} \eta_i s_i}$, $\mathcal{X} = \mathcal{A}\mathcal{B} + \mathcal{C}\mathcal{D}$,
- $s_i = p_i + \beta q_i$, $\beta = \frac{1}{2}(1-\alpha)$, $p_i = \hat{a}_i + \hat{\delta}_i c_i$, $q_i = \hat{\delta}_i d_i$.

6 Model Example

As an example, let's take a three-dimensional portfolio ($n = 3$) and the model specified in the previous section. Let's have

$$\begin{array}{lll} X_1(\gamma) \in Tr(-0.5,\ 1), & a_1(\omega) \in \mathcal{U}(1.065,\ 1.137), & \delta_1(\omega) \in \mathcal{U}(0.399,\ 0.401), \\ X_2(\gamma) \in Tr(-0.5,\ 1), & a_2(\omega) \in \mathcal{U}(1.172,\ 1.232), & \delta_2(\omega) \in \mathcal{U}(0.34,\ 0.46), \\ X_3(\gamma) \in Tr(0.5,\ 1), & a_3(\omega) \in \mathcal{U}(0.9,\ 0.91), & \delta_3(\omega) \in \mathcal{U}(0.3,\ 0.5), \end{array}$$

$$r(\gamma) \in Tr(1.07,\ 0.1), \quad \alpha = 0.6.$$

We will consider the portfolio model in a possibilistic context. For the specified input data, the vector of optimal portfolio shares will be equal to: $w^* = (0.501, 0.307, 0.192)$. Figure 1 shows a graphical interpretation of fuzzy random variables that determine the profitability of individual assets (on the left), and the structure of the optimal portfolio (on the right). For each distribution, the dotted line shows one of the possible triangular fuzzy values characterizing the asset return spread (initially set by an expert). In turn, the solid colored sides of the triangle show a random uniform fluctuation of the left and right shoulders of the distribution, as well as its modal value (due to market's stochastic uncertainty). The quasi-efficient frontiers of the found portfolio with different α-levels is shown in Fig. 2 (right parts of the parabolas).

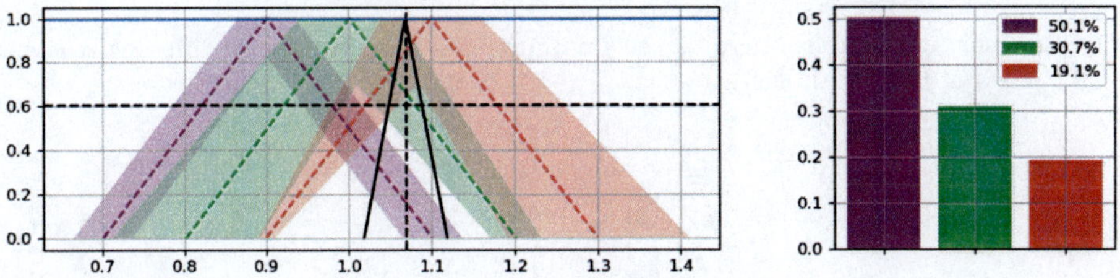

Fig. 1. Fuzzy random values of individual assets and the level of expected return (left) and the structure of the found minimum risk portfolio (right)

Let's illustrate the difference in the resulting models with crisp and fuzzy levels of expected return. With a fuzzy level of expected return the efficient frontier can be built in two ways: as a curve of risk dependence on the modal value of the fuzzy level and as a curve of risk dependence on the left boundary (in the possibility context) of the α-level set of the fuzzy level of expected return. Figure 3 shows both options.

As expected, the quasi-efficient frontier that depends on the left boundary of the α-level set of the fuzzy level of expected return behaves identically to the quasi-efficient frontier that depends on the crisp level of expected return. If the dependence is built on the modal value, then we observe a shift of the parabola by the amount of displacement of the boundary of the α-level set from the modal value. This spread will become wider if coefficient of fuzziness of the level of expected return becomes greater. We see that lowering the α level reduces the riskiness of the model.

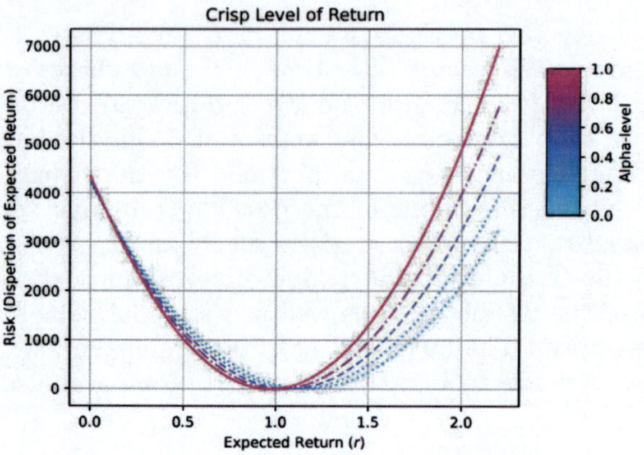

Fig. 2. Quasi-efficient frontiers of the minimum risk portfolio with different α-levels (right parts of the parabolas)

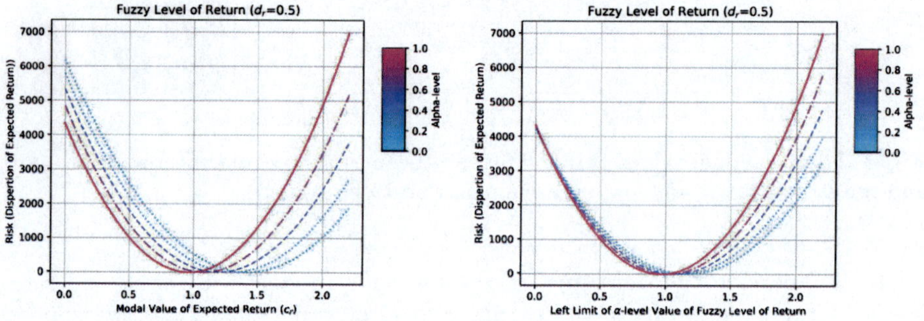

Fig. 3. The curve of risk dependence on the modal value of the fuzzy return level (left) and the left boundary of the α-level set of the fuzzy return level (right)

7 Summary

The article develops a method for constructing a quasi-efficient minimum risk portfolio frontier under conditions of hybrid uncertainty of the possibilistic-probabilistic type with allowed short sales and when an acceptable for the investor level of expected return is defined as crisp or fuzzy value. The formula for the dependence of the quasi-efficient frontier on the α-level is derived in analytical form. The results are illustrated by a numerical example.

The proposed method of analytical construction of a set of quasi-effective frontiers of a minimal risk portfolio simplifies the study of various mechanisms for managing the uncertainty of the model (various types of triangular norms, distributions of fuzzy and random components, etc.), which is an area for continuation of the presented work.

References

1. Markowitz, H.: Portfolio selection. J. Finan. **7**(1), 77–91 (1952). https://doi.org/10.2307/2975974
2. Markowitz, H.: Portfolio Selection: Efficient Diversification of Investments. Wiley, New York (1959)
3. Zadeh, L.A.: Fuzzy sets. Inf. Control **8**, 338–353 (1965)
4. Zadeh, L.A.: Fuzzy sets as a basis for a theory of possibility. Fuzzy Sets Syst. **1**, 3–28 (1978)
5. Nahmias, S.: Fuzzy variables. Fuzzy Sets Syst. **1**, 97–110 (1978)
6. Dubios, D., Prade, H.: Possibility Theory. Plenum Press, New York (1988)
7. Tanaka, H., Guo, P., Turksen, I.B.: Portfolio selection based on fuzzy probabilities and possibility distributions. Fuzzy Sets Syst. **111**(3), 387–397 (2000)
8. Inuiguchi, M., Tanino, T.: Portfolio selection under independent possibilistic information. Fuzzy Sets Syst. **115**(1), 83–92 (2000)
9. Parra, M.A., Terol, A.B., Uría, M.V.R.: A fuzzy goal programming approach to portfolio selection. Eur. J. Oper. Res. **133**, 287–297 (2001)
10. Ammar, E., Khalifa, H.A.: Fuzzy portfolio optimization a quadratic programming approach. Chaos, Solitons Fractals **18**(5), 1045–1054 (2003)
11. Zhang, W.G., Nie, Z.K.: On admissible efficient portfolio selection problem. Appl. Math. Comput. **159**, 357–371 (2004)
12. Bilbao-Terol, A., Pérez-Gladish, B., Arenas-Parra, M., Victoria, M., Uría, R.: Fuzzy compromise programming for portfolio selection. Appl. Math. Comput. **173**(1), 251–264 (2006)
13. Giove, S., Funari, S., Nardelli, C.: An interval portfolio selection problem based on regret function. Eur. J. Oper. Res. **170**(1), 253–264 (2006)
14. Vercher, E.: Portfolios with fuzzy returns: selection strategies based on semi-infinite programming. J. Comput. Appl. Math. **217**, 381–393 (2008)
15. Kwakernaak, H.: Fuzzy random variables. Inf. Sci. **15**, 1–29 (1978)
16. Nahmias, S.: Fuzzy variables in a random environment. In: Gupta, M.M., Ragade, R.K., Yager, R.R. (eds.) Advances in Fuzzy Sets Theory and Applications, pp. 165–180. NHCP, Amsterdam (1979)
17. Yazenin, A.V.: Linear programming with random fuzzy data. Sov. J. Comput. Syst. Sci. **30**, 86–93 (1992)
18. Liu, B.: Theory and Practice of Uncertain Programming. Physica, Heidelberg (2002)
19. Liu, B.: Uncertainty Theory: An Introduction to Its Axiomatic Foundations. Springer, Berlin (2004)
20. Luhandjula, M.K.: Optimisation under hybrid uncertainty. Fuzzy Sets Syst. **146**, 187–203 (2004)
21. Li, J., Xu, J.P., Gen, M.: A class of fuzzy random multiobjective programming problem. Math. Comput. Model. **44**, 1097–1113 (2006)
22. Yazenin, A.V.: Optimization with fuzzy random data and its application in financial analysis. In: Batyrshin, I., Kacprzyk, J., Sheremetov, L. (eds.) International Conference on Fuzzy Sets and Soft Computing in Economics and Finance. SPb (2004)
23. Yazenin, A.V.: Possibilistic-Probabilistic models and methods of portfolio optimization. In: Batyrshin, I., Kacprzyk, J., Sheremetov, L., Zadeh, L.A. (eds.) Perception-based Data Mining and Decision Making in Economics and Finance, Studies in Computational Intelligence, pp. 241–259. Springer, Berlin (2007). https://doi.org/10.1007/978-3-540-36247-0_9

24. Huang, X.: Two new models for portfolio selection with stochastic returns taking fuzzy information. Eur. J. Oper. Res. **180**(1), 396–405 (2007)
25. Yazenin, A., Soldatenko, I.: A portfolio of minimum risk in a hybrid uncertainty of a possibilistic-probabilistic type: comparative study. In: Kacprzyk, J., Szmidt, E., Zadrozny, S., Atanassov, K., Krawczak, M. (eds.) Advances in Fuzzy Logic and Technology 2017, vol. 643, pp. 551–563 (2017). https://doi.org/10.1007/978-3-319-66827-7_51
26. Soldatenko, I.S., Yazenin, A.V.: On one problem of portfolio analysis under soft constraints. Nechetkie Sistemy i Myagkie Vychisleniya [Fuzzy Syst. Soft Comput.] **15**(1), 64–76 (2020). (in Russian), https://doi.org/10.26456/fssc71
27. Yazenin, A.V., Soldatenko, I.S.: Model of a minimal risk portfolio under hybrid uncertainty. Control. Cybern. **50**(2), 315–332 (2021)
28. Barbaumov, V.E., Gladkikh, I.M., Chujko, A.S.: Finansovye investitsii: Uchebnik [Financial Investments: Textbook]. Finance and Statistics, Moscow (2003). (in Russian)
29. Feng, Y., Hu, L., Shu, H.: The variance and covariance of fuzzy random variables and their applications. Fuzzy Sets Syst. **120**(3), 487–497 (2001). https://doi.org/10.1016/S0165-0114(99)00060-3
30. Yazenin, A.V., Wagenknecht, M.: Possibilistic Optimization. A measure-based approach. Aktuelle Reihe 6/96, Brandenburgische Tecnische Universitat, Cotbus, Germany (1996)
31. Yazenin, A.V.: Osnovnye ponyatiya teorii vozmozhnostej. Matematicheskij apparat dlya prinyatiya reshenij v usloviyakh gibridnoj neopredelennosti [Basic concepts of the theory of possibilities. Mathematical decision-making apparatus in a hybrid uncertainty]. Fizmatlit Publ., Moscow (2016). (in Russian)

Applications

Fuzzy Mechanism for Data Transmission Adaptation

Łukasz Apiecionek(✉), Dawid Ewald, Jacek M. Czerniak, and Jan Baumgart

Kazimierz Wielki University, Bydgoszcz, Poland
lukasz.apiecionek@ukw.edu.pl

Abstract. This paper presents the possibility of application mechanisms of artificial intelligence and the protocol of parallel data transmission TCP for increasing the stability of information transfer in 5G networks. Protocol MPTCP running basics are presented as well as the idea of using artificial intelligence mechanisms to adapt the choice of data transmission path depending on the current error rate.

Keywords: MPTCP · artificial intelligence · 5G

1 Introduction

The 5G networks technology is currently implemented by services suppliers or potential beneficiaries of this technology, such as electricity supply. However, many GSM/3G/LTE network operators have invested in their infrastructure and want to get a return on this investment first. In terms of technology, 5G networks are defined by the following parameters [6,32,37]:

- 1 ms transmission delay for end connections of type end-to-end,
- 1000 times the bandwidth for an area,
- 10–100 times more connected devices,
- availability of services at the level of 99.999%,
- 100% area coverage,
- 90% reduction of energy consumption,
- possibility of devices operation with low power consumption up to 10 years.

Only networks that meet the above parameters can be called 5G networks. An important new aspect is the new radio interface [4,5,32,33]. It has the ability to co-work with more than one transmitter/receiver, known as MIMO (ang. multiple-input, multiple-output) [2,3,6,32,33]. It is worth noting that the 5G network should co-work together with the previous generation networks. In addition, 5G networks can use Ethernet protocols as backup links [32]. The fact of using multiple interfaces allows the use of protocols for multiplying transmission paths, such as Multi-Path Transmission Control Protocol (MPTCP) [43,44,47–49]?. The article presents the possibilities of using the MPTCP protocol and artificial intelligence mechanisms to increase the quality of data transmission.

K. T. Atanassov et al. (Eds.): BOS/SOR/IWIFSGN 2022, LNNS 793, pp. 283–295, 2023.
https://doi.org/10.1007/978-3-031-45069-3_26

2 MPTCP

In order to understand the operation\section of MPTCP technology it is necessary the knowledge of TCP concept [28,41,42,45], which is used for data transmission between devices. TCP can send data between two devices in both directions. TCP uses a unique number to identify a connection that consists of two values: the IP address and the port. The use of a checksum and sequence number allows TCP to provide data exchange for the application layer. However, before the application can start transmitting data, the transmission must be initialized. To this end, the first device sends the data segment with a SYN flag, then the receiver sends back a packet with a SYN and ACK flag. Finally, the initiating device responds with the data segment only with the ACK flag [28]. The TCP connection cannot be transferred after changing the IP address to another one. When a device switches from e.g. Ethernet to a Wi-Fi wireless network, it then gets a different IP address. Therefore, existing open connections must be closed and reestablished. This way, it is possible to manage the connection in terms of its quality, e.g. error rate or link performance.

MPTCP is a series of extensions to the TCP specification that allows the client to open multiple connections to a second host using different network cards (interfaces). In this way, it is possible to manage the connection in terms of its quality, e.g. error rate or link performance.

The main purpose of the solution is to establish multiple data transmission paths through the network for a single connection, as shown in Fig. 1. The advantage of this solution is to increase the connection bit rate. This solution also allows for controlling network congestion efficiently. The application of MTCP connection does not block standard TCP transmission in the network [18,26,31].

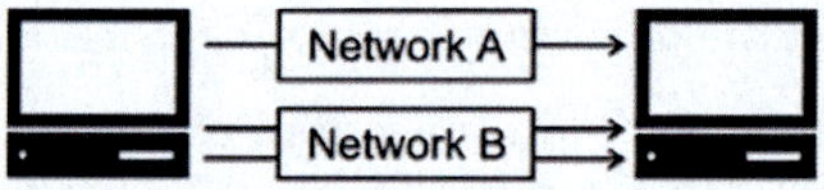

Fig. 1. Several TCP connections represented as one logical connection.

MTCP operates in the transport layer and is transparent to the upper and lower layers, as shown in Fig. 2. It is an additional functionality for the upper layer.

The new MTCP connections are set up in the same way as standard TCP connections as shown in Fig. 3. However, the protocol introduces a new function, allowing the MP_CAPABLE option to be sent, which informs both sides of the transmission that MPTCP is supported and can be used. Today, making connections is complicated by the use of intermediate devices (switches, routers). The pair of data: IP address and port of source and destination are not sufficient to identify the connection. Therefore, MPTCP extends these capabilities by providing an additional option - MP-JOIN, which allows you to add data streams that together form a single link. In the first step, the MP_JOIN option delivers a token that is generated together with the key created during connection initiation. Then the HMAC identification key is exchanged (hash-based message authentication code). Then the transmission stream is established and MPTCP can use it for data transmission. Each side of the transmission transmits data through open streams.

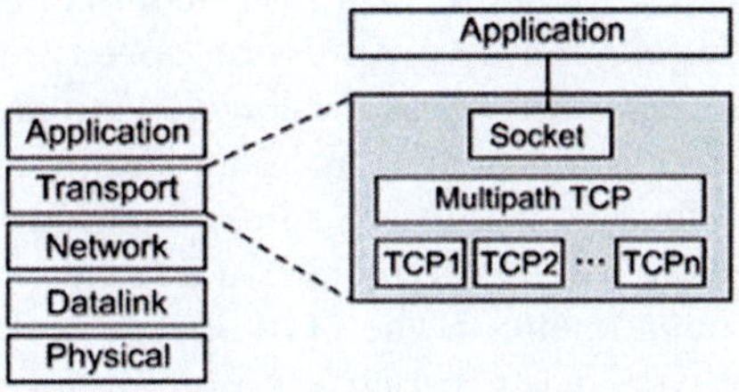

Fig. 2. MPTCP protocol stack.

The standard TCP number 'subflow sequence number' allows for the identification of the lost data in course of their transmission. The MPTCP uses the sequence number "data sequence number" to organize the received data and transfer it to the application layer [26]. When the device wants to inform the recipient that there is no more data to send, it sends the "Data FIN" flag. This flag has the same meaning as the TCP FIN flag on a standard TCP connection. Generally, users connect themselves via phones/tablets to the Internet via Wi-Fi or 3G connection, but not at the same time. In case they lose their connection to the network, they have to switch and reconnect. MPTCP avoids such a situation, because it can transfer data through both connections simultaneously and in the case of failure of one of them, dynamically switch the entire transfer to the other, without losing the connection. It also allows to choose the optimal connection speed.

The first mobile system that supports MPTCP [31,38,40,46] is iOS 7. It ensures uninterrupted data transmission in case the connection does not work. Currently, MPTCP is used for Siri data transmission. Siri is an intelligent assistant that supports many actions triggered by the operator. It allows, for example, to send messages, add meetings, and make calls using voice commands.

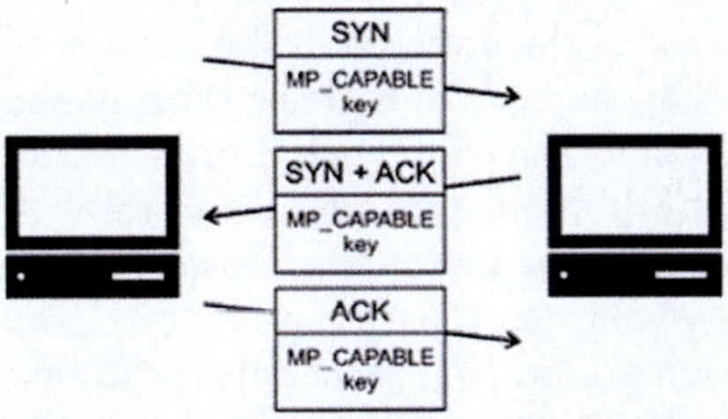

Fig. 3. Establishing MPTCP connection.

3 Adaptation Mechanism of Artificial Intelligence for 5G Networks

In order to adapt MPTCP to 5G networks, it is possible to use artificial intelligence mechanisms. Such an algorithm will allow for the adaptation of data transmission depending, for example, on the transmission error rate in the lines, the price of the connection or its transmission speed. Such an algorithm will allow for the adaptation of data transmission depending, for example, on the transmission error rate in communication links, the price of the connection or its transmission speed. For this purpose, you can be used, for example, Artificial Neural Networks or Fuzzy Logic. Artificial Neural Networks can be taught to recognize undesirable phenomena in the 5G network and adjust the method of data transmission. However, it is necessary to prepare a sufficiently large base of training data. Therefore, their application is currently difficult, as there are no existing data bases on which the learning process can be performed. At present, it is possible to use Fuzzy Logic. The term fuzzy corresponds to a situation where there are no fully precise data, such as defined metrics or object parameters. Instead of them, the concepts like more, less, and a little are used. Such concepts are used precisely in fuzzy logic. The concept of this field of knowledge was defined by professor at Columbia University in New York and at Berkeley University in California, Lofti A. Zadeh. He published the paper "Fuzzy sets" in Information and Control in 1965 [35]. He defined the term fuzzy set, which allows for the determination of imprecise data as a value from the interval (0, 1). The number assigned to the file defines the level of membership to the file. In his theory, L. Zadeh was based on three-valued logic, defined 45 years earlier by the Polish scientist Jan Łukasiewicz [19]. Another important step in the development of the theory was the introduction of the fuzzy number representation as L-R notation, proposed by D. Dubois and H. Prade [15–17]. One of the methods of generalizing fuzzy numbers is the concept invented by Professor Witold Kosiński [20] and his cooperators, called Ordered Fuzzy Numbers, which allows to combine a trend change with a fuzzy number. The first attempt to identify new operations on fuzzy files by prof. Kosiński has presented in the early 1990s [25]. Later papers published with colleagues (P. Prokopowicz, D. Ślęzak) allowed to define the arithmetic of fuzzy numbers [21–23].

In the case of 5G networks and data transmission adaptation mechanism, the algorithm may use the fact of the increase in the number of data retransmissions. It is worth noting, that an increase in data retransmission causes an increase in energy consumption. In this case, the algorithm may decide to change the data transmission path, which will reduce the number of retransmissions. The Ordered Fuzzy Numbers can speed up the decision to change the data transmission channel, which will allow to avoid quickly the necessary retransmissions.

The data transmission adaptation mechanism can operate as follows:

- the algorithm measures the number of TCP retransmission packet in all used data transmission links,

- retransmission errors values are recounted into values specifying the error percentage for a given link,
- the calculations are made in a continuous mode in assumed time periods - time slots.

On the basis of four consecutive measurements, a Ordered Fuzzy Number is created, specifying the number of errors for a given link, and at the same time their trend - increasing or decreasing:

- a positive order of the number means that the number of retransmissions is increasing,
- a negative order of the number means that the number of retransmissions is decreasing

The algorithm does not require complicated calculations, which will not cause the necessity to increase the computing power of devices or demand the energy increase. A special aspect is the fact that each retransmission increases the power consumption of the network. It should be noted that the data retransmission does not increase the power consumption only of the transmitter. The system consists of many devices, there are base stations, switches and routers on the way, each of them consumes energy. In the event that the packet is retransmitted, the power consumption is increased by the fact that the receiver has to detect a data problem, so it consumes energy to calculate the CRC, and then the receiver has to inform the sender that it has not received the data correctly. Finally, the sender has to resend the data. The application of the MPTCP mechanism with an adaptive data transmission control algorithm will allow for better use of the 5G network with lower energy consumption.

This algorithm was presented while analyzing the reduction in energy consumption for signal transmission [36]. The algorithm measures a TCP retransmission in all the used channels during the transmission and provides it in a percentage value of the transmitted packets. This measure is made both in a given period of time – a timeslot – and continuously. Four timeslots of the continuous measure could be defined in time:

$$t_i, t_{i-1}, t_{i-2}, t_{i-3} \tag{1}$$

where t_i is a current timeslot

These four results together give a fuzzy number in OFN notation, where

- $f_A(0)$ respond to t_{i-3}
- $f_A(1)$ respond to t_{i-2}
- $g_A(1)$ respond to t_{i-1}
- $g_A(0)$ respond to t_i

This fuzzy number in OFN notation is presented in Fig. 4.

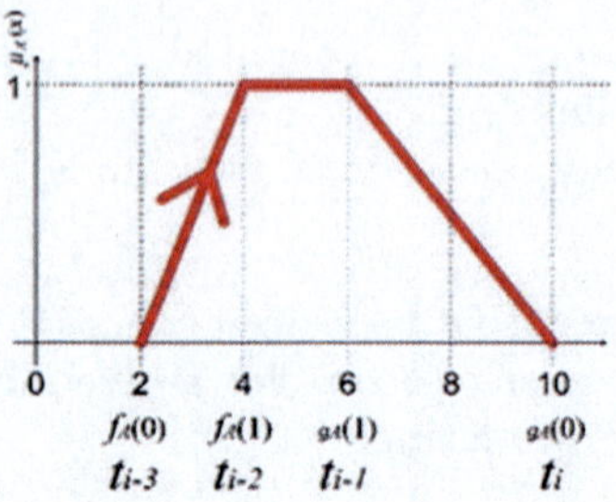

Fig. 4. Fuzzy number in OFN notation.

The definition of fuzzy observation of the connection used:

Definition 1. Fuzzy observation of C connection in timet_i is a set

$$C/t_i = \{f_c(0)/t_{i-3}, f_c(1)/t_{i-2}, g_c(1)/t_{i-1}, g_c(o)/t_i\} \tag{2}$$

where

$$t_i > t_{i-1} > t_{i-2} > t_{i-3} \tag{3}$$

$$\mid t_i - t_{i-1} \mid = \mid t_{i-1} - t_{i-2} \mid = \mid t_{i-2} - t_{i-3} \mid = t_n \tag{4}$$

timeslot of the measurement

$$f_c(0) \leq f_c(1) \leq g_c(1) \leq g_c(0) \tag{5}$$

This provides the Lemma 1.

Lemma 1

$$C_{positive} = \begin{cases} f_c(0) \leq f_c(1) \leq g_c(1) \\ or \\ f_c(1) \leq g_c(1) \leq g_c(0) \end{cases} \tag{6}$$

n other situation $C_{negative}$

According to this definition, the counters of retransmissions in the links during the connection observation should give:

- positive order of OFN when the packet retransmission count **increases**,
- negative order of OFN when the packet retransmission count **decreases**.

The four measurements performed during the data transfer allow to prepare fuzzy number in OFN notation. This gives a fuzzy observation of the MPTCP connections, defined as follows:

Definition 2. Fuzzy observation of the MPTCP connections is described by the following formula:

$$s_m = \sum_{i-1}^{n} \begin{Bmatrix} C_{positive} \mid C_{negative} \\ C_u * W_i \mid -C_i * W_i \end{Bmatrix} \tag{7}$$

Where

$w_i \in \{w_1, ..., w_i\}$ describes the impact on all connections.

According to definition 2, it is possible to define the MPTCP scheduler using OFN.

3.1 The Algorithm of OFN Scheduler

The algorithm with application of OFN for predicting transmission errors in MPTCP transmission consists of the four steps:

Step 1 - the administrator declares the start value for w_i and L_i for all MPTCP links, where w_i describes the impact of errors on all connections, and L_i describes the load of the whole data which should be sent by the i connections when the transmission starts. The L_i is expressed in percentage rate.
Step 2 - the amount of packets P_i which will be transferred over each connection during each timeslot is calculated using the following formula:

$$P_i = \frac{L_i}{\sum_{i=1}^{n} L_i} * Data \tag{8}$$

Step 3 - during the transmission the fuzzy observation of connection C_i is calculated for each connection accordingly to data retransmissions and fuzzy observation of MPTCP S_i is calculated accordingly to the given definition.
Step 4 - when the calculated S_i is positive and it is higher than the acceptance level AL, it means that the number of errors is increasing on this connection. In this situation the L_i for the given connection will be changed accordingly to the following formula:

$$L_i = \frac{L_i}{ErrorCorector} \tag{9}$$

When the calculated S_i is negative – the number of errors is decreasing on this connection. In this situation the L_i for given connection will be changed accordingly to the following formula:

$$L_i = L_i * ErrorCorector \tag{10}$$

The ErrorCorector is a value which describes how fast the system should stop using the connection in which the amount of errors increases. This value should be also declared by the network administrator.

Table 1. Fuzzy observation of the connections

Time	% Error on connections		S [p-positive, n-negative]			
	C1	C2	S1		S2	
1	1	**3**				
2	2	**4**				
3	2	6				
4	1	9	[1, 2, 2, 1]	p	[3, 4, 6, 9]	p
5	2	13	[2, 2, 1, 2]	n	[4, 6, 9, 13]	p
6	1	12	[2, 1, 2, 1]	p	[6, 7, 12, 12]	p
7	1	11	[1, 2, 1, 1]	p	[9, 13, 12, 11]	p
8	2	12	[2, 1, 1, 2]	p	[13, 12, 11, 12]	n
9	1	8	[1, 1, 2, 1]	p	[12, 11, 12, 8]	n
10	2	7	[1, 2, 1, 2]	p	[11, 12, 8, 7]	n
11	1	6	[2, 1, 2, 1]	p	[12, 8, 7, 6]	N
12	2	5	[1, 2, 1, 2]	p	[8, 7, 6, 5]	n
13	1	3	[2, 1, 2, 1]	p	[7, 6, 5, 3]	n
14	**2**	2	[1, 2, 1, 2]	p	[6, 5, 3, 2]	n
15	1	1	[2, 1, 2, 1]	p	[5, 3, 2, 1]	n

3.2 Simulation Test Results

n order to check a MPTCP scheduler with OFN, a simulation was performed.

The system was designed with two connection links: Connections 1 and 2, labeled C1 and C2, with maximum speed 100 Gbit/s. The parameters of the algorithm were declared as follows:

- the corrector for the links ErrorCorector = 2;
- acceptance level
- load balance on start for the connection C1 was $L_1 = 66$;
- load balance on start for the connection C2 was $L_2 = 34$;
- the timeslots used were 60 s.

Table 1 presents the results of the used algorithm dependently on the number of errors during the data transfer. The fuzzy observation of the link gives OFN numbers S1 and S2.

As it can be observed, after the 4th measure the OFN numbers are generated and the S1 exceeds the acceptance level, so OFN scheduler algorithm changes the load balance for the transmission, as it is presented in Table 2. The total number of packets passed to the connections changes during the timeslots accordingly to the load balance set at the beginning of the test. This affects the number of packets transferred through the connections, and depending on the error level on the link, provides different number of errors. In the simulation there were 40000 packets passed to the link. Table 3 presents the numbers of errors in the

Table 2. Load balance for the connections

Time	L - load balance	
	L1	L2
1	50	50
2	50	50
3	50	50
4	50	50
5	50	25
6	50	12, 5
7	50	6, 25
8	50	12, 5
9	50	25
10	50	50
11	50	50
12	50	50
13	50	50
14	50	50
15	50	50

connections with and without OFN scheduler in MPTCP. As it can be noticed, the total number of errors without OFN scheduler is higher than in the variant using this solution.

Table 3. Number of errors in packet transmission

Time	Errors count					
	MPTCP with OFN			MPTCP without OFN		
	L1	L2	L1+L2	L1	L2	L1+L2
1	400	1200	1600	400	1200	1600
2	800	1600	2400	800	1600	2400
3	800	2400	3200	800	2400	3200
4	400	3600	4000	400	3600	4000
5	1067	3467	4533	800	5200	6000
6	640	1920	2560	400	4800	5200
7	711	978	1689	400	4400	4800
8	1280	1920	3200	800	4800	5600
9	533	2133	2667	400	3200	3600
10	800	2800	3600	800	2800	3600
11	400	2400	2800	400	2400	2800
12	800	2000	2800	800	2000	2800
13	400	1200	1600	400	1200	1600
14	800	800	1600	800	800	1600
15	400	400	800	400	400	800

The analysis of the proposed algorithm allows drawing conclusions that simple calculation and optimal decision-making will reduce energy consumption in the entire IT system of the 5G network.

4 Conclusions

This article presents a special adaptive algorithm for 5G networks, which, using MPTCP mechanisms, allows for better use of network resources and to reduce energy consumption.

In this article, a special algorithm for mobile connections, applicable in 5G network, was introduced. For this algorithm a MPTCP concept was used and the energy consumption as well as the security aspect were developed by a special usage of fuzzy observation of the errors in the transmissions. An important aspect is the fact that the application of these mechanisms does not require hardware changes in the network infrastructure or end devices. Known mechanisms used in standard TCP protocol are used.

The proposed solution can be used not only in 5G networks, but also for critical infrastructure solutions. For this purpose, the use of MPTCP and fuzzy ordered numbers will increase the speed and security of the transmitted data [7,11,30].

References

1. 5G and e-Health, 5G-Infrastructure-Association (2015). www.5g-ppp.eu. Accessed 15 Aug 2016
2. 5G and Energy, 5G PPP Architecture Working Group. https://5g-ppp.eu/white-papers/. Accessed 15 Aug 2015
3. 5G Automotive Vision, 5G PPP Architecture Working Group. https://5g-ppp.eu/white-papers/. Accessed 15 Aug 2015
4. 5G Automotive Vision, 5G PPP Architecture Working Group. https://5g-ppp.eu/. Accessed 15 Aug 2015
5. Maternia, M., El Ayoubi, S.E. (eds.): 5G PPP Use Cases and Performance Evaluation Models, Version: 1.0. https://5g-ppp.eu. Accessed 15 Aug 2015
6. 5G radio access, Ericsson white paper, Uen 284 23-3204 Rev C (2016)
7. Andrysiak, T., Saganowski, L., Choras, M., Kozik, R.: Proposal and comparison of network anomaly detection based on long-memory statistical models. Log. J. IGPL **24**(6), 944–956 (2016)
8. Apiecionek, Ł, Romantowski, M.: Secure IP network model. Comput. Method Sci. Technol. **19**(4), 209–213 (2013). https://doi.org/10.12921/cmst.2013.19.4.209-216
9. Apiecionek, Ł., Romantowski, M.: Security solution for cloud computing. J. Inf. Control Manag. Syst. **11**(2) (2013). ISSN: 1336-1716
10. Apiecionek, Ł., Sobczak, M., Makowski, W., Vince, T.: Multi path transmission control protocols as a security solution. In: IEEE 13th International Scientific Conference on Informatics, INFORMATICS 2015, Poprad, Slovakia, pp. 27–31 (2015). ISBN: 978-1-4673-9867-1

11. Byrski, J., Byrski, W.: A double window state observer for detection and isolation of abrupt changes in parameters. Int. J. Appl. Math. Comput. Sci. **26**(3), 585–602 (2016). https://doi.org/10.1515/amcs-2016-0041
12. Chapman, B., Zwicky, E.D.: Building Internet Firewalls. O'Reilly & Associates, Inc. (1995). ISBN: 1-56592-124-0
13. Cheswick, W.R., Bellovin, S.M.: Firewalls and Internet Security: Repelling the Wily Hacker. Addison-Wesley Publishing Company (1994). ISBN: 0-201-63357
14. Czerniak, J.: Evolutionary approach to data discretization for rough sets theory. Fund. Inform. **92**(1–2), 43–61 (2009)
15. Dubois, D., Prade, H.: Operations on fuzzy numbers. Int. J. Syst. Sci. **9**(6), 613–626 (1978)
16. Dubois, D., Prade, H.: Fuzzy elements in a fuzzy set. In: Proceedings of IFSA, vol. 5, pp. 55–60 (2005)
17. Dubois, D., Prade, H.: Gradual elements in a fuzzy set. Soft. Comput. **12**(2), 165–175 (2008). https://doi.org/10.1007/s00500-007-0187-6
18. Ford, A., Raiciu, C., Handley, M.: TCP extensions for multipath operation with multiple addresses. RFC 6824 (2013)
19. Łukasiewicz, J.: O logice trójwartościowej (1988)
20. Kosiński, W.: On fuzzy number calculus. Int. J. Appl. Math. Comput. Sci. **16**(1), 51–57 (2006)
21. Kosiński, W., Prokopowicz, P., Frischmuth, K.: On algebra of ordered fuzzy numbers. In: Krassimir, J.K., Atanassow, T., Hryniewicz, O. (eds.) Soft Computing Foundations and Theoretical Aspects. EXIT (2004)
22. Kosiński, W., Prokopowicz, P., Rosa, A.: Defuzzification functionals of ordered fuzzy numbers. IEEE Trans. Fuzzy Syst. **21**(6), 1163–1169 (2013)
23. Kosiński, W., Prokopowicz, P., Ślęzak, D.: On algebraic operations on fuzzy numbers. In: Kłopotek, M.A., Wierzchoń, S.T., Trojanowski, K. (eds.) Intelligent Information Processing and Web Mining. AINSC, vol. 22, pp. 353–362. Springer, Heidelberg (2003). https://doi.org/10.1007/978-3-540-36562-4_37
24. Kosiński, W., Prokopowicz, P., Ślęzak, D.: Ordered fuzzy numbers. Bull. Pol. Acad. Sci. Ser. Sci. Math. **51**(3), 327–338 (2003)
25. Kosiński, W., Słysz, S.: Fuzzy numbers and their quotient space with algebraic operations. Bull. Pol. Acad. Sci. Ser. Tech. Sci. **41**, 285–295 (1993)
26. Krieger, U.: Evolution of transport protocols in high-speed networks. Lectures materials Multimedia-Kommunikation in Hochgeschwindigkeitsnetzen (KTR-MMK-M) (2014)
27. Paasch, C., Detal, G., Duchene, F., Raiciu, C., Bonaventure, O.: Exploring mobile/WiFi handover with multipath TCP. In: Proceedings of the 2012 ACM SIGCOMM Workshop on Cellular Networks: Operations, Challenges, and Future Design (2012)
28. Postel, J.: Transmission control protocol. RFC 793, IETF (1981)
29. Raiciu, C., et al.: How hard can it be? Designing and implementing a deployable multipath TCP. In: USENIX Symposium on Networked Systems Design and Implementation 2012 (2012)
30. Saganowski, L., Andrysiak, T., Kozik, R., Choras, M.: DWT-based anomaly detection method for cyber security of wireless sensor networks. Secur. Commun. Netw. **9**(15), 2911–2922 (2016)
31. Schneider, K.M., Mast, K., Krieger, U.R.: Adapting content-centric networking to the characteristics of multihomed mobile terminals (2014)
32. Understanding 5G: Perspectives on future technological advancements in mobile (2014). https://www.gsmaintelligence.com/. Accessed 15 Aug 2016

33. View on 5G Architecture, 5G PPP Architecture Working Group. https://5g-ppp.eu/white-papers/. Accessed 15 Aug 2015
34. Vokorokos, L., Ennert, M., Hartinger, M., Radusovský, J.: A survey of parallel intrusion detection on graphical processors. In: Proceedings of International Scientific Conference, INFORMATICS 2013, SpišskáNováVes, Slovakia, 5–7 November 2013 (2013)
35. Zadeh, L.: Fuzzy sets. Inf. Control **8**(3), 338–353 (1965)
36. Apiecionek, Ł: Limiting energy consumption by decreasing packets retransmissions in 5G network. Mob. Inf. Syst. (2017). https://doi.org/10.1155/2017/4291091
37. Jacko, P., et al.: Remote IoT education laboratory for microcontrollers based on the STM32 chips. Sensors **22**(4), 1440 (2022). https://doi.org/10.3390/s22041440. https://www.mdpi.com/1424-8220/22/4/1440
38. Baumgart, J., Sangho, B.: A case study of the effectiveness of new methods of swarm optimization compared to known methods. Stud. Mater. Appl. Comput. Sci. **13**(1), 47–50 (2021). ISSN: 1689-6300
39. Ewald, D., Zarzycki, H., Czerniak, J.M.: Certain aspects of the OFNBee algorithm operation for different fuzzifiers. In: Atanassov, K.T., et al. (eds.) IWIFSGN BOS/SOR 2020. LNNS, vol. 338, pp. 241–256. Springer, Cham (2022). https://doi.org/10.1007/978-3-030-95929-6_19
40. Galas, K.: Drive unit as a replacement for the platform. Stud. Mater. Appl. Comput. Sci. **12**(1), 10–14 (2020). https://doi.org/10.5281/zenodo.4362649. ISSN: 1689-6300
41. Namli, Ö.B., Tyburek, K.: Research of the efficiency of the reach of fire services to accidents in the city of Kocaeli on the basis of statistical data for the years 2013–2020. Stud. Mater. Appl. Comput. Sci. **14**(3), 1–5 (2022)
42. Piszcz, A., Mikolajewski, D.: Application for the Android platform with the system as a solution to the right problem horse nutrition. Stud. Mater. Appl. Comput. Sci. **12**(1), 5–9 (2020). https://doi.org/10.5281/zenodo.4362647. ISSN: 1689-6300
43. Prokopowicz, P., Mikołajewski, D., Mikołajewska, E., Kotlarz, P.: Fuzzy system as an assessment tool for analysis of the health-related quality of life for the people after stroke. In: Rutkowski, L., Korytkowski, M., Scherer, R., Tadeusiewicz, R., Zadeh, L.A., Zurada, J.M. (eds.) ICAISC 2017. LNCS (LNAI), vol. 10245, pp. 710–721. Springer, Cham (2017). https://doi.org/10.1007/978-3-319-59063-9_64
44. Rojek, I., Macko, M., Mikolajewski, D., Saga, M., Burczynski, T.: Modern methods in the field of machine modelling and simulation as a research and practical issue related to Industry 4.0. Bull. Pol. Acad. Sci.-Tech. Sci. **69**(2) (2021). https://doi.org/10.24425/bpasts.2021.136717
45. Sangho, B.: Comparison of selected wolf pack algorithms used in solving optimization problems. Stud. Mater. Appl. Comput. Sci. **13**(1), 17–32 (2021). https://doi.org/10.5281/zenodo.4362647. ISSN: 1689-6300
46. Tyburek, K., Bora, Ö.: Comparison of the efficiency of time and frequency domain descriptors for the classification of selected wind instruments. Stud. Mater. Appl. Comput. Sci. **14**(3), 6–13 (2022)
47. Zarzycki, H.: Comparative study of the firefly algorithm and the whale algorithm. In: Kahraman, C., Tolga, A.C., Cevik Onar, S., Cebi, S., Oztaysi, B., Sari, I.U. (eds.) INFUS 2022. LNNS, vol. 504, pp. 999–1006. Springer, Cham (2022). https://doi.org/10.1007/978-3-031-09173-5_114

48. Zarzycki, H., Dobrosielski, W.T., Czerniak, J.M., Ewald, D.: Use of OFN in the short-term prediction of exchange rates. In: Atanassov, K.T., et al. (eds.) IWIFSGN 2019 2019. AISC, vol. 1308, pp. 289–301. Springer, Cham (2021). https://doi.org/10.1007/978-3-030-77716-6_25
49. Zarzycki, H., Skubisz, O.: A new artificial bee colony algorithm approach for the vehicle routing problem. In: Kahraman, C., Cebi, S., Cevik Onar, S., Oztaysi, B., Tolga, A.C., Sari, I.U. (eds.) INFUS 2021. LNNS, vol. 307, pp. 562–569. Springer, Cham (2022). https://doi.org/10.1007/978-3-030-85626-7_66

Using the Hub and Spoke Structure to Improve the Organization of Public Transport in Warsaw

Barbara Mażbic-Kulma[2], Jan W. Owsiński[1](✉), Jarosław Stańczak[1,2], Aleksy Barski[1,2], and Krzyszof Sęp[1,2]

[1] Systems Research Institute Polish Academy of Sciences, Newelska 6, 01-447 Warsaw, Poland
{jan.owsinski,jaroslaw.stanczak,aleksy.barski,sep}@ibspan.waw.pl
[2] Warsaw School of Information Technology, Newelska 6, 01-447 Warsaw, Poland
kulma@wit.edu.pl
https://www.wit.edu.pl/en, https://www.ibspan.waw.pl/en/home/

Abstract. Large urban agglomerations require an efficient transport system. This is important at a time when city life requires frequent trips to work, school, for shopping, recreation and rest. Unfortunately, you cannot rely on individual transport here, as this, although very convenient, quickly leads to city congestion. A good solution here is properly designed, efficient public transport. However, public transport by itself does not even guarantee good communication, it is important to have adequate changeover options and fast transport to convenient transfer points. It seems that one of the ideas that would be particularly suitable for designing public transport connections is the idea of hub and spoke, developed since the 1970s, then especially in air transport [4].

It seems that this idea, at least in general terms, will be relatively easy to apply here. In general terms, all fast means of urban transport (metro, fast urban rail, suburban rail, high-speed trams) should be used as a kind of backbone, a structure of connections of high-speed and capacity communication hubs to which commuters are brought, and slower means of transport (buses, ordinary trams) should be used to a large extent to transport people to these communication hubs from other system components.

In this work we propose the location of such main changeover stations using some aspects of the hub and spoke method by applying a specialized evolutionary algorithm to analyze the actual city transportation system on the basis of publicly available timetable of Warsaw Public Transport vehicles.

Keywords: hub and spoke · evolutionary algorithm · transport system

1 Introduction

Although investments related to the improvement of the quality of public transport operation are highly recommended, they are usually very expensive. After

K. T. Atanassov et al. (Eds.): BOS/SOR/IWIFSGN 2022, LNNS 793, pp. 296–310, 2023.
https://doi.org/10.1007/978-3-031-45069-3_27

all, it is also possible to use what the city already has more effectively, of course not forgetting about further investments. One of possible ways to improve communication in the city may be application of the hub & spoke structure in urban transport to detect important stations and connections, see [9], or to detect points worth building Park and Ride facilities near communication hubs, [12].

The hub and spoke structure was discussed in [2,7,8] and [5]. This idea, at least in terms of its broadly understood principle, will be relatively easy to apply in a city and can be described as follows. All fast and high throughput means of urban transport such as metro, fast urban rail, suburban rail, high-speed trams should be used as a kind of backbone, a structure of connections of high-speed and capacity communication hubs to which commuters are brought, and slower means of transport (buses, ordinary trams, city bicycles, on foot, ...) should be used to a large extent to transport people to or from these communication hubs from other system components. In the case of the city this transport organization idea can be introduced slowly, partially with the maintenance of elements of the traditional system of connecting lines "every district to every district", transforming the existing one as the construction of fast connections between the districts progresses and the habits of commuters change.

For transportation system modeling we propose to use evolutionary methods to obtain solutions, because the problem of transforming the transportation connections graph into an instance of the hub and spoke structure according to problem-specific restrictions is very difficult. Our method, described in this paper, finds the candidates for communication hubs and attached to them spoke nodes using as an input information the timetable for the city communication system.

2 Definitions of Used Graph Structures

We treat the city communication system as a graph with stations as graph nodes and transport lines as edges, thus some basic notions from graph theory are presented here, following [14] and [1].

A ***graph*** is a pair $G = (V, E)$, where V is a non-empty set of *vertices* and E is a set of *edges*. Each edge is a pair of vertices $\{v_1, v_2\}$ with $v_1 \neq v_2$.

In our problem we can consider also a ***directed graph***, which is an ordered pair $G = (V, A)$ where:

- V is a set whose elements are called vertices or nodes;
- A is a set of ordered pairs of vertices, called arcs (directed edges).

A ***degree*** of a vertex is the number of edges to which this vertex belongs.

A ***clique*** (a *complete subgraph*) $Q = (V_q, E_q)$ in a graph $G = (V, E)$ is a graph such that $V_q \subseteq V$ and $E_q \subseteq E$ and $Card(V_q) = 1$ or each pair of vertices $v_1, v_2 \in V_q$ fulfils the condition $\{v_1, v_2\} \in E_q$, see [3].

It is important to notice that each subgraph of a clique is also a clique, and this can be utilized by greedy algorithms looking for the maximum clique in graph.

An **α-clique** (see [5,11]) can be defined as follows: let $A = (V',E')$ be a subgraph of graph $G = (V_q, E_q)$, $V' \subseteq V$, $E' \subseteq E$, $k = Card(V')$ and let k_i be the number of vertices $v_j \in V'$ such that $\{v_i, v_j\} \in E'$.

1. For $k = 1$ the subgraph A of graph G is an ***α-clique(α)***.
2. For $k > 1$ the subgraph A of graph G is an ***α-clique*** (α) if for all vertices $v_i \in V'$ the condition $\alpha = \frac{k_i+1}{k}$ is fulfilled, where $\alpha \in (0, 1]$.

The notion of *α-clique* is then used in the sense of *α-clique* (α) for established value of α. A subgraph of an *α-clique* may not be an *α-clique* for the established α (for $\alpha < 1.0$), which makes the properties of this structure differ from those of the *clique* or, in this case, also *α-clique* (1.0) or *α-clique* for $\alpha = 1.0$. It means that it is more difficult to build an approximate algorithm that seeks the maximum *α-clique* than a *clique* in graph. These properties are illustrated in Fig. 1.

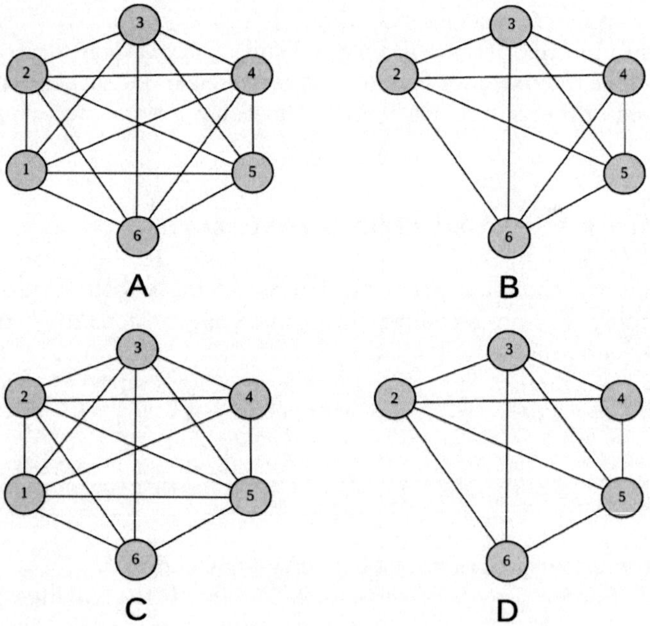

Fig. 1. An example of: A - clique or *α-clique* for $\alpha = 1.0$; B - a subgraph of clique from picture 2.1.A, which is also a clique; C - *α-clique* for $\alpha = 0.8333(3)$; D - a subgraph of *α-clique* from picture 2.1.C, which is *α-clique* with $\alpha = 0.8$.

A practical formula that can be used to find the value of parameter α in undirected and unweighted graph or subgraph is as follows:

$$\alpha = \frac{1}{n}\min_{j}\left(\sum_{i=1}^{n} a_{ij}\right) = \frac{1}{n}\min_{j}\left(1 + degree(v_j)\right), j = 1...n \quad (1)$$

where:

n - the number of graph (subgraph) nodes;
a_{ij} - entries of symmetric neighborhood matrix with indices i, j, describing the connection between vertices v_i and v_j of the graph, $a_{ij} \in \{0, 1\}$, 0 - no connection, 1 - presence of connection, we assume that a_{ii}=1 for all i =1, ..., n;
degree(v_j) - as mentioned earlier, the number of edges to which vertex v_j belongs.

For the weighted graph the formula used to compute α_w is:

$$\alpha_w = \frac{1}{n}\min_{j}\left(\sum_{i=1}^{n} \frac{1}{2a_{max}}(a_{ij} + a_{ji})\right), j = 1...n \quad (2)$$

n - the number of graph (subgraph) nodes;
a_{ij},a_{ji} - values of the elements of the generalized neighborhood matrix, describing the connections between nodes i and j in both directions (weights), we assume that $a_{ii} = 1$ for all i =1, ..., n;
a_{max} - value of the maximal weight - element of the neigborhood matrix.

The α-*clique* structures detected in a graph can be the basis for finding the connected graph blocks, which are very similar to the hub and spoke structure, proposed by [7,8]. The original clique structure in a graph imposes too strong constraints to use it in the majority of practical cases, where graphs connections are rather sparse. The α-*clique* structure is more flexible and its properties can be controlled using the α parameter. Besides, for $\alpha = 1$ it can also become a classic clique.

A hub and spoke structure is a graph $H_s = (G_h \cup G_s, E)$, where the subset G_h corresponds to α-*clique* with as big value of α as possible or at least to a connected graph (of hubs) with the relevant subset of the set E, each vertex of subset G_s (of spokes) has degree 1 and is connected exactly with one vertex from the subset G_h. This structure is illustrated in Fig. 2.

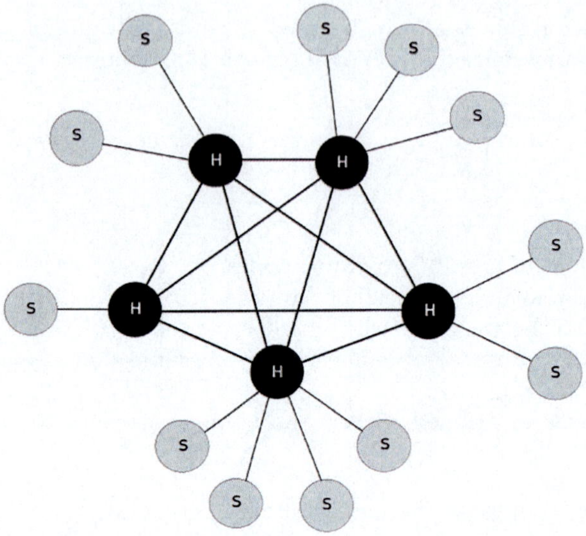

Fig. 2. An example of the Hub and Spoke structure (**H** - denotes hub nodes, **S** - spoke nodes).

3 Warsaw Urban Transport System

The actual data describing the public transport network in Warsaw come from https://www.wtp.waw.pl/rozklady-jazdy/ and ftp://rozklady.ztm.waw.pl.

A graphical diagram of the network is shown in Fig. 3. As you can see, this network is quite well developed and consists of:

- metro - 2 lines,
- high-speed city rail (SKM), suburban railway (WKD) and rail (KM) - 12 lines,
- trams - 26 lines,
- city, suburban and night buses - 303 lines.

WTP (Warszawski Transport Publiczny, Warsaw Public Transport, former ZTM) serves most of the public transport lines in Warsaw, only a small part is served by private carriers, who were not included here due to the lack of available data. Taxis, long-distance buses and long-distance trains were also not taken into account, because the former do not have a regular timetable, and long-distance lines are used only marginally as urban transport, similarly as the private lines, mainly serving the traffic between the city suburbs and its closer and more distant areas.

The entire urban transport system uses around 10,000 physical stops, but for calculation purposes it has been simplified by reducing it to 2,840 stops (graph nodes) as a result of combining into one stop the stops in opposite directions and stops divided into "sub-stops" in places with heavy passenger traffic, i.e.

stops, at which a large number of public transport lines stop, at loops or large intersections. Sometimes this leads to paradoxes, which in fact should lead to distinguishing some "sub-stops", as there are "sub-stops" of one "aggregated stop" so far apart that some communication lines stop at subsequent "sub-stops", causing problems in the description of traffic: loops with automatic unification or the need for manual interference and maintaining the division of such stops. As it can be seen, preparing of such real data for computations is not a simple task and requires a lot of manual corrections.

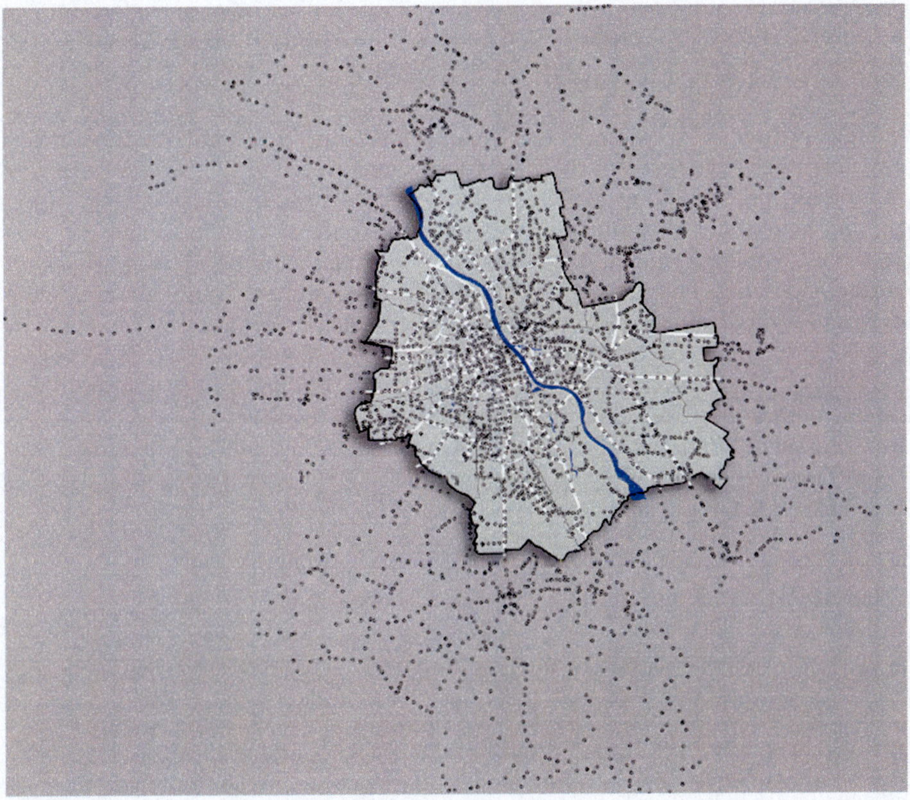

Fig. 3. The transport system of Warsaw (the state from the timetable as of November 25, 2021).

Automatic unification of stops is based on their numbering introduced by the WTP/ZTM. Each stop is identified by a four-digit number, the last two digits (minor number) of which identify the sub-stop and the first two (major number) its name. Only stops with the same names (major numbers), can be unified automatically.

The analysis of the stops graph also showed some drawbacks of the layout, namely it turned out that the graph was inconsistent. This was due to the lack of direct connections between the various modes of transport, in particular trains and other modes of urban transport. Usually, it is then necessary to walk several dozen or several hundred meters on foot to pass between stops, which prevents automatic (they have different major numbers) and rather excludes their manual unification. For this reason, it was necessary to add several dozen additional connections on foot in order to allow the consideration of many obvious and fastest routes in the city, using rail. Additional connections among close stops have the same parameters: maximal possible capacity and last 2 minutes. Of course, this is a simplification of the real conditions.

4 Evolutionary Based Method to Obtain Hub Nodes in Warsaw Communication System

Our graph model of Warsaw communication system was built on the basis of the timetable data from WTP/ZTM. In the constructed graph, we assumed that its vertices, i.e. stops, have a direct connection, as long as there is at least one running communication line (bus, tram, metro, SKM, suburban and ordinary rails) that connects them, i.e. it is possible to travel from one to the other stop without changing the transporting vehicle (transport line). Therefore, the resulting connection graph is not a sparse structure where only consecutive stops are connected but it is much denser, because the vertices, which are stops served by one line, usually form a full subgraph (sometimes it is not a full subgraph when the line does not follow exactly the same route both ways). Thus, the graph consists of overlapping blocks, because different communication lines have common stops. The data obtained from the timetable may present several properties of the transportation system:

- presence or not of the direct connections between the stops
- frequency or the number of courses in a certain unit of time
- travel time
- potential capacity of means of transport in a certain unit of time.

In the presented further results of computer simulation we used weighted graphs representing the travel time and potential capacity of means of transport.

In the case of stops and connections common to many communication lines, the collected values are, in our case, appropriately modified (aggregated), so that e.g. frequencies or capacity are added, the travel time is averaged in order to take into account a greater number of connections from more lines.

In the discussed example we decided to distinguish a set of several stops, called central stops, and present this set as a virtual destination for most travelers, for the sake of simplification, but also to make the analysis more specific. Of course, this is a simplification of the problem, because it is known that the real destination of commuters can be quite diversified. However, firstly, the selected stops are undoubtedly much more popular than others, and secondly, it is difficult to take all possible stops into account at the same time, so we considered a set of several very important stops located around the city center to be virtual destination of most trips. The choice of these stops is arbitrary, but they can be changed quite easily, especially if we have adequate data on the actual most popular travel aims of commuters or distribution of journeys for commuters in Warsaw, especially during the morning rush hour.

In the connection graph developed, each stop can become a communication hub, but it should meet several basic criteria:

- each hub should be strongly connected with a group of selected central stops (the first sum in parentheses in formula (3)),
- each hub should be strongly connected to the other selected hubs (the second sum in parentheses in formula (3)),
- the largest possible number of strong connections with stops from the spoke subgraph the given hub serves is also very important (the third sum in parentheses in formula (3)),
- similar numbers of spoke nodes for all hubs will be an additional advantage (the part $\frac{k-n}{n} - s_i$ of the formula (3)).

These requirements can be turned into the quality function used in computer simulations for both considered problems (time- and capacity-wise):

$$Q_1 = \sum_{i=1}^{n} \left(\sum_{j=1}^{m} w_{ij}^{C} + \sum_{l=i+1}^{n} w_{il} + \sum_{p=1}^{s_i} w_{ip} + \frac{k-n}{n} - s_i \right)$$
$$maxQ = \begin{cases} \frac{Q_1 \cdot \alpha_{WK}}{v \cdot n} & if\ \ Q_1 \geqslant 0 \\ \frac{Q_1 \cdot v \cdot n}{\alpha_{WK}} & if\ \ Q_1 < 0 \end{cases} \tag{3}$$

where:

Q - main optimized objective function,
Q_1 - auxiliary designation of a fragment of the calculated objective function,
n - number of hubs,
v - number of disconnected nodes + number of connected components of the graph (ideally 1),
m - number of central stops,
k - total number of stops in the connection graph,
s_i - number of stops served by the hub i (spoke stops),
w_{ij}^{C} - weight of the connection between the ith hub and the jth stop from the set of central stops highlighted,
w_{il} - weight of the connection between hubs i and l from the hub subgraph,
w_{ip} - weight of the connection between hub i and stop p from the subgraph it serves.
α_{WK} - value of the coefficient α_W for the *α-clique* of hub nodes (can be computed using formula (2)).

Of course, different weights are used in both problems. For the problem regarding travel time the weight between two stops is computed as: 1440 (number of minutes in a day) minus the average travel time in minutes (taken from the timetable) to turn it into maximization problem.

In the case of capacity, the weight means the average number of passengers that can be carried on a given route in an hour. Average capacities of individual means of transport can be taken from the ZTM website with information about the fleet of vehicles owned.

Before the evolutionary computations had been carried out, the weights were scaled to the interval (0, 1).

The evolutionary method used is a dedicated method with several specific improvements based on [10]. The standard evolutionary algorithm (Algorithm 1) is not an efficient tool in its pure form and requires many problem specific improvements to work efficiently [6].

Algorithm 1. The basic version of evolutionary algorithm

```
Input:
  Input data
Output:
  Output data
begin
  Random initialization of the population of solutions.
  while stop condition is not satisfied do
  begin
    Reproduction and modification of solutions using genetic operators
    Evaluation of obtained solutions
    Selection of individuals for the next generation
  end;
end;
```

The specialization of the evolutionary algorithm includes the following aspects: proper encoding of solutions, specialized genetic operators, a method of managing the execution of genetic operators and a selection method. We used a modification of approach, presented in [9]. In this approach, particular emphasis was placed on ensuring the consistency of the obtained connection graph in the form of hub & spoke, thus the fact of consistency of the graph is checked during mixed random and greedy generation of initial solutions:

- as a first hub, a node is drawn with probability proportional to its degree in the graph,
- spoke nodes are drawn and accepted if they are connected with the currently chosen hub,
- if no new spokes can be found for the currently chosen hub, one unattached so far node is drawn, if it is connected with the rest of the hub subgraph, it becomes a new hub, if not, the procedure of selecting a new hub is repeated 20 times,

- after that a new hub is accepted even if it is not connected with the subgraph of hubs (if so, its quality function will be significantly worse) and the procedure of selecting new spokes and hubs repeats until all graph nodes are processed.

Also a special new genetic operator is added that tries to connect the not connected elements of the modified graph. The set of genetic operators consists of:

- mutation - exchange of randomly chosen nodes in different sets of spokes;
- relocation 1 - if a spoke node of not connected hub is consistent with the biggest and consistent part of the hub subgraph, it is exchanged with the previous hub and becomes a new hub, possibly unattached spoke nodes are moved to different hubs' subgraphs of spokes if they can be connected to their hubs;
- relocation 2 - a randomly chosen node (spoke) is moved to a different hub's set of spokes;
- exchange of a randomly selected hub for randomly selected spoke.

The aspect of consistency is also taken into account in the evaluation of solutions (see formula (3)).

Genetic operators (more than the standard two), are managed using a method that is based on reinforcement learning, called Q-learning [13], where an agent (problem solution) performs action (executes a genetic operator) and gains a reward or penalty for it - solution becomes better, worse or unchanged. This reward enables computation of the operator quality (formula (4)), which is stored in a solution and inherited by its children. Stored values (evaluations of operators) are normalized and become probabilities of operator selection and execution independently by the individuals in the population.

$$V(s_{t+1}) = V(s_t) + \alpha \left(r_{t+1} + \gamma \left(V^{\star}(s_{t+1}) - V(s_t) \right) \right) \tag{4}$$

where:

t - time step,
$V(st)$ - quality factor or discounted cumulative reward,
$V * (st{+}1)$ - estimated value of the best quality factor (in conducted experiments the value attained by the best operator is used),
S_t, S_{t+1} - states of the agent in consecutive time steps,
α - the learning factor,
γ - the discount factor,
r_{t+1} - reward for the action, equal to the improvement of the quality of solution after execution of the genetic operator.

5 Results of Computer Simulations

In the study [9] we presented the results obtained for the imposed numbers of hub nodes, without the possibility of passenger transfers on foot between the very close and important stops connecting different means of city transport, and for old data (from 2016) without over 300 new stops and over 100 new bus lines and a new metro line. Hubs were located in those results mainly at the metro and SKM stations.

The new results are obtained for newer data (from 2021) with new quality functions, enabling the algorithm to find hubs independently without any assumptions on their expected number. Additional modification is introduction of pedestrian connections between the selected nearby transfer stops. Unfortunately, introduction of possible pedestrian crossings between all stops is not easy, as it requires manual intervention in the data, and preferably also the data on how commuters travel the distance between stops. Unfortunately, such data are not generally available, and so only in places where such connections are evident they were accounted for. It is possible that access to accurate data about frequent transport mode changes would alter the solutions.

The results obtained for the new method are as follows. Our new method proposes 51 hubs considering times of travels as the weights of the graph of public transport connections. The results are presented in Fig. 4.

In the case of capacity of connections, the resulting subgraph of hubs consists of 58 stops, which are presented in Fig. 5.

The indicated hubs are shown as bigger red dots in both pictures. As it can be seen, they are mainly located at the crossroads or forks of the main communication routes. The results are quite similar but not identical (29 stops are identical), their common part being shown in Fig. 6. Some of the different hubs-stops are geographically very close.

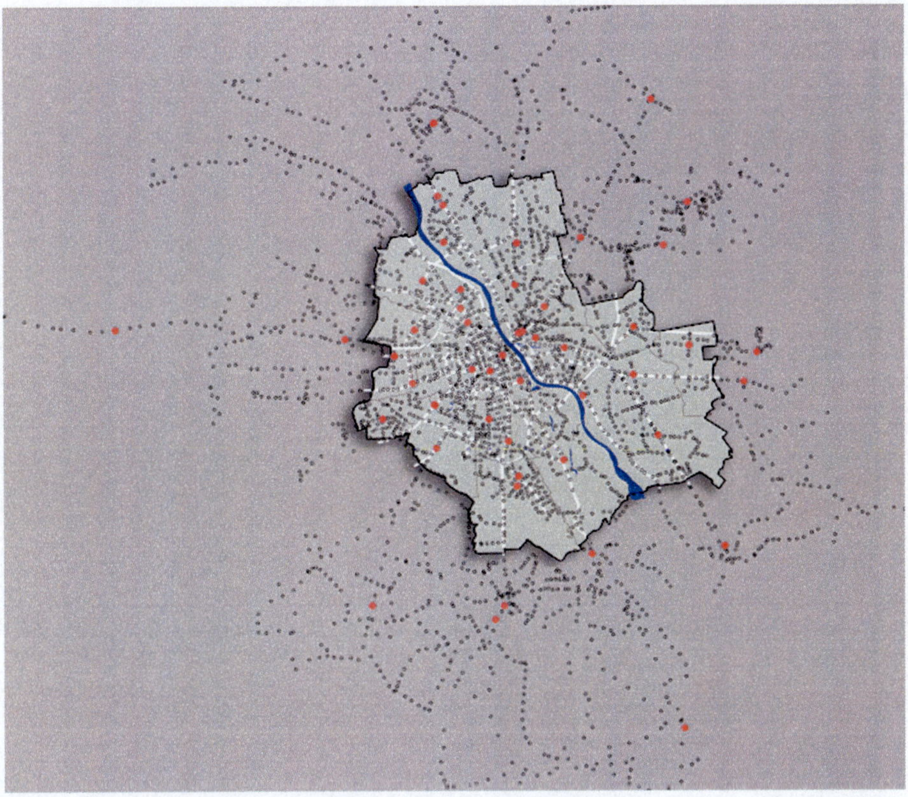

Fig. 4. Warsaw transport system with hubs (51) computed on the basis of time of travels (the state of the timetable as of November 25, 2021).

The results here shown are much more sophisticated than those obtained in 2016, because we obtained many hubs at metro and SKM stations (18 for the time criterion, 13 for the capacity criterion) as previously, but also in some additional, important communication points, scattered all over the city and even in its suburbs. It seems that the new method found some additional important communication routes apart from the largest communication lines related to railways, and also placed there proposals for the creation of communication hubs.

Since the evolutionary algorithm used to solve the problem is a heuristic method and largely depends on randomness, the results obtained in different simulations may vary. It is quite difficult to thoroughly analyze all the solutions obtained, these are presented as quite representative and useful in the case of practical implementation of its elements. In addition, the table below presents simple statistics of the obtained solutions on the example of 10 simulations of both versions of the problem.

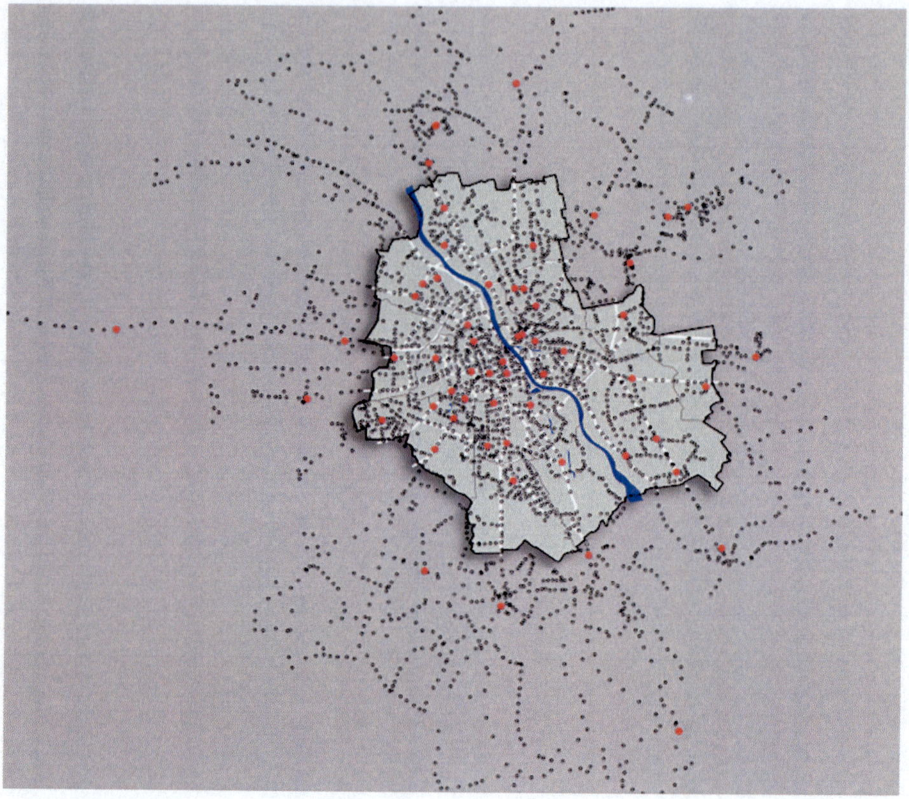

Fig. 5. Warsaw transport system with hubs (58) computed on the basis of capacity of connections (the state of the timetable as of November 25, 2021).

Table 1. Minimum, average and maximum numbers of hubs obtained during 10 simulations of both problems.

Problem	Average number of hubs	Minimum number of hubs	Maximum number of hubs
Time	56.3	51	63
Capacity	71.8	56	105

As it can be seen in Table 1, the problem with capacity criterion requires more hubs and the results obtained are more varied than in the case of the time criterion.

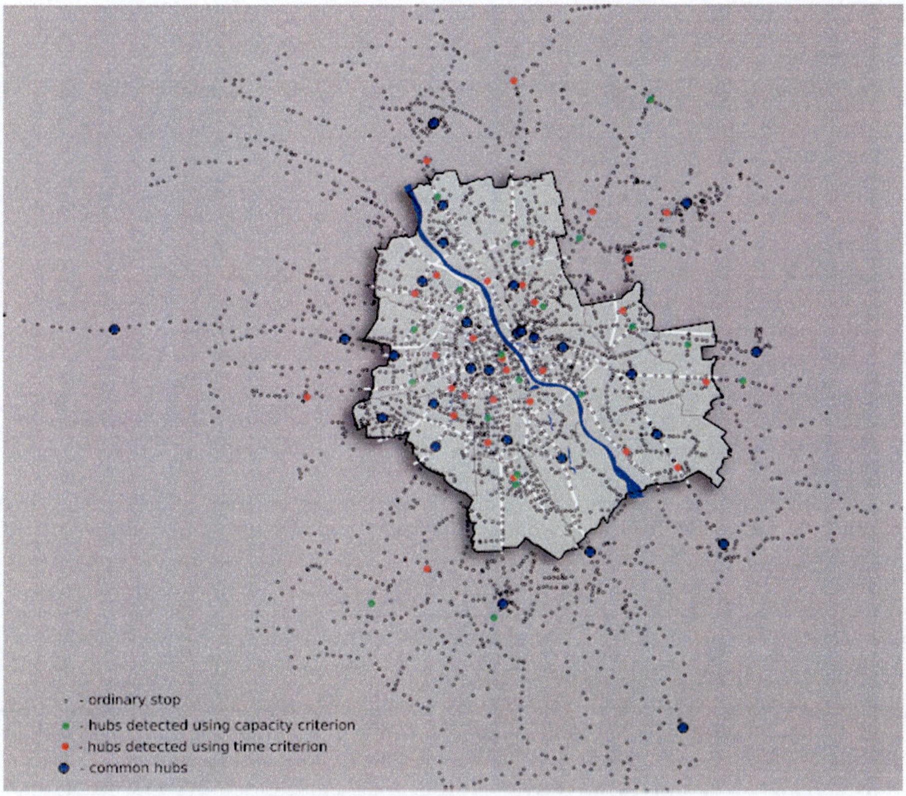

Fig. 6. Warsaw transport system with marked 29 common hubs selected by both criteria used (the state of the timetable as of November 25, 2021).

6 Conclusions

This work deals with the possibility of improving the public transport system by using elements of the hub & spoke concept. The proposed specialized evolutionary method of the transport data processing is quite efficient and can deal with large sets of stops, characterizing big cities.

As a result we obtained two quite similar solutions, although we used two different criteria: capacity and time of travel. In both solutions hubs are located mainly at crossings of main communication routes, which is rather in line with our expectations, apart from the indication of the largest communication lines related to rail transport that we obtained in the previous approach, our new, improved method also placed proposals for creation of communication hubs in suburbs with important bus and tram lines. This approach seems to be more universal and flexible.

References

1. Cormen, T., Leiserson, C., Rivest, R., Stein, C.: Introduction to Algorithms. MIT Press, Cambridge (2022)
2. Coyle, J.J., Bardi, E.J., Novack, R.A.: Transportation, 4th edn. West Publishing Company, New York (1994)
3. Hansen, P., Mladenovic, N., Urosevic, D.: Variable neighborhood search for the maximum clique. Discret. Appl. Math. **145**(1), 117–125 (2004)
4. Matisziw, T.C., Lee, Ch.-L., Grubesic, T.H.: An analysis of essential air service structure and performance. J. Air Transp. Manage. **18**(1), 5–11 (2012)
5. Mażbic-Kulma-Kulma, B., Potrzebowski, H., Stańczak, J., Sęp, K.: Evolutionary approach to solve hub & spoke problem using α-cliques. Evol. Comput. Glob. Optim. Prace naukowe PW Warszawa **2008**, 121–130 (2008)
6. Michalewicz, Z.: Genetic Algorithms + Data Structures = Evolution Programs. Springer, Berlin Heidelberg (1996). https://doi.org/10.1007/978-3-662-03315-9
7. O'Kelly, M.E., Bryan, D.: Interfacility interaction in models of hubs and spoke networks. J. Reg. Sci. **42**(1), 145–165 (2002)
8. O'Kelly, M.E.: A quadratic integer program for the location of interacting hub facilities. Eur. J. Oper. Res. **32**, 392–404 (1987)
9. Owsiński, J.W., Stańczak, J., Barski, A., Sęp, K.: Identifying main center access hubs in a city using capacity and time criteria. Evol. Approach Control Cybern. **45**(2), 207–223 (2016)
10. Stańczak, J.: Biologically inspired methods for control of evolutionary algorithms. Control. Cybern. **32**(2), 411–433 (2003)
11. Stańczak, J., Potrzebowski, H., Sęp, K.: Evolutionary approach to obtain graph covering by densely connected subgraphs. Control. Cybern. **41**(3), 80–107 (2011)
12. Stańczak, J., Barski, A., Sęp, K., Owsiński, J.W.: The problem of distribution of park-and-ride car parks in Warsaw. Int. J. Inf. Manage. Sci. **27**, 179–190 (2016). https://doi.org/10.6186/IJIMS.2016.27.2.6
13. Sutton, R.S., Barto, A.G.: Reinforcement Learning: An Introduction. MIT Press, Cambridge (1998)
14. Wilson, R.J.: Introduction to Graph Theory. Addison Wesley Longman (1996)

Evaluation Metrics for a Hybrid Classification System Based on the Distributivity Equation and the UNSW-NB15 Cyberattack Dataset

Ewa Rak(✉) and Jaromir Sarzyński

College of Natural Sciences, University of Rzeszów, Rzeszów, Poland
{erak,jsarzynski}@ur.edu.pl

Abstract. The aim of the study was to apply and evaluate the usefulness of the hybrid classifier to detect network intrusion threats on a comprehensive dataset. Our approach is established on ensemble classifiers involving the distributivity law which aggregates classifiers accordingly. The paper includes results of experiments performed on the UNSW-NB 15 dataset i.e. a collection of network packets exchanged between hosts. A five-fold cross-validation technique was used for the performance evaluation of classifiers on the dataset using the Scikit-learn tool. We present the results of classification measures of the performance of our hybrid algorithm created from aggregations associated with the distributivity equation of selected classification algorithms (Multilayer Perceptron Network, k Nearest Neighbors and Naive Bayes) and compare them with the results of these individual algorithms on raw data.

Keywords: Combined classifier · Machine learning · Cybersecurity · Accuracy · Precision · Aggregation · Distributivity law

1 Introduction

One of the interesting and dynamically developing trends in Machine Learning is combined classification (known also as hybrid, multiple or ensemble method). The algorithms developed within its framework try to build a model of the classification system based on component classifiers in order to use their best features and competencies needed to solve a given decision problem. During the construction of such systems, we encounter two types of problems: how to choose a valuable set of classifiers and how to obtain the final decision based on the answers of the members of the above-mentioned group of classifiers. In this case, the aggregation related to the distributivity law proved to be very effective. Various methods proposed over the years use different strategies to derive such a combination (see [9,15,16]).

K. T. Atanassov et al. (Eds.): BOS/SOR/IWIFSGN 2022, LNNS 793, pp. 311–321, 2023.
https://doi.org/10.1007/978-3-031-45069-3_28

This paper continued research with a novel hybrid approach (see [13,14]) to increase the main classification measures: accuracy, sensitivity, and precision, while minimizing the number of classifiers by applying the distributivity law that aggregates classifiers appropriately. Ensemble (hybrid) methods have been shown to be useful and effective solutions to improve classification efficiency. Despite being able to achieve the highest classification accuracy, ensemble methods have suffered greatly from a large number of base classifiers. Nevertheless, we can solve this problem by combining some classifiers by the appropriate aggregation operators for which the distributivity law holds. Both mean functions and triangular norms perform well.

The selection of appropriate classifiers for the most predictive hybrid classifier is extremely important. Each classification method shows different efficiency and precision, also depending on the type of dataset [8]. Three different classification algorithms were used in this paper - Multilayer Perceptron Network, k-Nearest Neighbors and Naive Bayes from Scikit-learn [26], to detect a protocol-based attack. Classification algorithms were used to determine which attack was better classified. In particular, the UNSW-NB15 dataset was applied to a non-standard aggregation method involving the distributivity law, and its quality on standard performance measures was compared with the quality obtained on raw data of selected algorithms.

The rest of the paper is organized as follows. Section 2 describes the background of hybrid classification systems. In Sect. 3, concepts connected with aggregation functions and distributivity between them are recalled. Section 4 explains the dataset used. Section 5 describes the experimental setting, evaluates the results of applying the ensemble classification method based on the distributivity law (with three selected classifiers) on a cyber-attacks dataset, and indicates the accuracy, sensitivity, and precision of classifiers for each type of attack. Finally, Sect. 6 summarizes this work.

2 A Hybrid Classification Model

Hybrid classification systems are becoming popular as they offer many possibilities for solving increasingly complex real-world problems. They provide the opportunity to use both a priori knowledge and raw data to solve problems in an innovative and promising way. Therefore, this multidisciplinary field of research is constantly evolving in the Artificial Intelligence research community. Hybridization occurs in many areas of human activity. Hybrid approaches seek to take advantage of the strengths of individual components, achieving increased efficiency by combining them.

As there is no possibility of constructing a universal classifier, therefore, the so-called hybrid methods attempt to use the best features of base classifiers to construct a complex classifier. The basic idea is to combine the results of several data mining algorithms, each one applied to a diversified dataset. This approach has often been much more accurate than classic approaches [24].

The earliest paper of combined learning can date back to the last century. Dasarathy and Sheela [3] proposed to utilize component classifiers trained in different categories to constitute a composite classification system, thus improving the performance of identification systems. The past decades have witnessed combined learning drawing increasing attention, and researchers have carried out a great amount of exploration and innovation. Dos Santos et al. [5] combined the optimization process with the dynamic selection strategy to select the most confident subsets of classifiers with high accuracy. Moreover, it is an important issue how to design and utilize effective methods to integrate results from basic models. There are also some other studies that introduce different types of basic models into ensemble classification, and these basic models include neural networks, support vector machines, etc. Zhou et al. [25] proposed a neural network-based ensemble method, where the weights of networks were updated by a genetic algorithm. In addition, ensemble classification methods have been applied to the intelligent transportation area ([12,18,23]), pattern recognition ([17,19]) and especially for network intrusion detection as one of the most critical network security problems in the world technology (see e.g., [6,21,22]). The intrusion detection system (IDS) has gained an essential role in computer and network security. IDS monitoring and analysis a network traffic and is used to classify different types of attacks. The network traffic action consists of many features collected in the form of a dataset to detect different types of attacks [20]. The increase of the massive amount of data that is generated daily through the Internet has caused the world of technology to face a big challenge [10].

Based on the above-mentioned papers, we find that there are still many challenges in the ensemble classification field that are worth further exploring in the future. Specifically, although most methods mainly consider improving the accuracy of the model at the architecture level of ensemble models, there is quite a lot of research on determining the appropriate model size and reducing the complexity of the model to increase the training speed.

3 Distributivity Equation of Aggregation Functions

The process of projection methods of multidimensional space input data to one dimension is usually carried out by the so-called aggregation functions (aggregation operators). They were formalized forty years ago [4], and since then have been extensively investigated (see e.g., [2] or [7]).

The selection of the aggregation function should be based on the properties dedicated to the framework in which the aggregation is performed. Among the huge number of classes of aggregation functions, averaging functions (means) and triangular operations are usually easily applied to the classification problem (see Table 1).

Distributivity specifies the relationship between two binary functions including aggregation functions.

Definition 1 (cf. [1], p. 318). *Let F and G be some binary functions in the nonempty set U, where F is symmetric.*

Table 1. Means and t-norms considered in this paper.

Mean	Name	T-norm	Name
$M_\wedge(X,Y) = \min(X,Y)$	Minimum		
$M_\vee(X,Y) = \max(X,Y)$	Maximum	$T_P(X,Y) = X \cdot Y$	Product
$M_A(X,Y) = \frac{X+Y}{2}$	Arithmetic	$T_E(X,Y) = \frac{X \cdot Y}{2-(X+Y-X \cdot Y)}$	Einstein
$M_H(X,Y) = \begin{cases} 0, & X=Y=0 \\ \frac{2XY}{X+Y}, & elsewhere \end{cases}$	Harmonic	$T_H(X,Y) = \begin{cases} 0, & X=Y=0 \\ \frac{XY}{X+Y-XY}, & elsewhere \end{cases}$	Hamacher
$M_P(X,Y) = \sqrt{\frac{X^2+Y^2}{2}}$	Power mean	$T_L(x,y) = \max(x+y-1,0)$	Łukasiewicz

We say that F is distributive over G if for all $X, Y, Z \in U$ the following equality is fulfilled

$$F(X, G(Y,Z)) = G(F(X,Y), F(X,Z)). \tag{1}$$

The lack of distributivity is a big problem in any algebraic transformations, and therefore also in computer modeling. In general, aggregations are not distributive from each other. Of all known pairs of aggregation functions for which the distributivity law (1) formally holds, we have selected (see Table 1) and used in the elaborated algorithm the following equalities

D1 $T_P(X, M_\wedge(Y,Z)) = M_\wedge(T_P(X,Y), T_P(X,Z))$
D2 $T_P(X, M_\vee(Y,Z)) = M_\vee(T_P(X,Y), T_P(X,Z))$
D3 $T_P(X, M_A(Y,Z)) = M_A(T_P(X,Y), T_P(X,Z))$
D4 $M_A(X, M_A(Y,Z)) = M_A(M_A(X,Y), M_A(X,Z))$
D5 $T_P(X, M_H(Y,Z)) = M_H(T_P(X,Y), T_P(X,Z))$
D6 $M_P(X, M_P(Y,Z)) = M_P(M_P(X,Y), M_P(X,Z))$
D7 $T_E(X, M_\wedge(Y,Z)) = M_\wedge(T_E(X,Y), T_E(X,Z))$
D8 $T_E(X, M_\vee(Y,Z)) = M_\vee(T_E(X,Y), T_E(X,Z))$
D9 $T_H(X, M_\wedge(Y,Z)) = M_\wedge(T_H(X,Y), T_H(X,Z))$
D10 $T_H(X, M_\vee(Y,Z)) = M_\vee(T_H(X,Y), T_H(X,Z))$
D11 $T_L(X, M_\wedge(Y,Z)) = M_\wedge(T_L(X,Y), T_L(X,Z))$
D12 $T_L(X, M_\vee(Y,Z)) = M_\vee(T_L(X,Y), T_L(X,Z))$

4 UNSW-NB15 Dataset

The dataset that we have used for this work is the UNSW-NB15 dataset [11] of the intrusion detection system (IDS) available from UNSW Canberra at ADFA (https://research.unsw.edu.au/projects/unsw-nb15-dataset).

The raw network packets of the UNSW-NB15 dataset was created by the IXIA PerfectStorm tool in the Cyber Range Lab of the Australian Centre for Cyber Security (ACCS) for generating a hybrid of real modern normal activities and synthetic contemporary attack behaviours. The tcpdump tool was utilised to capture 100 GB of the raw traffic (e.g., Pcap files). Researchers processed the raw network data using the Bro-IDS, and an additional security tool, called Argus, to create richer features. Since they generated synthetic traffic, they used reporting functionality in the network traffic generator to export the labels. In total, they generated nine different types of attacks in addition to the normal traffic (see Table 2). Twelve algorithms were developed to generate totally 49 features with the class label, where (1 to 5) are the flow features. The matched features are categorized into three sets, namely basic (6 to 18), content (19 to 26) and time (27 to 35). Features 36 to 40 and 41 to 47 are labeled as general-purpose features and connection features, respectively. However, not all 49 features are necessarily relevant for the class labels. 42 are numeric in nature.

Compared to the criteria for a good intrusion detection dataset, UNSW-NB15 has both audit logs and raw network data. It has a more complete repertoire of attacks. It includes realistic network activity, and it is well labeled. Since it is synthetic data, there are no privacy concerns. The complete datasets even include source and destination IP addresses, which are generally considered too sensitive to expose by most organizations.

Table 2. UNSW-NB15 dataset classes.

No.	Class	Cardinality	Card. of samples	Description
1	**Normal**	2218764	**59529**	Natural transaction data
2	**Generic**	215481	**59529**	A technique to clash the block-cipher configuration by using # functions
3	**Exploits**	44525	**44525**	An instruction that takes advantage of bugs/errors caused by unintentional behavior on the network
4	**Fuzzers**	24246	**24246**	An attack to crash the system by inputting a lot of random data
5	**DoS**	16353	**16353**	Attacks computer memory
6	**Reconnaissance**	13987	**13987**	A probe to evade network security controls by collecting relevant information
7	**Analysis**	2677	**2677**	An attack targets web applications through emails, ports, or web scripts
8	**Backdoors**	2329	**2329**	Using backdoor to secure remote access
9	Shellcode	1511	–	Code is used to exploit software vulnerabilities
10	Worms	174	–	A set of virus codes can add to a computer system or other programs
$\sum$		2540047	**223175**	

5 The Hybrid Classification Method Based on the Distributivity Law

We recall the novel approach (an ensemble method) proposed in our previous article [13] to increase the accuracy of a classification and, at the same time, minimize a group of classifiers by applying the distributivity law which will aggregate classifiers accordingly.

The proposed algorithm was subjected to further modifications in order to improve both the accuracy of the classification as well as the sensitivity, false positive rate (fall-out), and precision. These are the so-called classification quality measures based on a confusion matrix. The confusion matrix is a situation analysis table for summarizing the prediction results of classification models in ML. As shown in Table 3 (the case of binary classification), TP is the number of correctly predicted positive cases, which is the number of connection records correctly classified as attack class in network attack security detection. FN is the number of incorrectly predicted positive cases, FP is the number of incorrectly predicted negative cases, and TN is the number of correctly predicted negative cases. Our experimental evaluation metrics based on a confusion matrix are the following:

Table 3. A confusion matrix for IDS.

Real ↓ Predicted→	Attack	Normal
Attack	TP	FN
Normal	FP	TN

- **Accuracy (ACC)** $= \frac{TP+TN}{TP+TN+FP+FN}$,
- sensitivity (recall or true positive rate) **TPR** $= \frac{TP}{TP+FN}$,
- false positive rate **FPR** $= \frac{FP}{FP+TN}$,
- precision (positive predictive value) **PPV** $= \frac{TP}{TP+FP}$.

The pycm library (https://www.pycm.io/doc/index.html) was used for their determination, naturally for each decision class.

5.1 The Proposed Hybrid Approach

- Data pre-processing: choose 8 classes of attack types, including normal (no attack) denoted by d_0 and 7 classes of attacks denoted by d_j, $j = 1, ..., 7$ whose cardinality exceeds 2000. The whole dataset contains 2540047 objects (see these listed classes in Table 2). Features have been scaled (normalized) to a given range using MinMaxScaler[1]. It scales the values to a specific value range without changing the shape of the original distribution.

[1] https://scikit-learn.org/stable/modules/preprocessing.html.

Randomize 223175 samples on which the necessary calculations will be perform.

- The process of building classifiers and determining the decision values for the test objects is given by **Algorithm 1**.
- The method used for $\mathbf{P}, \mathbf{P_j}, \mathbf{C_j}$ classifiers construction: MultiLayer Perceptron, k-NN and Naïve Bayes in Python with Scikit-learn (https://scikit-learn.org/stable/modules/classes) - the most popular Python library for performing classification, regression, and clustering algorithms.
  For each methods used, the default options were left.
- When determining classifiers, symbolic attributes have been omitted (due to the distance in kNN).
- The algorithm proposed above has been optimized to improve the quality of the classification measures.

Algorithm 1:

1 **Step 1.** Construct classifiers as follows:

(1.) The **P** classifier - used to pre-classify an object as to whether it belongs to decision class d_0 (label "0") or to some other decision class $\neg d_0 = \{d_1, d_2, d_3, d_4, d_5, d_6, d_7\}$ (label "1");

(2.) Collection of classifiers $\mathbf{C_j}$ ($j = 1, 2, ..., 7$) - used to distinguish objects with decision value d_j (label "1") from objects from other classes $\neg d_j = \{d_1, ..., d_7\} \setminus \{d_j\}$ (label "0") without the class d_0;

(3.) Collection of classifiers $\mathbf{P_j}$ ($j = 1, 2, ..., 7$) - used to distinguish objects with decision value d_j (label "1") from objects of the class d_0 (label "0").

Step 2. Determine the probability of membership to decision class with label "1" (weight for a class labeled "1") for each instance. Next, construct the weight table of classifiers $\mathbf{P}, \mathbf{C_j}, \mathbf{P_j}$ for the class labeled "1" (classifiers as columns, objects as rows, weights (probabilities) for class "1" as values).

Step 3. For a given distributivity law $D1 - D12$ compute the value of its left side L (as $L = R$) considering the obtained classifiers X, Y, Z in the combination of their settings i.e. $X = P$, $Y \in C_j$, $Z \in P_j$ and $X = P$, $Y \in P_j$, $Z \in C_j$.

Step 4. Fix the parameter $\varepsilon \in (0, 1)$. It denote the threshold for classifying an object into one of the d_j classes.

Step 5. Determine the maximum for the previously calculated values of the left side of the distributivity law.

Step 6. For each tested object propose a decision value as follows: if the maximum value of the left side ($\max_{aggr}$) of distributivity was obtained for the j-th decision (after aggregating the weights of the classifiers $\mathbf{P}, \mathbf{C_j}, \mathbf{P_j}$) and $\max_{aggr} > \varepsilon$, then propose the j-th decision value for the given object; else if $\max_{aggr} \leq \varepsilon$ then propose a neutral decision - no attack (class "0").

5.2 Experimental Results and Discussion

The algorithm, along with the use of all three proposed classification methods, was implemented and tested in the Python programming language (using the Scikit-learn library for the construction of $\mathbf{P}, \mathbf{C_j}, \mathbf{P_j}$ classifiers). It uses examples selected above pairs of distributive aggregation functions denoted $D1 - D12$. An important fact is that 12 (due to the symmetry of aggregation) of 27 possible cases of permutations of these classifiers were taken into account. To test the quality of classifiers, we applied five-fold cross-validation. Experiments were carried out using 38 numeric features (there are 42 numerical attributes, but we discarded one '*sport*' and 3 with missing values '$ct - flw - http - mthd$', '$is - ftp - login$', '$ct - ftp - cmd$').

The performed classification experiments are designed to evaluate the performance of our hybrid intrusion detection model using four metrics: accuracy (ACC), sensitivity (TPR), false positive rate (FPR) and precision (PPV).

Comparative analysis of such a measurement results on the original raw data and newly created data by using a hybrid method based on the distributivity equation with three algorithms (MLP, kNN and NB) and the fixed $\varepsilon \in (0, 1)$ are presented in Table 4.

Due to the large size of the result tables and the limited space of this paper, we present only the best results for each type of attack. The complete list of results includes the folder **Experiments results** with files .xlsx placed on Google Drive (see [27]).

According tothe results presented, it is easy to notice that the best classification metrics with respect to raw data are obtained when we adopt globally distributivity equations D5, D7, D9 and D11 with $\varepsilon = 0.1$ (also with $\varepsilon = 0.2$) taking into account the order of algorithms used in it $X = kNN$, $Y = MLP$ and $Z = NB$. Similar, although sometimes slightly weaker results are obtained in the case of $X = Y = Z = MLP$, while the worst measures are included in the case of $X = Y = Z = NB$.

It can be certainly stated that the best value of results was also influenced by the selection of aggregation functions satisfying the (1) distributivity equation. Generally, the best results are for equations $D7$, $D9$ and $D11$, in which there is a combination of the Einstein, Hamacher and Łukasiewicz t-norm with minimum and also autodistributivity with the root mean square $D6$. The value of ε is also significant. In the case of the above distributivity equations, it is closer to the beginning of the interval $(0, 1)$. The detailed data obtained (placed in [27]) indicate a certain improvement in the accuracy of the classification for particular types of attack under consideration, mainly "*Exploits*", "*Fuzzers*", "*Generic*", "*Reconnaissance*", in the case of higher cardinality of a given class. For the attack classes "*Analysis*", "*Backdoors*" and "*Normal*" (no attack), the results for this hybrid method and raw data kNN are almost the same.

However, the result of TPR, FPR and PPV measurements is even better. There is a significant improvement in all results in relation to raw data, in particular for the precision of our classification. In general, this shows the enormous

Table 4. The best results in case $X = kNN$, $Y, Z \in \{MLP, NB\}$ obtained for D5, D7, D9 and D11 with $\varepsilon = 0.1$

Class	ACC %	Raw_MLP_ACC %	Raw_NB_ACC %	Raw_kNN_ACC %
Analysis	98,9	95,8	98,8	98,8
Backdoors	98,9	97,8	95,1	98,9
DoS	91,7	89,4	92,5	92,1
Exploits	87,6	71	80,1	85
Fuzzers	97,2	78,7	87,0	92,2
Generic	99,2	80,3	76,3	98,1
Normal	99,3	90,9	90	99,3
Reconnaissance	97,7	90,2	79,7	94,9

Class	TPR %	Raw_MLP_TPR %	Raw_NB_TPR %	Raw_kNN_TPR %
Analysis	15,4	7	0	0
Backdoors	2,4	2,3	20,9	0
DoS	21	3,1	0,3	10
Exploits	86,6	36,1	6,5	75,7
Fuzzers	86,4	36	22,2	77,7
Generic	97,4	61,2	95,3	96,5
Normal	98,3	70,3	67,3	98,1
Reconnaissance	74,1	10,2	40,5	57,8

Class	FPR %	Raw_MLP_FPR %	Raw_NB_FPR %	Raw_kNN_FPR%
Analysis	0,1	3,1	0	0
Backdoors	0,02	1,1	4,1	0
DoS	2,7	3,7	0,2	1,4
Exploits	12,2	20,3	1,5	12,7
Fuzzers	1,5	16,1	5,1	6,0
Generic	0,1	12,7	30,5	1,2
Normal	0,3	1,6	1,7	0,1
Reconnaissance	0,7	4,4	17,7	2,6

Class	PPV %	Raw_MLP_PPV %	Raw_NB_PPV %	Raw_kNN_PPV%
Analysis	61,8	2,6	0*	0*
Backdoors	53,4	2,1	5,1	0*
DoS	37,6	6,1	11,7	36,4
Exploits	63,9	30,8	52	59,7
Fuzzers	87,8	21,4	34,5	61,4
Generic	99,7	63,6	53,2	96,6
Normal	99	94,2	93,6	99,6
Reconnaissance	87,6	13,4	13,3	59,9

where 0* is in the case: true positive + false positive = 0.

potential of the proposed hybrid method. It can be successfully used to detect various network threats, and with additional modifications and enhancements, it can be an integral part of an Intrusion Detection System.

6 Conclusions

In Machine Learning, a combination of classifiers, known as an ensemble (hybrid) classifier, often outperforms individual ones. While many ensemble approaches exist, it remains, however, a difficult task to find a suitable ensemble configuration for a particular dataset. The aim of this work was to apply the hybrid method (classifiers aggregation based on the distributivity law) to detect network attacks. It presents the measurement results of an innovative method using various standard algorithms for supervised classification with the appropriate distributivity law. The base algorithms for the custom aggregation method that were connected and compared with raw data by accuracy, precision, TPR and FPR measures are namely Multilayer Perceptron Network, k-Nearest Neighbor, and Naive Bayes. Our experimental results suggest that the hybrid approach can generate ensembles that outperform traditional algorithms in terms of classification precision and accuracy. With the appropriate value of the ε parameter, we obtained results up to 5% (ACC) and 59,2% (PPV) better (than for raw data) for individual decision classes. Therefore, this approach can be helpful in creating new, better adapted Intrusion Detection Systems.

References

1. Aczél, J.: Lectures on Functional Equations and Their Applications. Academic press, New York, London (1966)
2. Beliakov, G., Bustince, H., Calvo, T.: A Practical Guide to Averaging Functions. Studies in Fuzziness and Soft Computing, vol. 329. Springer, Cham (2016). https://doi.org/10.1007/978-3-319-24753-3
3. Dasarathy, B.-V., Sheela, B.-V.: A composite classifier system design: concepts and methodology. Proc. IEEE **67**(5), 708–713 (1979)
4. Dombi, J.: Basic concepts for the theory of evaluation: the aggregative operator. Eur. J. Oper. Res. **10**, 282–293 (1982)
5. Dos Santos, E.M., Sabourin, R., Maupin, P.: A dynamic overproduce-and-choose strategy for the selection of classifier ensembles. Pattern Recogn. **41**(10), 2993–3009 (2008)
6. Golrang, A., Golrang, A.-M., Yayilgan, S.-Y., Elezaj, O.: A Novel Hybrid IDS Based on Modified NSGAII-ANN and Random Forest. Electronics **9**(4), 577 (2020)
7. Grabisch, M., Marichal, J.-L., Mesiar, R., Pap, E.: Aggregation Functions. Encyclopedia of Mathematics and Its Applications, vol. 127. Cambridge University Press, New York (2009)
8. Kim, Y.-S.: Comparision of the decision tree, artificial neural network, and linear regression methods based on the number and types of independent variables and sample size. J. Expert Syst. Appl. **34**(2), 1227–1234 (2008)
9. Kuncheva, L.-I.: Combining Pattern Classification, 2nd edn. Wiley, Hoboken (2014)
10. Lee, J., Kim, J., Kim, I., Han, K.: Cyber threat detection based on artificial neural networks using event profiles. IEEE Access **7**, 165607–165626 (2019)
11. Moustafa, N., Slay, J.: UNSW-NB15: a comprehensive data set for network intrusion detection systems (UNSW-NB15 network data set). In: Military Communications and Information Systems Conference (MilCIS), pp. 1–6. IEEE (2015). https://doi.org/10.1109/MilCIS.2015.7348942

12. Oliveira, L., Nunes, U., Peixoto, P.: On exploration of classifier ensemble synergism in pedestrian detection. IEEE Trans. Intell. Transp. Syst. **11**(1), 16–27 (2010)
13. Rak, E., Bazan, J.-G., Szczur, A., Rząsa, W.: The distributivity law as a tool of k-NN classifiers'aggregation: mining a cyber-attack data set. In: IEEE International Conference on Fuzzy Systems (FUZZ-IEEE), pp. 1–8 (2020). https://doi.org/10.1109/FUZZ48607.2020.9177686
14. Rak, E., Szczur, A.: A comparative assessment of aggregated classification algorithms with the use to mining a cyber-attack data set. In: IEEE International Conference on Fuzzy Systems (FUZZ-IEEE), pp. 1–6. (2021). https://doi.org/10.1109/FUZZ45933.2021.9494487
15. Rokach L.: Pattern Classification Using Ensemble Methods. Series in Machine Perception and Artificial Intelligence. World Scientific (2010)
16. Seni G., Elder J.: Ensemble Methods in Data Mining: Improving Accuracy Through Combining Predictions. Morgan and Claypool Publishers (2010)
17. Su, Y., Shan, S., Chen, X., Gao, W.: Hierarchical ensemble of global and local classifiers for face recognition. IEEE Trans. Image Process. **18**(8), 1885–1896 (2009)
18. Sun, S., Zhang, C.: The selective random subspace predictor for traffic flow forecasting. IEEE Trans. Intell. Transp. Syst. **8**(2), 367–373 (2007)
19. Zhang, P., Bui, T.-D., Suen, C.-Y.: A novel cascade ensemble classifier system with a high recognition performance on handwritten digits. Pattern Recogn. **40**(12), 3415–3429 (2007)
20. Singh, R., Kalra, M., Solanki, S.: A hybrid approach for intrusion detection based on machine learning. Int. J. Secure. Netw. **15**(4), 233–242 (2020)
21. Ugochukwu, C.-J., Bennett, E.-O., Harcourt, P.: An Intrusion Detection System Using Machine Learning Algorithm. LAP LAMBERT Academic Publishing (2019)
22. Shashank, K., Balachandra, M.: Review on network intrusion detection techniques using machine learning. In: 2018 IEEE Distributed Computing, VLSI, Electrical Circuits and Robotics (DISCOVER), pp. 104–109. IEEE (2018). https://doi.org/10.1109/DISCOVER.2018.8673974
23. Zhang, B.: Reliable classification of vehicle types based on cascade classifier ensembles. IEEE Trans. Intell. Transp. Syst. **14**(1), 322–332 (2013)
24. Zhou, Z.-H.: Ensemble Methods: Foundations and Algorithms. Chapman and Hall/CRC (2012)
25. Zhou, Z.-H., Wu, J., Tang, W.: Ensembling neural networks: many could be better than all. Artif. Intell. **137**(1–2), 239–263 (2002)
26. https://scikit-learn.org/stable/
27. https://drive.google.com/drive/u/0/folders/1i-3igmAs6Y088oHSP2ItpXeVYqwF0i2H

Hybrid Sensor System for Robot Control with Nonlinear Autoregressive Network with Exogenous Inputs

Sotir Sotirov[1] and Simeon Ribagin[1,2](✉)

[1] University "Prof. D-r Asen Zlatarov", Burgas, Bulgaria
ssotirov@btu.bg, simribagin@gmail.com

[2] Department of Bioinformatics and Mathematical Modelling Institute of Biophysics and Biomedical Engineering, Bulgarian Academy of Sciences, Sofia, Bulgaria

Abstract. In this paper, nonlinear autoregressive exogenous input (NARX) neural network has been used for modeling a sensor based hybrid system for possible control of robotic devices. The signals for the model are obtained from the MyoWare surface muscle sensors and the so proposed hybrid sensor system represents a sensor-based control for various task execution and direct human-robot interaction.

Keywords: NARX neural network · Muscle sensor · Robot control

1 Introduction

The need to control complex systems like robotic devices and manipulators under strong uncertainties has led to a reevaluation of the adaptive control methods currently available. Robot control is already being heavily commercialized in industry but the methods required to regulate interaction and collaboration between humans and robots have not been fully established yet [1]. With the emergence of Artificial neural networks (ANNs), adaptive control schemes can be produced that possess stronger robustness and high level of intelligence. In general, a neural network is a system that is modeled after the design of organic nerve systems. It is made up of linked neurons arranged in a layering system that includes an input layer, one or more intermediate hidden layers, and an output layer. Among all kinds of artificial neural networks, the nonlinear autoregressive network with exogenous inputs (NARX) neural network usually has a higher level of accuracy, superior generality and practicability for time series prediction because of various advantages such as the ability to keep a memory of past events to predict future trends [2, 3]. Moreover, the NARX network has the ability to provide a high degree of flexibility to address features as including a large number of exogenous variables in time-series models, in other words, the ability to accept almost any number of exogenous inputs, which relieves the neural network builder from the task of picking a small number of informative input variables [4]. Learning of NARX networks is more effective than in other neural networks and these networks converge much faster and generalize better

K. T. Atanassov et al. (Eds.): BOS/SOR/IWIFSGN 2022, LNNS 793, pp. 322–326, 2023.
https://doi.org/10.1007/978-3-031-45069-3_29

than other networks. NARX are recurrent dynamic networks that correlate the current value assumed by the output parameters y(t) in a time series to the past values of the same parameters and of the driving parameters. Thus, this network allows us to use as an input not only the series $x(t)$ but also the values of the series $y(t)$ in the past times. The fundamental equation of the NARX model is as follows:

$$y(t) = f(x(t-1), x(t-d), y(t-1), y(t-d)),$$

where d is a delay parameter.

On the other hand, different bio signals have been explored for use as control signals for different robotic, orthotic or rehabilitation devices. In the present study we propose the use of muscle signals as an input signals for the NARX model in order to predict the possibility of using such signals for robot control. Electromyography (EMG) is an experimental technique concerned with the development, recording and analysis of myoelectric signals which represents the sum of the motor unit action potentials (MUAPs) that combine to initiate muscular contraction. There are two distinct modalities of electro-physiological techniques for neuromuscular investigation. Electromyography can be performed with either surface or needle electrodes. Surface electromyography (sEMG) is a non-invasive method of measuring muscle activity in which conductive gel electrodes are placed on the skin to measure overall activity of a large region of muscle fibers. The measured voltage represents the amount of force that a muscle (or group of muscles) is exerting in real-time. sEMG is commonly used in biomedical, engineering applications and human computer interaction. The MyoWare is a small surface sEMG sensor developed by Advancer Technologies [5]. It automatically analyzes the incoming raw voltage signal and outputs a smooth posi-tive waveform of integer values. It is suitable for producing raw sEMG signals and analog output signals for microcontroller based applications. The sensor is designed for reliable sEMG output with low power consumption. It operates with single power supply (+2.9 V to +5.7 V) with polarity reversal protection. An additional feature in this sensor is that the sensitivity gain can be adjusted.

2 The NARX Model

In the present paper we will use nonlinear autoregressive network with exogenous inputs (NARX) is a one of the recurrent dynamic network, with feedback connections enclosing several layers of the network. The autoregressive network with exogenous inputs model is based on the linear ARX model, which is usually used in time-series modeling. The equation for the model is:

$$y(t) = f\left(y(t-1), y(t-2), \ldots, y\left(t-n_y\right), u(t-1), u(t-2), \ldots, u(t-n_u)\right),$$

where the next value of the output signal $y(t)$ is regressed on previous values of the output signal and previous values of an independent input signal and n_{y} and n_{u} are the number of the delay blocks. A figure of the neural network is shown below (Fig. 1), where a two-layer feedforward network is used for the recognition and prediction, in cases where the input and output can be multidimensional.

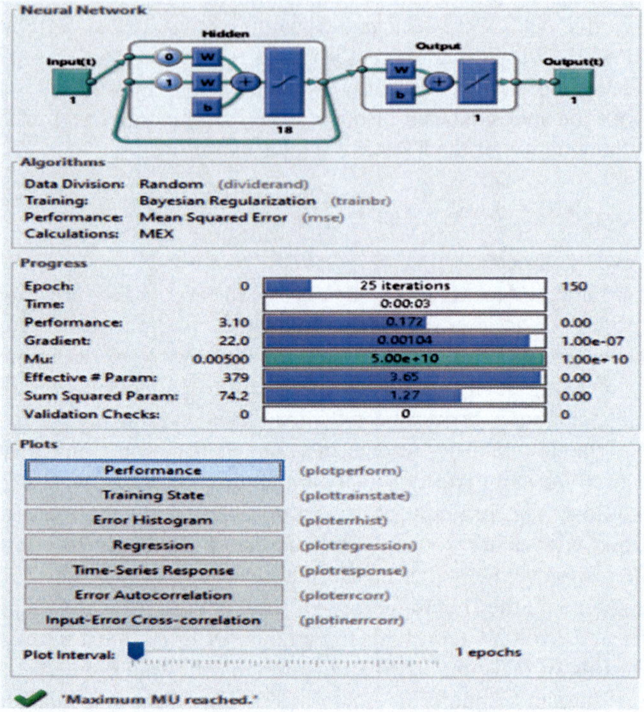

Fig. 1. The structure of the NARX model

3 Results

In this case, a signal from the MyoWare muscle sensor (AT-04-001) is fed to the input of the neural network. MyoWare sensor signal is Envelope Output (ENV). At the output of the neural network, a voltage is obtained to control a selected robotic manipulator motor. The structure of the neural network with one input 18 neurons in the middle layer and one output. The training set is 1428 values measured with an experimental model of the robotic manipulator, and it is divided as previously described into three parts. The first part is the largest and is 70% of all data. The second part is the testing data. They are 20% of all data. The last data is the verification data. It is characteristic of them that they are so-called external data and do not participate in the training of the neural network. They are 10% of all data. It is important to say that the data allocation of training/testing/verification data is randomized by the system. For the learning process, we set the following parameters: Performance (MSE) = 0.00001; Validation check = 15. The input vector is divided into three different parts: Training (70/100); Validation (10/100) and Testing (20/100).

The learning process of the neural network and the error distribution is shown on (Fig. 2), as can be seen the neural network is trained at step 25 and the best root mean square error (0.17237) is achieved at step 9.

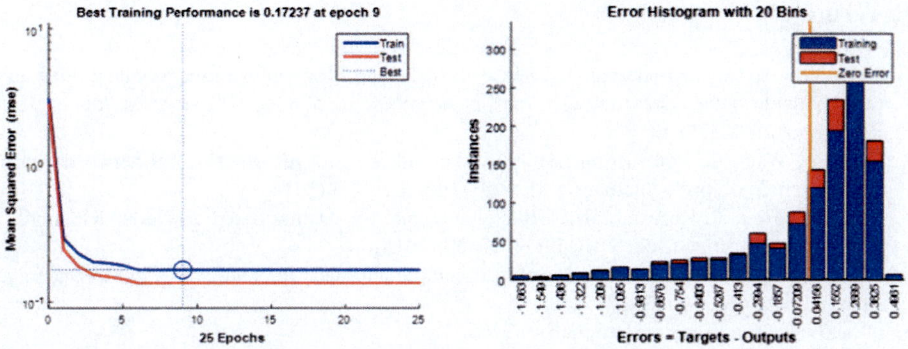

Fig. 2. Visualization of the learning process of the neural network and the distribution of the mean squared error of the neural network

As seen in the Fig. 3 shown, the regression coefficient 0.67135 for training, 0.68146 for testing and 0.67241 for all.

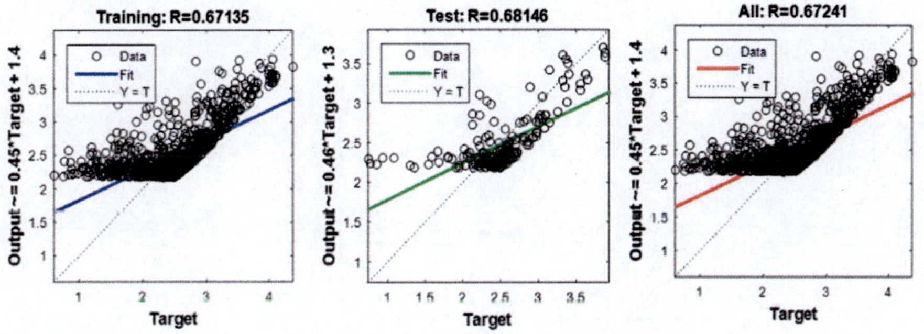

Fig. 3. The regression coefficient

4 Conclusions

The so proposed hybrid sensor system may be used for control of different types of robotic devises and particularly robotic manipulators and robotic arms. The proposed approach combines a robotic device, MyoWare surface muscle sensors and a neural network. The individual signals of the MyoWare surface muscle sensors are processed by the nonlinear autoregressive exogenous input (NARX) neural network, thus allowing for the performance of various tasks and direct human-robot interaction.

Acknowledgements. This work is partially supported by the grant: BG05M20P001-1.002-0011 "Centre of competence MIRACle - Mechatronics, Innovation, Robotics, Automation, Clean technologies".

References

1. Cherubini, A., Navarro-Alarcon, D.: Sensor-based control for collaborative robots: fundamentals, challenges, and opportunities. Front. Neurorobot. **14**, 576846 (2021). https://doi.org/10.3389/fnbot.2020.576846
2. Wang, J., Wang, J.: Forecasting energy market indices with recurrent neural networks: case study of crude oil price fluctuations. Energy **102**, 365–374 (2016)
3. Wang, N., Hou, T.: Research of RMB exchange rate forecasting based on the NARX model. J. Northeastern Univ. (Soc. Sci.), **17**(1), 32–37 (2015)
4. Zhao, Y., Li, J., Yu, L.: A deep learning ensemble approach for crude oil price forecasting. Energy Econ. **66**, 9–16 (2017)
5. https://www.sparkfun.com/products/13723

Generalized Net Model of a Collective Work of a DOBOT Magician Robot System with a Centralized Control

Simeon Ribagin[1,2(✉)], Iasen Hristozov[3], and Krassimir Atanassov[1,2]

[1] Department of Bioinformatics and Mathematical Modelling, Institute of Biophysics and Biomedical Engineering, Bulgarian Academy of Sciences, Acad. G. Bonchev Str., Block 105, 1113 Sofia, Bulgaria
simribagin@gmail.com, krat@bas.bg

[2] Prof. Asen Zlatarov University, 1,"Prof. Yakimov" Blvd, 8010 Burgas, Bulgaria

[3] European Soft Design, LTD., Sofia, Bulgaria
iasen_hristozov@abv.bg

Abstract. Applying the theory of Generalized Nets (GN), proven as successful tool for parallel processes modeling and analysis of complex systems, a GN model of the DOBOT Magician robot system with centralized control is developed. The GN model describes the collective work of five robots (DOBOT Magician serial manipulators) and two conveyor belts which joint operation is controlled by a centralized system. Presented model may be considered as a basis of a more complex GN model aiming to describe a collective work of a group of self-controlled system of robots.

Keywords: Generalized net · Manipulator · Model · Robotic arm

1 Introduction

The control of the collective work of a group of the robots depends on the features of the robots itself, and as such, is not a trivial task. DOBOT Magician is one of the core products in DOBOT's K12 customized artificial intelligence education ecosystem presents itself a multi-functional lightweight intelligent robotic arm. DOBOT Magician integrates mechanics and electronics, automation and programming. Dobot Magician has multiple tool heads and the capability to work with the Arduino platform. It also has a graphical programming environment called "DobotBlockly". All aforementioned features present the Dobot Magician as a great STEAM [5] teaching device aiming to strengthen knowledge across multiple subjects by a high precision and user-friendly user interface [4,6,7].

The aim of this study is to be developed a model of the collective work of a system of robots with centralized control applying the theory of Generalized Nets (GN). The GN model presents the use of five serial manipulators and two conveyor belts to demonstrate robot cooperative work in a production line. The

K. T. Atanassov et al. (Eds.): BOS/SOR/IWIFSGN 2022, LNNS 793, pp. 327–334, 2023.
https://doi.org/10.1007/978-3-031-45069-3_30

serial manipulators are DOBOT Magician robots. Each robot has four degrees of freedom and replaceable end effector. The end effector can be 3D printer, laser engraver, suction cup, gripper, or a pen. This allows robot to be used for different purposes in different educational applications. In this case three of the robots are used with gripper or suction cup as end effector in order to place on or remove objects from the conveyor belts. Two of them will be used for processing of the objects, i.e. engraving. Developed GN model represents feed forward control of the process, based on the estimation of the time needed for each operation.

2 A GN Model

In the present paper, we will use a minimal reduced GN (see, [1,2]), i.e., it contains only a set of transitions, a set of tokens, a set of initial tokens characteristics and a characteristic function that determines, the next token characteristic that each token will obtain after its transfer from input to output transition places. Each transition will contain only sets of an input and an output places and an index matrix (see [3]) with transition condition predicates. Developed here GN model of a collective work of a system of robots contains (see Fig. 1) 10 transitions, 35 places and 14 types of tokens that represent:

- $\alpha_1, \alpha_2, \ldots$ - the sequential objects for machining
- κ - the controller (in place l_{24})
- $\kappa_1^1, \kappa_2^1, \ldots, \kappa_1^2, \kappa_2^2, \ldots, \kappa_1^9, \kappa_2^9, \ldots$ - the sequential signals from the controller to the separate components of the system
- μ^1 - the first robot with an engraving tool (in place l_{12})
- μ^2 - the second robot with an engraving tool (in place l_{27})
- ν^1 - the first conveyor belt (in place l_6)
- ν^2 - the second conveyor belt (in place l_{33})
- σ^1 - a storage for raw objects (in place l_3)
- $\sigma_1^1, \sigma_2^1, \ldots$ - the signals from the storage for raw objects
- σ^2 - a storage for ready objects (in place l_{36})
- $\sigma_1^2, \sigma_2^2, \ldots$ - the signals from the storage for ready objects
- χ^1 - the first robot with a gripping tool (in place l_9)
- χ^2 - the second robot with a gripping tool (in place l_{15})
- χ^3 - the third robot with a gripping tool (in place l_{30})

Below, for brevity, we will denote the α-, σ^i- (for $i = 1, 2$), and κ^i- (for $i = 1, 2, \ldots, 9$) tokens without their concrete indices, but as α_{cu}, σ_{cu}^i, and κ_{cu}^i, where cu is a natural number, corresponding to the current number of an object or a signal.

In the present (simpler) model we will assume that the transfer of any object over the first conveyor belt has duration t_1 and over the second conveyor belt - duration t_7; the work of the first, second and the third robot with a gripping tool - durations t_2, t_4, t_6, respectively; and the work of the first and second robot with an engraving tool - durations t_3 and t_5, respectively. Let

$$t_{proc} = \sum_{i=1}^{7} t_i.$$

At the first time-moment T, token κ splits to two tokens - the same token κ, that continues to stay in place l_{20} and a token κ_1^1 that enters place l_{12} and on the next time-step $T + t^o$ (where t^o is the elementary time-step) - place l_3, where it unites with a token σ^1. At time-moment $T + 2t^o$, the token σ^1, that has an initial characteristic

"list of the raw objects, numbered by 1, 2, ..., that stay in the storage",

splits to three tokens - the same token σ^1 that continues to stay in place l_3 with a current characteristic

"current list of the raw objects in the storage",

a token α (i.e., α_1) that enters place l_1 and a token σ_1^1 that enters place l_2 with a characteristic

"raw object with number 1 is put over the first conveyor belt".

The GN transitions (see Fig. 1) have the following forms (for the index matrices, used in the paper, see [3]).

$$Z_1 = \langle \{l_3, l_{12}\}, \{l_1, l_2, l_3\}, \begin{array}{c|ccc} & l_1 & l_2 & l_3 \\ \hline l_3 & W_{3,1} & W_{3,2} & true \\ l_{12} & false & false & true \end{array} \rangle,$$

where
$W_{3,1} = W_{3,2}$ –"there is a signal from κ for placing a new object".

When the predicate $W_{3,1}$ is "*true*" token σ^1 splits to three tokens: the same token σ^1 that continues to stay in place l_3, token α_{cu} that enters place l_1 and obtains a characteristic

"the current task of the collective robot work has started",

and token σ_{cu}^1 that enters place l_2 and obtains a characteristic:

"an object leaves the storage".

$$Z_2 = \langle \{l_1, l_5, l_{13}\}, \{l_4, l_5\}, \begin{array}{c|cc} & l_4 & l_5 \\ \hline l_1 & W_{1,4} & false \\ l_5 & false & true \\ l_{13} & false & true \end{array} \rangle,$$

where
$W_{1,4}$ ="there is a signal from κ for transferring of the current object".

When the predicate $W_{1,4}$ is "*true*" token α_{cu} enters place l_4, and obtains a characteristic:

"the first conveyor belt starts to transfer the current object".

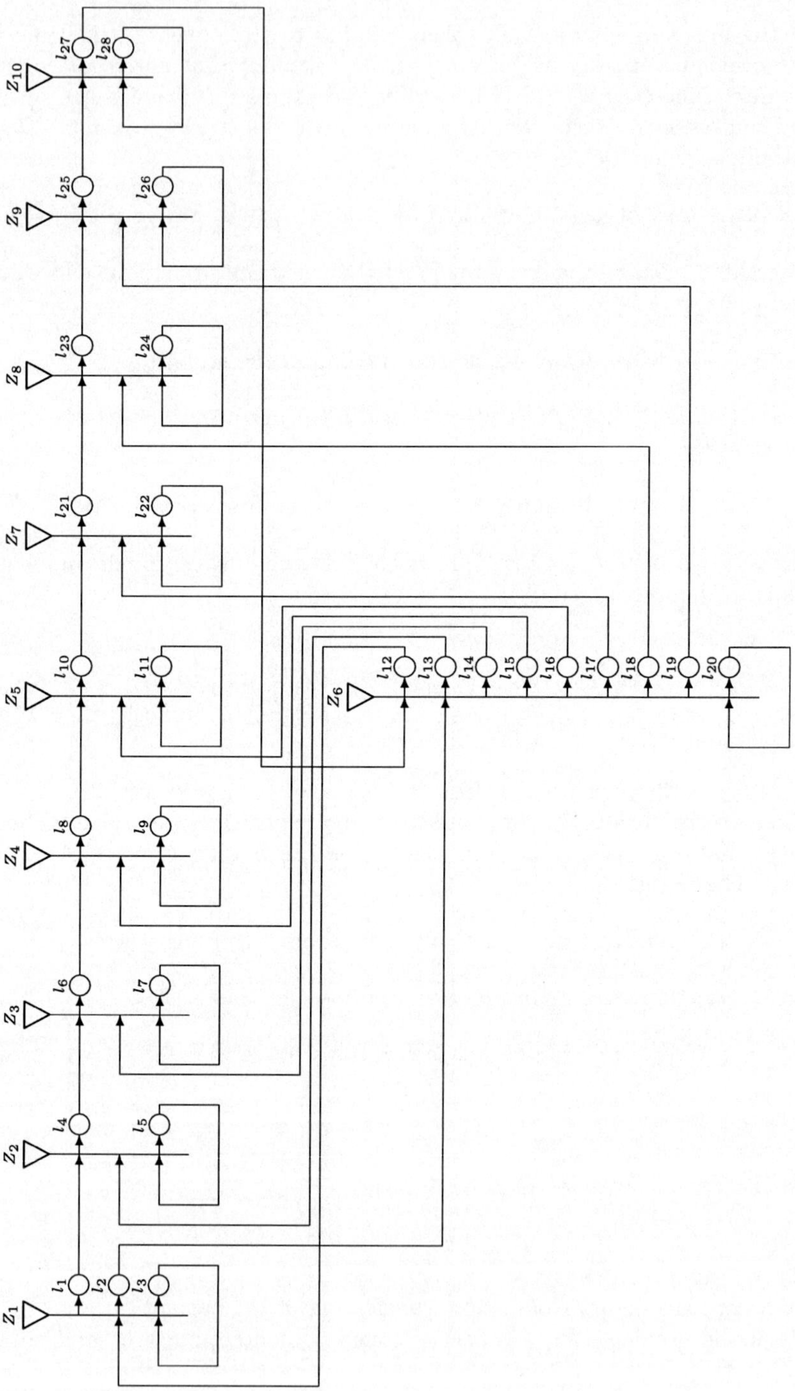

Fig. 1. A GN model

$$Z_3 = \langle \{l_4, l_7, l_{14}\}, \{l_6, l_7\}, \begin{array}{c|cc} & l_6 & l_7 \\ \hline l_4 & W_{4,6} & false \\ l_7 & false & true \\ l_{14} & false & true \end{array} \rangle,$$

where
$W_{4,6}$ = "there is a signal from κ for transferring of the current object".

When the predicate $W_{4,6}$ is "*true*" token α enters place l_6, and obtains a characteristic:

"the first robot with a gripping tool starts to transfer the current object from the first conveyor belt".

$$Z_4 = \langle \{l_6, l_9, l_{15}\}, \{l_8, l_9\}, \begin{array}{c|cc} & l_8 & l_9 \\ \hline l_6 & W_{6,8} & false \\ l_9 & false & true \\ l_{15} & false & true \end{array} \rangle,$$

where
$W_{6,8}$ = "there is a signal from κ for manipulation over the current object".

When the predicate $W_{6,8}$ is "*true*" token α enters place l_8, and obtains a characteristic:

"the first robot with an engraving tool starts to engrave the current object".

$$Z_5 = \langle \{l_8, l_{11}, l_{16}\}, \{l_{10}, l_{11}\}, \begin{array}{c|cc} & l_{10} & l_{11} \\ \hline l_8 & W_{8,10} & false \\ l_{11} & false & true \\ l_{16} & false & true \end{array} \rangle,$$

where
$W_{8,11}$ = "there is a signal from κ for transferring of the current object".

When the predicate $W_{8,10}$ is "*true*" token α enters place l_{10}, and obtains a characteristic:

"the second robot with a gripping tool starts to transfer the current object after the manipulation".

$$Z_6 = \langle \{l_2, l_{20}, l_{27}\}, \{l_{12}, l_{13}, l_{14}, l_{15}, l_{16}, l_{17}, l_{18}, l_{19}, l_{20}\},$$

$$\begin{array}{c|ccccccccc} & l_{12} & l_{13} & l_{14} & l_{15} & l_{16} & l_{17} & l_{18} & l_{19} & l_{20} \\ \hline l_2 & false & false & false & false & false & false & false & false & true \\ l_{20} & W_{20,12} & W_{20,13} & W_{20,14} & W_{20,15} & W_{20,16} & W_{20,17} & W_{20,18} & W_{20,19} & true \\ l_{27} & false & false & false & false & false & false & false & false & true \end{array} \rangle,$$

where
$W_{20,i}$ = "there is a signal from κ to the i-th device of the robotic system", where $i = 12, 13, \ldots, 19$.

The first token κ_1^1 enters place l_{12} in time-moment $T+3t^o$. When the current token σ_{cu}^1 unites with token κ in place l_{20}, on the next time-step

$$T + 2t^0 + (cu - 1)(t_{proc} + t^o)$$

token κ splits to eight tokens - the same token κ that continues to stay in place l_{20} and tokens $\kappa_{cu}^2, \ldots, \kappa_{cu}^9$ that enter places $l_{13}, \ldots l_{19}$ with following characteristics:

"the first conveyor belt must start work immediately",

"the first robot with a grapping tool must start work after $t_1$time-steps",

"the first robot with an engraving tool must start work after$t_1 + t_2$

time-steps",

"the second robot with a grapping tool must start work after$t_1 + t_2 + t_3$

time-steps",

"the second robot with an engraving tool must start work after$t_1 + t_2$

$+t_3 + t_4$ time-steps",

"the third robot with a grapping tool must start work after $t_1 + t_2 + t_3$

$+t_4 + t_5$ time-steps",

"the second conveyor belt must start work after $t_1 + t_2 + t_3 + t_4 + t_5 + t_6$

time-steps",

respectively.

If at the time moment $T+2t^o+cu(t_{proc}+t^o)$ token σ_{cu}^2 does not enter place l_{27} and respectively in the next time-moment it does not enter place l_{20}, this will be a signal that there is a problem in the system and token κ will not send signals for the beginning of the work on the next job. Opposite, if token σ_{cu}^2 enters place l_{20} in time, token κ will send signals for continuation of the work.

$$Z_7 = \langle \{l_{10}, l_{17}, l_{22}\}, \{l_{21}, l_{22}\}, \begin{array}{c|cc} & l_{21} & l_{22} \\ \hline l_{10} & W_{10,21} & false \\ l_{17} & false & true \\ l_{22} & false & true \end{array} \rangle,$$

where

$W_{10,21}$ = "there is a signal from κ for manipulation over the current object".

When the predicate $W_{10,21}$ is "*true*" token α enters place l_{22}, and obtains a characteristic:

"the second robot with a engreaving tool starts to manipulate

the current object".

$$Z_8 = \langle \{l_{18}, l_{21}, l_{24}\}, \{l_{23}, l_{24}\}, \begin{array}{c|cc} & l_{23} & l_{24} \\ \hline l_{18} & false & true \\ l_{21} & W_{21,23} & false \\ l_{24} & false & true \end{array} \rangle,$$

where
$W_{21,23}$ ="there is a signal from κ for transferring of the current object".
When the predicate $W_{21,23}$ is "*true*" token α enters in place l_{28}, and obtains a characteristic:

"the third robot with a gripping tool starts to manipulate

the current object".

$$Z_9 = \langle \{l_{19}, l_{23}, l_{26}\}, \{l_{25}, l_{26}\}, \begin{array}{c|cc} & l_{25} & l_{26} \\ \hline l_{19} & false & true \\ l_{23} & W_{23,25} & false \\ l_{26} & false & true \end{array} \rangle,$$

where
$W_{23,25}$ ="there is a signal from κ for transferring of the current object".
When the predicate $W_{23,25}$ is "*true*" token α enters place l_{25}, and obtains a characteristic:

"the second conveyor belt starts to transfer the current object".

$$Z_{10} = \langle \{l_{25}, l_{28}\}, \{l_{27}, l_{28}\}, \begin{array}{c|cc} & l_{27} & l_{28} \\ \hline l_{25} & false & true \\ l_{28} & W_{28,27} & true \end{array} \rangle,$$

where
$W_{28,27}$ = "the object successfully entered the storage".
When the predicate $W_{28,27}$ is "*true*" token σ^2 splits to two tokens: the same token σ^2 that continues to stay in place l_{28} and token σ^2_{cu} that enters place l_{27} and obtains one of the following two characteristics:

$$\begin{cases} \text{"the current object enters the storage for ready objects",} & \text{if the process finishes successfully} \\ \text{"the storage for the ready objects enters in a standby mode"} & \text{if the process does not finish in time} \end{cases}$$

3 Conclusion

The purpose of this study is to present the application of GNs in the field of robotics thus proving again the GN advantages as an alternative tool for modelling of parallel processes and processes in complex systems. The proposed GN model describes the collective work of five robots, representatives of the DOBOT Magician manipulators, and two conveyor belts. The GN model presents the results of their joint operation when centralized control has been applied. The model can be further enhanced and detailed in specifying the tasks of any of the robot manipulators, as well as in their collective work in case of self-controlled system of robots.

Acknowledgment. The authors acknowledge the support from the project UNITe BG05M2O P001-1.001-0004/28. 02.2018 (2018–2023).

References

1. Atanassov, K.: Generalized Nets. World Scientific, Singapore (1991)
2. Atanassov, K.: On Generalized Nets Theory. Prof. M. Drinov Academic Publishing House, Sofia (2007)
3. Atanassov, K.: Index Matrices: Towards an Augmented Matrix Calculus. Springer, Cham (2014). https://doi.org/10.1007/978-3-319-10945-9
4. Cai, J., Deng, J., Zhang, W., Zhao, W.: Modeling method of autonomous robot manipulator based on D-H algorithm. Mob. Inf. Syst. **2021**, 10 (2021). Article ID: 4448648. https://doi.org/10.1155/2021/4448648
5. Yakman, G.: STEAM education: an overview of creating a model of integrative education (2008)
6. https://www.robotlab.com/store/dobot-robotic-arm
7. https://www.alcom.no/wp-content/uploads/2019/11/DOBOT-Product-Catalog-Educational-Version.pdf

Optimizing the Operation of a Wireless Sensor Network Using a Swarm Intelligence Algorithm

Hubert Zarzycki[1](✉), Łukasz Apiecionek[2], Dawid Ewald[2], and Jacek Czerniak[2]

[1] General Tadeusz Kosciuszko Military Academy of Land Forces, Wroclaw, Poland
hubert.zarzycki@awl.edu.pl

[2] Institute of Technology, Department of Computer Science, Kazimierz Wielki University, Bydgoszcz, Poland
{lapiecionek,dewald,jczerniak}@ukw.edu.pl

Abstract. This paper presents the results obtained after testing and comparing the whale algorithm (WOA) and the artificial bee colony (ABC) algorithm in a practical problem. The task of the prepared solution was to support the operation of wireless sensor networks. As part of the experiments, several sets of solutions were generated for studying the number of active sensors and energy consumption. One of the sets of computational simulation results is presented. The research shows that the use of selected swarm intelligence algorithms can increase the effectiveness of wireless sensor networks.

Keywords: Whale algorithm · Artificial Bee Colony · optimization · wireless sensor network · swarm intelligence

1 Introduction

In the presented paper, an analysis of selected algorithms inspired by nature was carried out. There is no limit to the number of possible solutions to nature-based algorithms, because almost every phenomenon from the surrounding natural environment can be an inspiration in creating optimal solutions. These optimal algorithms have been refined by thousands of generations of diverse organisms for better adaptation and survival. The most famous algorithms inspired by nature include, among others ant colony optimization (ACO), bee algorithm (ABC), particle swarm algorithm (PSO), firefly (FA) or cat (CSO) algorithm. This study compiled and analyzed the results for the whale algorithm (WOA) and the bee algorithm.

1.1 Swarm Intelligence

Swarm intelligence algorithms are a subgroup of nature-inspired algorithms. Characteristic here is the phenomenon of cooperation of many swarm agents (e.g. bees) in order to achieve a specific goal related to the welfare of the swarm [3, 11]. In this approach, there is no individual managing all tasks or imposing the exact course of action of individuals.

K. T. Atanassov et al. (Eds.): BOS/SOR/IWIFSGN 2022, LNNS 793, pp. 335–342, 2023.
https://doi.org/10.1007/978-3-031-45069-3_31

Despite these apparent aberrations, cooperation results in the solution of complex problems. Commonly known examples of swarming algorithms are bee algorithm (ABC), ant (ACO), particle swarm optimization (PSO) and cat algorithm (CSO). In this study, solutions are analyzed using the relatively new whale algorithm (WOA) and the standard bee algorithm (ABC) [2, 9].

This article is a continuation of the authors work on various applications of artificial intelligence in the area of optimization methods [14], fuzzy numbers [4, 16, 17], and in particular on nature-inspired algorithms [20–22].

2 ABC Algorithm

The principle of the bee algorithm is based on the concept of mimicking the way of searching for food by bee colonies. Individual bees exchange information about places with abundant food [6, 7].

The ABC algorithm divides bees into three categories. In terms of their functions, bees are divided into scouts, observers and working bees. Observers are bees waiting for information about food sources. Scouts search for new food sources at random [8]. On the other hand, worker bees are those that go to a previously discovered food source. There are three steps in each of the food search cycles in the bee algorithm.

- in the first stage, female bees go to known food sources, where the amount of collected nectar is assessed.
- in the second stage, the bees, the observers, collect information from the worker bees and select the food source. The higher the amount of nectar found, the greater the chance of choosing this food source.
- in the third stage, the bees selected from among the colonies are sent to random places in order to find food sources. Upon arriving at a given place with food, the scout checks the adjacent solution and makes an assessment. If a food source is abandoned by a scout, it is replaced with another randomly found food source.

After the observers' bees return, the information found on the food is passed on to the observers. In the standard ABC algorithm, in each cycle, a maximum of one scout is sent to search for new sources. The same number of observing and hired bees is given. In this algorithm, the food source represents a possible solution to the optimization problem [20]. The amount of food in the source determines the suitability of the solution. On the other hand, the number of observing or employed bees determines the number of solutions in a given population.

3 Whale Algorithm

Seyedali Mirjalili and Andrew Lewis developed the Whale Optimization Algorithm (WOA) in 2016 [10]. It is one of the newest and most interesting algorithms in the area of swarm intelligence. It is based on observing the behavior of humpback whales. They are whales that are large in size, although they mainly feed on small fish and krill. Over the generations, they developed a method of encircling large numbers of captured organisms. This method inspired Mirjalili and Lewis to develop the WOA algorithm.

Closer observation of humpback whales revealed that they hunt close to the surface of the water in spiral formations. Meanwhile, these whales release significant amounts of air through their nostrils. Krill and small fish are so little that the mesh of air bubbles formed around them prevents them from escaping. Closed by ever narrower spirals with air bubbles, they are pushed upwards. Right at the surface, they become easy targets for humpback whales. No other type of whale uses this method in hunting (Fig. 1).

Fig. 1. A spiral network of air bubbles formed by humpback whales to encircle small organisms [10].

The method of hunting humpback whales is based on three phases. In the first phase, these whales surround the krill or small fish. In this phase, all humpback whales head towards the leader. The leader is the agent that has the best solution in the current iteration. Then a selection is made, a spiral is formed, or the loops around the organisms are narrowed. In nature, humpback whales perform both of these activities simultaneously. The algorithm draws one of these two behaviors with a probability of 50%.

The final phase is the search for the best possible solution by free whales (agents). After exceeding the set number of iterations or when a better solution is found, the algorithm is stopped and the results are displayed.

Phase I

The location of the prey is unknown. The whales go to the leader, ie the agent with the best solution.

$$D = \left|CX^*(t) - X(t)\right| \quad A = 2ar - a \tag{1}$$

$$X(t+1) = X^*(t) - AD \quad C = 2r \tag{2}$$

where

t - iterative number,

X* - vector of whale best position,

X - vector of current whale position,
r - random number [0,1]
a - convergence factor (a = 2–2t/tmax)

Phase II
The spiral position is updated in phase II. The preying area may be narrowed. Each of the two procedures is selected with a 50% probability.

$$D\prime = \left|X^*(t) - X(t)\right| \tag{3}$$

$$X(t+1) = D\prime e^{bl}cos(2\pi l) + X^*(t) \tag{4}$$

$$X(t+1) = \begin{cases} X^*(t) - AD & if\ p < 0.5 \\ D\prime e^{bl}cos(2\pi l) + X^*(t) & if\ p \geq 0.5 \end{cases} \tag{5}$$

where
D' - distance between the pray and the humbag
I - random number of range [−1, 1]
b – constant b is modifying the shape of a spiral
p - random number of range [0,1]

Phase III
The task of Phase III is to continue the global search of the results and avoid falling into the local minimum. In this phase, prey are randomly searched based on the location of individual humpback whales.

$$D = |CX_{rand}(t) - X(t)| \tag{6}$$

$$X(t+1) = X_{rand}(t) - AD \tag{7}$$

where
X_{rand} - random positional vector (random whale)

4 Optimal Use of Energy of Wireless Sensor Network

Multiple sensors are networked together to fulfill specific functions. Typically it is monitoring traffic, humidity, water level, temperature or many other phenomena, mainly natural [19]. An important issue related to the optimal operation of the sensor network is energy consumption [1]. Sensors consume energy regardless of whether they receive a signal at the moment or not. The work uses the whale (and bee) algorithm for optimal management of switching nodes on and off. This enables the minimization of energy consumption while maintaining the highest possible coverage of the monitored phenomenon.

In principle, each node is managed as a separate agent (whale) of our algorithm. Based on the position and signal, it is calculated which nodes should be running at the moment. The proposed solutions were verified and analyzed in a tool created in the

C# language [23]. After running the simulation, the nodes imitated typical sensors with standard power consumption by default. The obtained results are presented in the table below.

Both with a small and larger number of sensors, energy consumption was lower for the solution based on the whale algorithm. This confirms the high efficiency of the use of WOA in the area of applications related to the optimization of energy consumption of sensor networks.

5 Analysis of the Obtained Results

As can be seen in the table below regardless of the number of nodes, the results of the WOA algorithm are significantly better than ABC, which is characterized by a lower demand for energy and a lesser number of active sensors (Table 1).

Table 1. Table of results set number 1.

Network size	Number of active nodes		Energy Consumption	
	WOT	ABC	WOT	ABC
100	31	39	11607	16307
200	40	51	18074	26129
300	38	50	17805	23321
400	41	51	17740	23568
500	44	56	21746	28278
600	40	51	20441	27341
700	39	49	18685	26528
800	40	52	20349	28574
900	37	50	19070	26615
1000	42	53	20095	28631

The whale algorithm is known to be used in the optimization of sensor networks. The comparison of the number of active nodes actually shows that the WOA algorithm performs much better. The whale algorithm results are 25 to 35% more optimal than the ABC algorithm (Fig. 2).

The diagram shows that with the rapid percentage increase of sensors from 100 to 200, there is also a sharp increase in energy demand. Then, with a smaller percentage increase in the number of active sensors, the rate of increase in energy demand decreases. The amount of energy needed fluctuates slightly, going down at 700 knots and increasing at 800 and 1000 (Fig. 3).

In the initial size of the diagram above it can be seen that between the number of nodes 100 and 200 there is a significant increase in energy demand. The whale algorithm keeps the energy consumption in the range of 15,000–20,000 with slight deviations. A slight

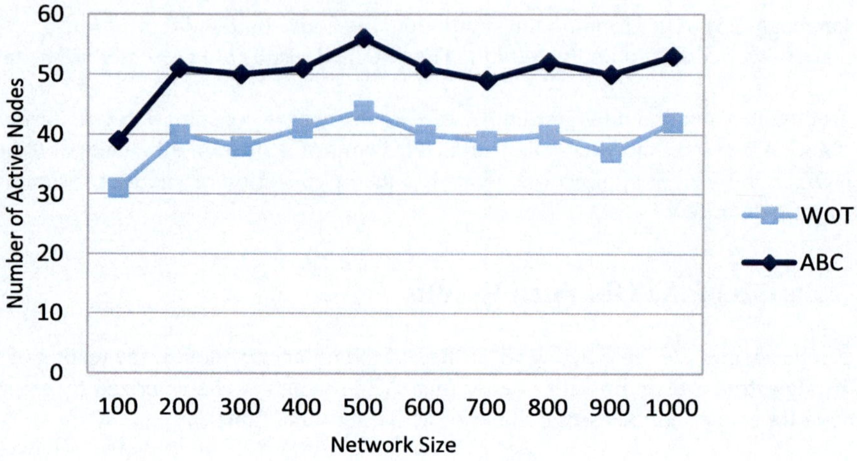

Fig. 2. Diagram of active nodes for the Artificial Bee Colony (ABC) and the WOA algorithm (WOT) [21]

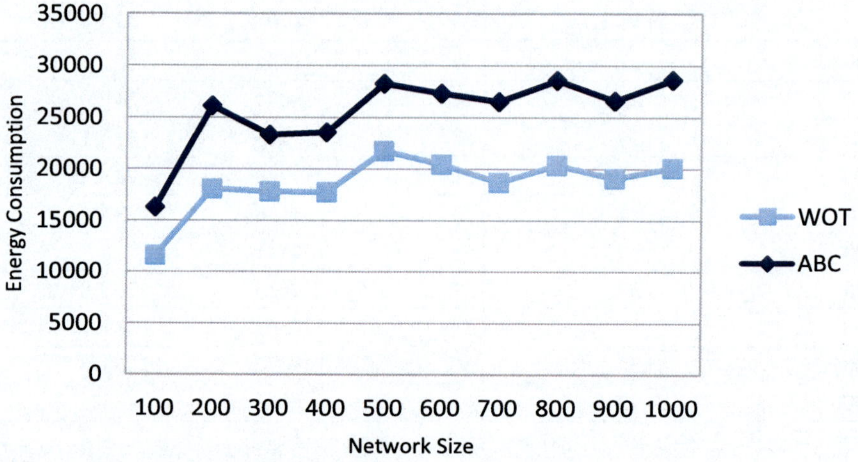

Fig. 3. Diagram of the energy consumption for the whale optimization algorithm ABC and Artificial Bee Colony (ABC) algorithm [21]

increase is only visible for the size of the 500 network. However, for the ABC algorithm, for the network size of more than 400 nodes, energy consumption is at relatively high levels, from 25,000 to 30,000. For any size of wireless sensor networks, the WOA algorithm is able to find a solution with significantly lower consumption energy.

Undoubtedly, it is worth looking for energy optimization solutions for wireless sensor networks based on nature-inspired algorithms [18, 21]. Contemporary approaches in the area of swarm intelligence bring a new quality and perhaps are the future of optimal applications of the sensor network.

6 Conclusions

Over millions of years, organisms living on earth have developed many optimal behaviors in order to survive in the natural environment. Researchers of these phenomena managed to describe some of these phenomena in the form of algorithms. There are more and more nature-inspired algorithms. They are developed and improved [5]. These algorithms are used in more and more problems encountered in practice [13]. Currently, many algorithms inspired by nature are the only practical solution to specific complex optimization tasks [12].

The article presents examples of modern WOT and ABC algorithms from the area of swarm intelligence. These algorithms were used to optimize the number of active nodes and energy consumption in the wireless sensor networks. The obtained and analyzed results indicate the advantage of the WOA algorithm in this application. The WOA results were more than 25% better for network sizes ranging from 100 to 1000.

Swarm intelligence algorithms perform well compared to with the deterministic and force-based approaches. Especially when it comes to the amount of computing power needed and the associated costs and the time needed for calculations. It is worth noting that further improvement in the search for optimal solutions can be obtained, among others by using better fine-tuned parameters in formulas or by adapting accurate and modern improvements in algorithms.

References

1. Ahmed M.M., Houssein, E.H., Hassanien, A.E., Taha, A., Hassanien, E.: Maximizing lifetime of wireless sensor networks based on whale optimization algorithm. In: Hassanien, A., Shaalan, K., Gaber, T., Tolba, M. (eds.) Proceedings of the International Conference on Advanced Intelligent Systems and Informatics 2017. AISI 2017. Advances in Intelligent Systems and Computing, vol. 639, pp. 724–733. Springer, Cham (2018). https://doi.org/10.1007/978-3-319-64861-3_68
2. Panigrahi, B.K., Shi, Y., Lim, M.H., Red., Handbook of Swarm Intelligence, t. 8, Adaptation, Learning, and Optimization. Springer, Heidelberg (2011)
3. Engelbrecht, A.P.: Fundamentals of Computational Swarm Intelligence. Wiley, Hoboken (2005)
4. Ewald, D., Czerniak, J.M., Zarzycki, H.: OFNBee method used for solving a set of benchmarks. In: Kacprzyk, J., Szmidt, E., Zadrożny, S., Atanassov, K., Krawczak, M. (eds.) Advances in Fuzzy Logic and Technology 2017. EUSFLAT IWIFSGN 2017 2017. Advances in Intelligent Systems and Computing, vol. 642, pp. 24–35. Springer, Cham (2018). https://doi.org/10.1007/978-3-319-66824-6_3
5. Ewald, D., Zarzycki, H., Apiecionek, Ł, Czerniak, J.M.: Ordered fuzzy numbers applied in bee swarm optimization systems. J. Univ. Comput. Sci. **26**(11), 1475–1494 (2020)
6. Karaboga, D.: An idea based on honey bee swarm for numerical optimization. Technical Report for Erciyes University, Kayseri, Turkey (2005)
7. Karaboga, D., Basturk, B.: A powerful and efficient algorithm for numerical function optimization: artificial bee colony (ABC) algorithm. J. Glob. Optim. **39**, 459–471 (2007)
8. Karaboga, D., Basturk, B.: On the performance of artificial bee colony (ABC) algorithm. Appl. Soft Comput. **8**, 687–697 (2008)
9. Merkle D.: Swarm Intelligence: Introduction and Application, Springer, Heidelberg (2008). https://doi.org/10.1007/978-3-540-74089-6

10. Mirjalili, S., Lewis, A.: The whale optimization algorithm. Adv. Eng. Softw. **95**, 51–67 (2016)
11. Agarwal, P., Mehta, S.: Nature-inspired algorithms: state-of-art, problems and prospects. Int. J. Comput. Appl. **100**, 14–21 (2014)
12. Slowik A.: Swarm Intelligence Algorithms. Modifications and Applications. CRC Press/Taylor and Francis, Boca Raton (2020)
13. Skubisz, O., Zarzycki, H., Wincewicz-Bosy, M., Ewald, D.: Applying improved bee-inspired algorithm for the vehicle routing problem. In: International Business Information Management Association, IBIMA publishing, Seville (2021)
14. Śmigielski, G., Dygdała, R., Zarzycki, H., Lewandowski, D.: Real-time system of delivering water-capsule for firefighting. In: Kobayashi, S., Piegat, A., Pejaś, J., El Fray, I., Kacprzyk, J. (eds.) ACS 2016. AISC, vol. 534, pp. 102–111. Springer, Cham (2017). https://doi.org/10.1007/978-3-319-48429-7_10
15. Zarzycki, H., Apiecionek, Ł., Czerniak, J.M., Ewald, D.: The proposal of fuzzy observation and detection of massive data DDOS attack threat. Atanassov, K., *et al.* (eds.) Uncertainty and Imprecision in Decision Making and Decision Support: New Challenges, Solutions and Perspectives. IWIFSGN 2018. Advances in Intelligent Systems and Computing, vol. 1081, pp. 363–378. Springer, Cham (2021). https://doi.org/10.1007/978-3-030-47024-1_34
16. Zarzycki, H., Czerniak, J.M., Dobrosielski, W.T.: Detecting Nasdaq composite index trends with OFNs. In: Prokopowicz, P., Czerniak, J., Mikołajewski, D., Apiecionek, Ł., Ślęzak, D. (eds.) Theory and Applications of Ordered Fuzzy Numbers. SFSC, vol. 356, pp. 195–205. Springer, Cham (2017). https://doi.org/10.1007/978-3-319-59614-3_11
17. Zarzycki, H., Dobrosielski, W.T., Vince, T., Apiecionek, Ł.: Center of circles intersection, a new defuzzification method on fuzzy numbers. Bulletin of the Polish Academy of Sciences. Technical Sciences (2020)
18. Zarzycki, H., Ewald, D., Skubisz, O., Kardasz, P.: A comparative study of two nature-inspired algorithms for routing optimization. In: Atanassov, K.T., *et al.* Uncertainty and Imprecision in Decision Making and Decision Support: New Advances, Challenges, and Perspectives. IWIFSGN BOS/SOR 2020 2020. Lecture Notes in Networks and Systems, vol. 338, pp. 215–228. Springer, Cham (2021). https://doi.org/10.1007/978-3-030-95929-6_17
19. Zarzycki, H., Czerniak, J.M., Lakomski, D., Kardasz, P.: Performance comparison of CRM systems dedicated to reporting failures to IT department. In: Madeyski, L., Śmiałek, M., Hnatkowska, B., Huzar, Z. (eds.) Software Engineering: Challenges and Solutions. Advances in Intelligent Systems and Computing, vol. 504, pp 133–146. Springer, Cham (2016). https://doi.org/10.1007/978-3-319-43606-7_10
20. Zarzycki, H., Skubisz, O.: A new artificial bee colony algorithm approach for the vehicle routing problem. In: Kahraman, C., Cebi, S., Cevik Onar, S., Oztaysi, B., Tolga, A.C., Sari, I.U. (eds.) INFUS 2021. LNNS, vol. 307, pp. 562–569. Springer, Cham (2022). https://doi.org/10.1007/978-3-030-85626-7_66
21. Zarzycki, H., Skubisz, O.: An application of ant algorithm for routing optimization problem. In: International Business Information Management Association, IBIMA Publishing (2021)
22. Zarzycki H.: Comparative study of the firefly algorithm and the whale algorithm. In: Kahraman, C., Tolga, A.C., Cevik Onar, S., Cebi, S., Oztaysi, B., Sari, I.U. (eds.) Intelligent and Fuzzy Systems. INFUS 2022. Lecture Notes in Networks and Systems, vol. 504, pp. 999–1006. Springer, Cham (2022). https://doi.org/10.1007/978-3-031-09173-5_114
23. https://docs.microsoft.com/pl-pl/dotnet/csharp. Accessed 21 Aug 2022

Author Index

K. T. Atanassov et al. (Eds.): BOS/SOR/IWIFSGN 2022, LNNS 793, pp. 343–344, 2023.
https://doi.org/10.1007/978-3-031-45069-3